Understanding the Electromagnetic Field

Understanding the Electromagnetic Field

Basil S Davis
University of Notre Dame, USA

NEW JERSEY • LONDON • SINGAPORE • BEIJING • SHANGHAI • HONG KONG • TAIPEI • CHENNAI • TOKYO

Published by

World Scientific Publishing Co. Pte. Ltd.
5 Toh Tuck Link, Singapore 596224
USA office: 27 Warren Street, Suite 401-402, Hackensack, NJ 07601
UK office: 57 Shelton Street, Covent Garden, London WC2H 9HE

Library of Congress Control Number: 2023012282

British Library Cataloguing-in-Publication Data
A catalogue record for this book is available from the British Library.

UNDERSTANDING THE ELECTROMAGNETIC FIELD

ISBN 978-981-127-481-7 (hardcover)
ISBN 978-981-127-536-4 (paperback)
ISBN 978-981-127-482-4 (ebook for institutions)
ISBN 978-981-127-483-1 (ebook for individuals)

For any available supplementary material, please visit
https://www.worldscientific.com/worldscibooks/10.1142/13368#t=suppl

Printed in Singapore

Preface

Electromagnetism — or electrodynamics, as it is more commonly called — is an elegant science that is founded on a few experiments and the mathematical apparatus of vector calculus. The subject was largely developed in the 19th century by a small number of physicists, among whom James Clerk Maxwell holds a preeminent position. Maxwell succeeded in uniting all the known phenomena of electricity and magnetism into a set of four elegant equations. Maxwell also provided the inspiration and foundation for Einstein's Special Theory of Relativity which arose in the early 20th century. This past century also witnessed the birth of quantum theory, and the integration of the latter with electromagnetism, which was largely the work of Richard P. Feynman. Some basic problems of classical electromagnetism have been solved by quantum electrodynamics. The challenge of the 21st century is to integrate electromagnetism with Einstein's General Theory of Relativity, in which gravity plays the central role. This integration needs to be done at both macroscopic and microscopic levels, and no consensus has been reached in these areas.

In this book the focus is on the field itself, and so interactions of the electromagnetic field in material media — important as these are — are omitted. There are excellent texts that cover these areas.

No new physics is introduced in this book. But the pedagogy is original. The study of the electromagnetic field is related to special relativity and quantum mechanics, and in a final chapter the relationship between electromagnetism and general relativity is introduced. Much effort has been made to simplify derivations of important formulas and theorems.

This book diverges from most textbooks on electrodynamics in the detailed attention given to quantum mechanics and relativity, both special and general. In the 21st century the study of electromagnetism cannot be isolated from a knowledge of quantum mechanics and relativity. A firm basic knowledge of these topics is essential for a graduate student specializing in any area of physics. The author offers this book as evidence that it does not require extraordinary intellectual abilities to gain a mastery of electromagnetism, acquire a basic knowledge of quantum mechanics, and obtain a real grasp of special and general relativity.

The inspiration for this book developed over the years I taught physics at St. Edmund's College (Shillong), Tulane University (New Orleans), and Xavier University of Louisiana (New Orleans).

I dedicate this book to my wife Shyla, who worked hard so that I could take time off to write this book, and to our children Melinda, Jessica and Peter, who have always been my cheerleaders.

Basil S. Davis

South Bend, Indiana, 2023.

Contents

Chapter 1

The Study of Empty Space

1.1 Introduction

The word *field* is a multivalent term capable of a variety of connotations. In its primary literal sense, a field is a piece of land, an agricultural space upon which a crop such as corn, rice or wheat is cultivated. Another meaning — perhaps as literal as the preceding — suggests its usage in sports, such as a soccer field, a hockey field, or a football field. Used metaphorically, it could refer to an area of one's expertise, interest, or specialization — be it medicine, mathematics or music. But in physics the word *field* has a particular meaning, one that is both concrete and abstract. In this book we will study one particular field that plays an extremely important role in physics — the electromagnetic field. And whereas some introductory textbooks might convey the impression that the electromagnetic field is just a mathematical device that is useful for understanding the real forces between charges, in this book we will show that the electromagnetic field is just as real as anything else in physics.

The use of the word *field* in physics does have parallels in the world of sports. Just as there are different kinds of fields upon which different sports are played — soccer, football, cricket, etc. so in physics one studies different sorts of fields — gravitational, electric, magnetic, nuclear, etc. in which different interactions occur. The objects also differ from field to field. A soccer ball differs from a cricket ball, and an electron differs from a neutron. And just as the rules of soccer are different from the rules of cricket, so too, the laws that govern interactions in an electromagnetic field are different from those in a gravitational field. There is yet another detail in this parallelism that is worthy of particular attention. In sports the field is not merely the background upon which the action takes place, but the field plays an active part in the game itself. So, in American football, a reception or an interception is completed only if the ball thrown by the passer did not touch the ground before it was caught. And in physics, charged particles exchange momentum not only with each other, but also with the electromagnetic field. Thus both in physics and in sports the field is not a lifeless surface upon which the action unfolds, but it is something that actively participates in the action itself. So, if we were to adapt the well-known Shakespearean quote "all the world's a stage, and all the men and women merely players," we could say the stage is also a player, albeit one without an entrance or an exit.

The geometry of the ground upon which the field is marked is important for the specific sport. Soccer and football fields are built on level ground. Golf, however, requires uneven terrain, and that is perhaps why one speaks of a golf *course*, but not a golf field. In physics the ground upon which the field exists is called *space*. The electromagnetic field can be thought of as something created by electric charges — either stationary, or in motion — and superimposed on or embedded in physical space. In classical physics — which we shall deal with in most of this book — this electromagnetic field does not change the space in which it operates. It is an altogether different matter with the gravitational field of general relativity. Material objects having mass create a gravitational field, and the gravitational field has an effect on the space itself. Unlike the gravitational field, the electromagnetic field exists in a flat three-dimensional space. (This is not strictly true for an extremely strong electromagnetic field, but it is true enough for electromagnetic fields we can currently generate on earth, and which we encounter in outer space.) By "flat" we mean that one can draw straight lines in such a space, and the three angles made by the straight line triangles drawn in this space will always add up to two right angles. Such a space is called a Euclidean space. Thus, to begin the study of the electromagnetic field we need to study three-dimensional Euclidean space.

We begin our analysis of three-dimensional space by using rectangular or Cartesian coordinates, which are defined by three mutually perpendicular straight lines called coordinate axes. Each one of the axes is an infinite straight line. The fundamental assumption we make here is that there is a one-to-one correspondence between the set of points on the line and the set of real numbers. This implies two things: first, that every point in space can be described by three real numbers, and second, there is no set of three real numbers that does not have a corresponding point in three-dimensional space.

1.2 Real Numbers

To understand the characteristics of real numbers it may help to examine some of the more familiar subsets of the real numbers. The commonest subset is the set of natural numbers 1, 2, 3.... Since there is no such thing as the largest possible natural number, the set of natural numbers $\mathbb{N} = \{1, 2, 3...\}$ is said to be a countably infinite set. The cardinality or cardinal number of a finite set is simply the number of elements of the set. Cardinalities of infinite sets have also been defined, though these cardinal numbers are certainly not obtained by counting the elements. The cardinal number or cardinality of $\mathbb{N}$ is written as a symbol

$$n(\mathbb{N}) \equiv |\mathbb{N}| = \aleph_0 \tag{1.1}$$

where the symbol $\aleph_0$ (pronounced aleph nought, aleph zero or aleph null) is what we conventionally label as "infinity." But mathematicians have found more than one infinity, so that $\aleph_0$ is actually the smallest of the possible infinities. Let us check this claim by trying to find a set having a greater cardinality than $|\mathbb{N}|$. Now, $+3 \in \mathbb{N}$ (read "+3 is an element of $\mathbb{N}$") though $-3 \notin \mathbb{N}$. But negative numbers are also real (since a negative potential energy is associated with an attractive force), and so we create a wider set which includes negative numbers and the number 0, together called the set of integers. The set of integers is written as

$$\mathbb{Z} = \{... - 3, -2, -1, 0, 1, 2, 3...\} \tag{1.2}$$

But though the set of integers contains many — indeed infinitely many — more elements than the set of natural numbers, paradoxically $\mathbb{Z}$ has the same cardinality as the set of natural numbers. When comparing two infinite sets, cardinality is compared by pairing off elements in the two sets. If we can establish a one-to-one correspondence (also called a "bijection") between the elements of the two sets, we could then say that they have the same cardinality. So, if we were to define the set $\mathbb{Z}_2$ as the set of even integers $\{... - 6, -4, -2, 0, 2, 4, 6...\}$ it is easy to see that $|\mathbb{Z}_2| = |\mathbb{Z}|$.

We will now prove that the set of integers has the same cardinality as the set of natural numbers. First, we separate out the natural numbers into even and odd numbers. Clearly, there is a one-to-one correspondence between the positive elements of $\mathbb{Z}_2$ defined above and the even elements of $\mathbb{N}$. Next, we define the function $1 - n$ where n is an odd natural number. Now, for odd values of the natural number n the function $1 - n$ takes on the values $0, -2, -4, -6$, etc. Now, it is evident that for every n there is a unique $1 - n$ and vice versa, and so it follows that there is a one-to-one correspondence between the odd natural numbers and the set consisting of 0 and the negative even numbers. Putting all this together it follows that $|\mathbb{Z}| = |\mathbb{N}|$.

Let us picture a straight line that extends to infinity in either direction. If we were to mark off points at equal distances along the line, we would find that there is a one-to-one correspondence between the integers and the marked points on the line. But there are also points on the line lying between those that have been marked. These can also be identified with numbers, but these numbers will not be integers.

If we obtain the ratio of any two integers, with the restriction that the denominator should not be 0, we would get a number that could be an integer, or a positive or negative fraction. The ratios so obtained from integers are called rational numbers. $3, -2.5, 1/4, -5$, etc. are examples of rational numbers. The set of rational numbers is sometimes written as

$$\mathbb{Q} = \{x : x = p/q \ \forall \ p, q \in \mathbb{Z}, q \neq 0\} \tag{1.3}$$

With a little more effort we can prove that $|\mathbb{Q}| = |\mathbb{N}|$.

Exercise:
Prove that $|\mathbb{Q}| = |\mathbb{N}|$.

We could depict each rational number as a point on a straight line. Such points representing rational numbers may coincide with the points marking the integers, but in general they would lie between neighboring integer points.

Exercise:
Prove that for every rational number there is a corresponding point on a number line. Is the reverse true?

But is there a one-to-one correspondence between the rational numbers and the points on a line? The answer is in the negative, because there are points on a line that would not correspond to any rational numbers. If we draw a right angled triangle whose equal sides are exactly 1 centimeter, the hypotenuse would have length $\sqrt{2}$ cm. Thus, if we can draw a line of length

1 cm, then we can draw a line of length equal to $\sqrt{2}$ cm, and hence the number $\sqrt{2}$ corresponds to a point on a line. But $\sqrt{2}$ cannot be written as a ratio of two integers, and hence it is an irrational number. Other numbers that can be depicted on a line, but which are also irrational numbers, are π and the natural logarithm base e. The union of the set of irrational numbers and the set of rational numbers is the set of real numbers $\mathbb{R}$. The square root of 2 can be written as a decimal expansion using an algorithm for finding square roots of numbers that are not perfect squares. This method uses nothing but algebraic operations, and for any finite number of decimal places we can obtain the square root using a finite number of operations. Thus the decimal expression of an irrational number can never be 100% accurate. But every real number can be assigned a geometrical point on a line with total accuracy. This is an important mathematical principle that is foundational to all of physics, and in particular to electromagnetic field theory.

We will assume that there is a one-to-one correspondence between the real numbers and the points on an unbroken line. This correspondence can be thought of as a mapping between the algebraic numbers and the geometrical points. The cardinality of the set of real numbers, or the set of points in a line, is written as $n(\mathbb{R}) \equiv |\mathbb{R}| = \aleph_1$. There is also a one-to-one mapping between the numbers in a finite line segment and those in a line of infinite length, indicating that the cardinality of the set of points in a line is independent of the length of the line. There is a mathematical principle called the *continuum hypothesis* which states that there is no infinity that is intermediate between $\aleph_0$ and $\aleph_1$.

Exercise:
Show that every real number that can be constructed by using algebra or calculus corresponds to a unique point on a straight line.

The cardinality of the set of points on a line is the same as the cardinality of the set of real numbers $n(\mathbb{R}) = |\mathbb{R}|$ and this cardinality is different from and greater than the cardinality of rational numbers. $|\mathbb{R}| = \aleph_1$ is a higher order of infinity than $\aleph_0$. Indeed, one could say that $\aleph_1$ is infinitely greater than $\aleph_0$. One expresses the relationship between these infinities by the equation $\aleph_1 = 2^{\aleph_0}$ suggesting that the former is both exponentially and infinitely greater than the latter. That the cardinality of real numbers is infinitely greater than the cardinality of rational numbers (or integers or natural numbers) has a profound bearing on experimental physics. Quantities in physics are expressed in the decimal system, following the scientific notation, e.g. 2.4078×10^{-4}. In this example the quantity is expressed to five significant figures, where the number of significant figures represents the level of accuracy of the experiment. The decimal expansion of a rational number would eventually run into repeating strings of digits which could also be zeros. So, $4/5 = 0.80000$, $14/13 = 1.076923076923...$ and $3/7 = 1.857142857142....$ An irrational number cannot be expressed in these terms. There is no repetitious pattern in the digits of irrational numbers. The result of any experimental measurement is always expressed as a rational number with a small number of digits. But if we were to measure some property of a solid sphere of radius 1.000 m, and find the experimental value of this property to be 3.142, then we could reasonably guess that the property is related to the radius r by the formula πr. The challenge then would be to derive this formula theoretically.

Roulette wheels, and the wheels on television shows such as *Wheel of Fortune* or *The Price is Right* have a finite number of slots. There is therefore a non-zero probability of landing on any one of the slots. But if we have a circle with a continuum of points, and each point represents a real number, the probability that we will land exactly on any one number that we choose beforehand is always zero. Volumes, areas, speeds, etc. in nature are real numbers, which cannot be written exactly in a terminating or recurring decimal expansion. The speed of light in space is often expressed as $c = 3.00 \times 10^8$ m/s, but it can be written more precisely as 299792458 m/s, and even this is an approximation, because c is a real number.

Measurements of ratios between two quantities can tell us if there is any fundamental relationship between the quantities. In Arthur C. Clarke's *2001: A Space Odyssey* the monolith that was uncovered on the moon was identified as a device manufactured by intelligent beings, and not as a natural occurrence, because the ratios of its dimensions appeared to be exactly 1 : 4 : 9. James Clerk Maxwell found that the two numbers measured from experiments on electricity and magnetism — equivalent to our modern $\epsilon_0 = 8.85 \times 10^{-12}$ and $\mu_0 = 1.26 \times 10^{-6}$ — when multiplied together by the square of the measured speed of light $c = 3.00 \times 10^8$ — yields a number that is remarkably close to 1. Indeed, when we carry out this product using these same numbers to three significant figures we obtain 1.00. Maxwell knew this was no coincidence, and correctly inferred that light is a form of electromagnetic radiation. Atomic spectroscopy revealed surprising patterns in the wavelengths of radiation emitted by gases such as hydrogen and helium. For instance, when the wavelengths of the different light waves emitted by hydrogen were compared, it was found that their ratios could be expressed as simple functions of small natural numbers. A new law of physics — the quantum theory — had to be brought in to explain this phenomenon. In sum, we expect the results of experiments to be real numbers, and when rational numbers seem to pop up we seek an explanation.

1.3 Scalar Fields

Three-dimensional space is analyzable using a set of three real numbers. The real numbers associated with each point are the coordinates of the point. The simplest way of analyzing space is with the Cartesian or rectangular coordinates, though in many real life situations spherical and cylindrical polar coordinates are easier to use. We shall discuss such situations later. But for now we will work with rectangular coordinates.

As we saw in the previous section, a field is an abstract entity that is superimposed on space. As such, the field is subject to the geometry of the space. Now, the electromagnetic field does not alter the geometry of the space in which it is embedded. And that is why the study of the geometry of three-dimensional flat space is important for electromagnetism.

The simplest kind of field is a scalar field. If to every point within a particular region of space we associate a real number, such a set of numbers is called a scalar point function, or a scalar field, which we may write as

$$\phi = f(x, y, z) \tag{1.4}$$

where the coordinates x, y and z are real numbers. A real physical scalar field must be single-valued, so that ϕ cannot have more than one value at any point in space. Moreover, at every point in space ϕ must have a definite value, though exceptions are permitted at certain points called singularities or singular points. Thus, apart from these singularities, a scalar field must be continuous. A scalar field also should not have discrete jumps, i.e., it must also be differentiable everywhere, except possibly at certain boundary points.

To be differentiable at a point having the coordinate x, the field must have a single derivative when approached from the left or from the right of x.

This means $\lim_{\Delta x \to 0} \frac{\Delta \phi}{\Delta x} \equiv \frac{\partial \phi}{\partial x}$ must be a single valued function at the point x. Similar requirements are made for y and z.

In a region where such a field is both continuous and differentiable, we can define the gradient of the field, which is a vector point function or vector field,[1] written as

$$\nabla \phi = \frac{\partial \phi}{\partial x}\hat{\imath} + \frac{\partial \phi}{\partial y}\hat{\jmath} + \frac{\partial \phi}{\partial z}\hat{k}$$

In electromagnetism we study a scalar field of particular importance called the scalar potential or the electric potential, written as ϕ or sometimes V.

Scalar point functions can also be defined in regions where the space is filled with material particles, like a solid or a liquid or a gas. So, the temperature or the pressure at a point can be a scalar field whose value is a function of the coordinates of that point.

Whereas scalar fields of interest to physicists do vary in general from point to point, there may be several points where they have the same value. So, it is possible that the electric potential has the same value at every point on the surface of an imaginary sphere, or on the flat surface of a conductor, etc. Such a surface along which the potential is constant is called an equipotential surface, which can be expressed algebraically by the equation $\phi(x, y, z) = c$. Let us consider two points A and B separated by a small displacement $d\mathbf{r} = dx\hat{\imath} + dy\hat{\jmath} + dz\hat{k}$ on this equipotential surface.

Clearly, the potential is the same at both points A and B. And so the change in potential from A to B is zero: $d\phi = 0$. Since ϕ is a function of the three coordinates (x, y, z), we can write the total differential $d\phi$ in terms of the partial derivatives

$$d\phi = \frac{\partial \phi}{\partial x}dx + \frac{\partial \phi}{\partial y}dy + \frac{\partial \phi}{\partial z}dz = 0 \tag{1.5}$$

So, for any small displacement $d\mathbf{r}$ along the equipotential surface

$$\nabla \phi \cdot d\mathbf{r} = 0 \tag{1.6}$$

Therefore $\nabla \phi$ is perpendicular to any displacement along the equipotential surface. Suppose now that $d\mathbf{r}$ is an infinitesimal displacement in an arbitrary direction. Then $\nabla \phi \cdot d\mathbf{r} = |\nabla \phi||d\mathbf{r}| \cos \theta$, which has maximum value

[1] In some books the gradient of a scalar field is called a one-form, not a vector.

when $\theta = 0$, and becomes zero when $\theta = 90^0$, i.e. when $d\mathbf{r}$ is perpendicular to the equipotential surface. This means that $d\phi$ is maximum along a displacement that is perpendicular to the equipotential surface.

Example:

A plane intersects the coordinate axes at $x = 3$, $y = 4$, and $z = 3$ units respectively. Find the components of a unit vector $\hat{u}$ that is perpendicular to the plane and directed into the first octant.

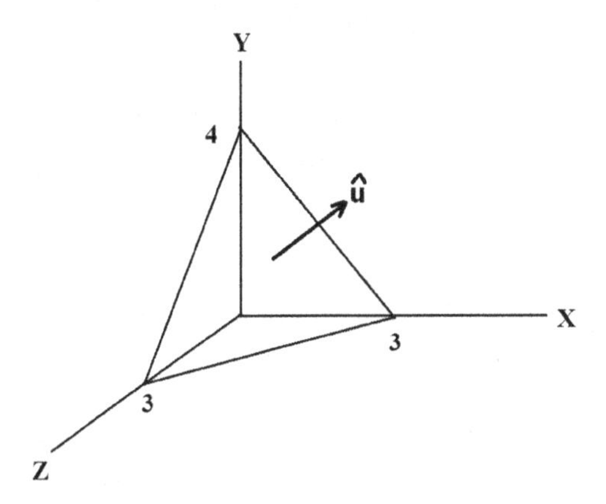

If we think of the plane in this example as an equipotential surface, then the normal would be perpendicular to the equipotential surface and thus would be parallel to the gradient of the potential.

First, we need to obtain the potential as a function of the coordinates.

Now, the equation of a plane intersecting the coordinate axes can be written easily as

$$\frac{x}{3} + \frac{y}{4} + \frac{z}{3} = 1 \tag{1.7}$$

Or $4x + 3y + 4z = 12$. So, the plane can be written as an equipotential surface $\phi = 12$ where $\phi = 4x + 3y + 4z$. We know that the value of ϕ is constant all along the plane and increases (or decreases) most rapidly perpendicular to the plane. From the diagram it is evident that the line of increase points in the first octant. So, we need to find a unit vector that is parallel to $\nabla\phi$. This is simply

$$\hat{u} = \frac{\nabla\phi}{|\nabla\phi|} = \frac{4\hat{i} + 3\hat{j} + 4\hat{k}}{\sqrt{4^2 + 3^2 + 4^2}} = \frac{4}{\sqrt{41}}\hat{i} + \frac{3}{\sqrt{41}}\hat{j} + \frac{4}{\sqrt{41}}\hat{k} \tag{1.8}$$

Exercise:
The potential due to a point charge q placed at the origin of a coordinate system has the form $\phi = \frac{a}{\sqrt{x^2+y^2+z^2}}$. Find the components of the unit vector perpendicular to the equipotential surface at the point (−3, 2, 4) and pointing away from the origin.

1.4 Delta Function

The volume charge density ρ appears frequently in electromagnetic equations. Two-dimensional surface charge densities are generally represented by the symbol σ and one-dimensional linear density by the symbol λ. In the study of electromagnetism one encounters functions that are not continuous or differentiable. For example, an electron does not have a volume, and so its charge is not considered to be distributed over a volume, like a charged spherical conductor. The charge density of a point charge requires a unique mathematical function to describe it. Such a function is called the Dirac delta function $\delta(x)$ (in one-dimensional space) which satisfies these two conditions, which may be taken as the definition of the function:

1. $\int_a^b \delta(x)dx$ has the value 1 if $a < 0 < b$, and has the value 0 otherwise.

2. $\int_a^b f(x)\delta(x)dx$ has the value $f(0)$ if $a < 0 < b$, and has the value 0 otherwise.

The definition of the delta function indicates that the delta function has a dimension that is the inverse of length. Now, linear charge density is defined as charge per unit length. So, a point charge q located at the point $x = a$ is described by the charge density $\lambda(x) = q\delta(x - a)$.

A charge located at the point (a, b, c) is described by the three-dimensional charge density $\rho = q\delta(x - a)\delta(y - b)\delta(z - c)$.

There are other functions that occur commonly in physics and in the study of the electromagnetic field, which are related to the delta function.

Consider a thin conducting plate whose thickness is very small in comparison with its other dimensions. Suppose this plate has a very large surface area A, and is given a positive charge Q which is equally distributed throughout the plate. So the plate has charge per unit area given by $\sigma = \frac{Q}{A}$.

Let us name a line perpendicular to the plate as the x axis, and let us situate the plate itself at $x = 0$. Since the zero of electric potential can be defined arbitrarily, we will take the potential to be zero on the plate. Then, by the laws of electrostatics which we will study presently, the potential on the x axis on either side of this plate would be given by $V(x) = -E|x|$ where E is a positive number, and it can be readily seen that $V(0) = 0$.

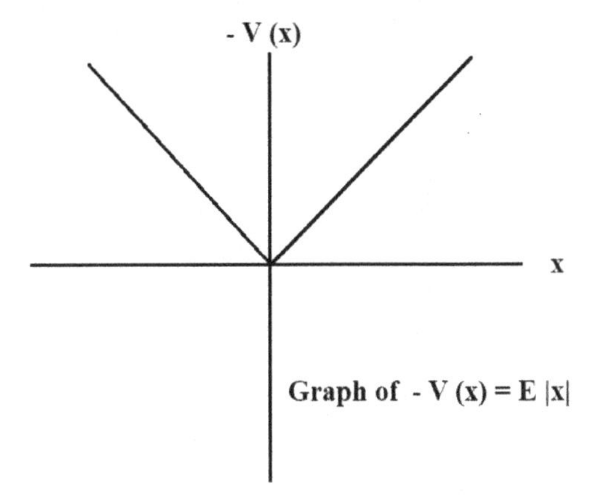

The function $-V(x) = E|x|$ is continuous for all values of x, but it is not differentiable at $x = 0$.

$$\frac{d(|x|)}{dx} = +1 \text{ for } x > 0 \text{ and } -1 \text{ for } x < 0$$

But we can make $|x|$ differentiable at $x = 0$ by redefining it as a limit

$$|x| = \lim_{\epsilon \to 0^+} \sqrt{x^2 + \epsilon} \tag{1.9}$$

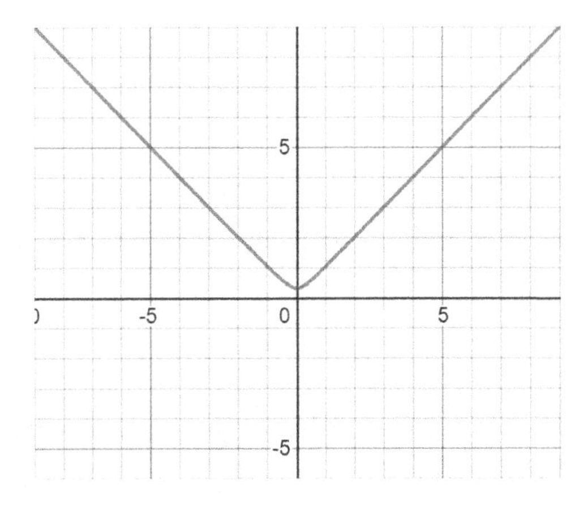

The curve shown above is a graph of $y = \sqrt{x^2 + 0.1}$, which is a rectangular hyperbola, and has a continuous derivative for all real values of x. Using this definition for the absolute value function, we obtain its derivative at $x = 0$ as:

$$\frac{d|x|}{dx} = \lim_{\epsilon \to 0^+} \frac{x}{\sqrt{x^2 + \epsilon}} = 0 \text{ at } x = 0 \tag{1.10}$$

So far our discussion is mathematical. But a physical plate has a finite thickness, however narrow it may be. And at every point on the conducting plate the potential is the same. So the derivative of the potential at $x = 0$ is actually equal to 0. So our definition of the absolute value in Eq. (1.9) is physically useful.

Now, if $V(x)$ is the electric potential, the derivative of $-V(x)$ with respect to x is an important quantity, and as we shall see later, it equals the component of the electric field along the x direction:

$$E_x = -\frac{dV(x)}{dx} = E\frac{d|x|}{dx}$$

which is $-E$ for $x < 0$, 0 at $x = 0$ and E for $x > 0$. These values are consistent with the physical fact that there is no electric field inside a conductor.

$$E_x = \begin{cases} -E, & \text{if } x < 0. \\ 0, & \text{if } x = 0. \\ +E, & \text{if } x > 0. \end{cases} \tag{1.11}$$

E_x can be written in closed form as a function in terms of the Heaviside step function $\theta(x)$,which is defined as

$$\theta(x) = \begin{cases} 0, & \text{if } x < 0. \\ \frac{1}{2}, & \text{if } x = 0. \\ +1, & \text{if } x > 0. \end{cases} \tag{1.12}$$

It is evident that $E_x = E[2\theta(x) - 1]$.

The step function is clearly discontinuous, but it can be defined as the limiting case of a continuous function:

$$\theta(x) = \lim_{\epsilon \to 0} \left[\frac{x}{2\sqrt{x^2 + \epsilon^2}} + \frac{1}{2} \right] \quad . \tag{1.13}$$

Spatial derivatives of the electric field are also extremely important in the study of the electromagnetic field.

The field E_x that we are examining varies only in the x direction.

$$\frac{dE_x}{dx} = 2E\frac{d\theta(x)}{dx} \tag{1.14}$$

It is not hard to show that $\frac{d\theta(x)}{dx} = \delta(x)$.

> **Exercise:**
> Show that $\frac{d\theta(x)}{dx} = \delta(x)$.

In the case of the uniformly charged plate, we found that $E_x = E[2\theta(x) - 1]$. Hence

$$\frac{dE_x}{dx} = 2E\delta(x) \tag{1.15}$$

We have thus shown the relationship between three irregular functions: $|x|, \theta(x)$, and $\delta(x)$. It is also instructive to consider the dimensions of each of these functions. $|x|$ has the dimension of length L, $\theta(x)$ has no dimension or L^0, and $\delta(x)$ has the dimension of L^{-1}.

We shall see in a later chapter that for a large uniformly charged plate carrying surface charge density σ the electric field on either side of the plate has the magnitude $E = \frac{\sigma}{2\epsilon_0}$ where ϵ_0 is a constant number whose significance we will discuss later. Since $\sigma = \frac{Q}{A}$, the quantity $\frac{Q}{A}\delta(x)$ is equivalent to charge Q divided by volume, which is volume charge density, a quantity that is usually expressed as ρ. So we can write

$$\frac{dE_x}{dx} = \frac{\rho}{\epsilon_0} \tag{1.16}$$

where ϵ_0 is a constant number.

For the sake of completeness we will also obtain the second x derivative of the electric field

$$\frac{d^2E_x}{dx^2} = 2E\frac{d(\delta(x))}{dx}$$

> **Exercise:**
> Show that $\frac{d(\delta(x))}{dx} = -\frac{\delta(x)}{x}$.

We have thus seen three "abnormal" scalar functions which do have mathematical and physical meaning.

1.5 Curvilinear Coordinates

1.5.1 *Spherical Coordinates*

Problems having spherical symmetry — such as the field due to a point charge or a spherical conductor — are more easily solved using spherical polar coordinates. In the graph shown below a point (x, y, z) has the spherical polar coordinates (r, θ, φ).

The transformation equations between these sets of coordinates are as follows:

$$r = \sqrt{x^2 + y^2 + z^2}; \quad \theta = \cos^{-1} \frac{z}{\sqrt{x^2+y^2+z^2}}; \quad \varphi = \tan^{-1} \frac{y}{x}$$

$$x = r \sin\theta \cos\varphi; \quad y = r \sin\theta \sin\varphi; \quad z = r \cos\theta$$

When we work in Cartesian coordinates we express a volume element as $dxdydz$ which is the volume of a rectangular block of sides dx, dy and dz. In spherical polar coordinates it is convenient to define the volume element as the volume $r^2 \sin\theta d\theta d\varphi dr$ of the rectangular block of sides $dr, rd\theta$ and $r \sin\theta d\varphi$ shown in the figure below:

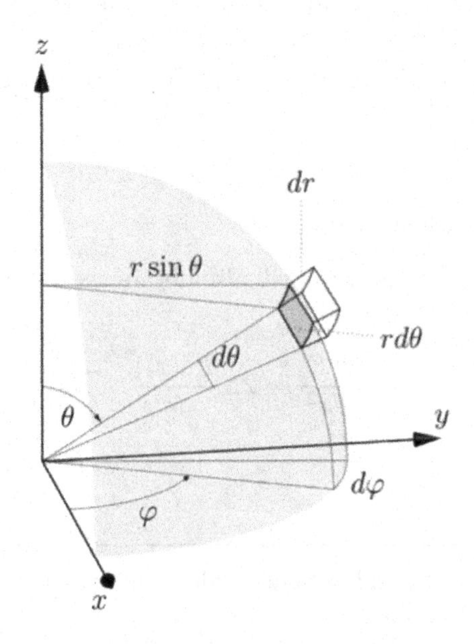

An infinitesimal displacement vector is expressed in Cartesian coordinates as

$$d\mathbf{s} = \hat{\imath}dx + \hat{\jmath}dy + \hat{k}dz \tag{1.17}$$

The same displacement is expressed in spherical polar coordinates as

$$d\mathbf{s} = \hat{e}_r dr + \hat{e}_\theta r d\theta + \hat{e}_\varphi r \sin\theta d\varphi \tag{1.18}$$

where the unit vectors are defined as $\hat{e}_r = \frac{\partial \mathbf{s}}{\partial r}, \hat{e}_\theta = \frac{1}{r}\frac{\partial \mathbf{s}}{\partial \theta}$, and $\hat{e}_\varphi = \frac{1}{r\sin\theta}\frac{\partial \mathbf{s}}{\partial \varphi}$.

Unlike the Cartesian unit vectors, these polar unit vectors are not constants, but vary in direction at different points. $\hat{e}_r$ is everywhere radial, directed away from the origin. $\hat{e}_\theta$ is tangential to a longitude passing through a point, and $\hat{e}_\phi$ is tangential to a latitude passing through a point. It can be seen from the graph that these unit vectors are orthonormal: $\hat{e}_i \cdot \hat{e}_j = \delta_{ij}$ where the subscripts i, j stand for r, θ, φ. Now, the difference *df* in the value of a scalar function f between two closely spaced points is a scalar and hence independent of the coordinate system. In its general form, it can be written as

$$df = \nabla f \cdot d\mathbf{s} \tag{1.19}$$

Since $df = \frac{\partial f}{\partial r}dr + \frac{\partial f}{\partial \theta}d\theta + \frac{\partial f}{\partial \varphi}d\varphi$, it follows that the gradient of a scalar field can be expressed as

$$\nabla f = \hat{e}_r \frac{\partial f}{\partial r} + \hat{e}_\theta \frac{1}{r}\frac{\partial f}{\partial \theta} + \hat{e}_\varphi \frac{1}{r\sin\theta}\frac{\partial f}{\partial \varphi} \tag{1.20}$$

Thus the del operator can be written in spherical polar coordinates as

$$\nabla = \hat{e}_r \frac{\partial}{\partial r} + \hat{e}_\theta \frac{1}{r}\frac{\partial}{\partial \theta} + \hat{e}_\varphi \frac{1}{r\sin\theta}\frac{\partial}{\partial \varphi} \tag{1.21}$$

The divergence of a vector field $\mathbf{A}$ can be obtained by carrying out the operation

$$\mathrm{div}\mathbf{A} = \nabla \cdot (\hat{e}_r A_r + \hat{e}_\theta A_\theta + \hat{e}_\varphi A_\varphi) \tag{1.22}$$

The unit vectors are not independent of the coordinates. When we carry out the dot product and evaluate the derivatives we obtain

$$\nabla \cdot \mathbf{A} = \frac{1}{r^2}\frac{\partial}{\partial r}\left(r^2 A_r\right) + \frac{1}{r\sin\theta}\frac{\partial}{\partial \theta}\left(\sin\theta A_\theta\right) + \frac{1}{r\sin\theta}\frac{\partial A_\varphi}{\partial \varphi} \tag{1.23}$$

We can also write expressions for the curl and the Laplacian operations.

$$\nabla \times \mathbf{A} = \hat{e}_r \frac{1}{r\sin\theta}\left[\frac{\partial}{\partial\theta}(\sin\theta A_\varphi) - \frac{\partial A_\theta}{\partial\varphi}\right] + \hat{e}_\theta\left[\frac{1}{r\sin\theta}\frac{\partial A_r}{\partial\varphi} - \frac{1}{r}\frac{\partial}{\partial r}(rA_\varphi)\right] + \hat{e}_\varphi \frac{1}{r}\left[\frac{\partial}{\partial r}(rA_\theta) - \frac{\partial A_r}{\partial\theta}\right] \tag{1.24}$$

$$\nabla^2\phi = \left(\underbrace{\frac{\partial^2\phi}{\partial r^2} + \frac{2}{r}\frac{\partial\phi}{\partial r}}_{\frac{1}{r^2}\frac{\partial}{\partial r}\left(r^2\frac{\partial\phi}{\partial r}\right)} + \frac{1}{r^2\sin\theta}\frac{\partial}{\partial\theta}\left(\sin\theta\frac{\partial\phi}{\partial\theta}\right) + \frac{1}{r^2\sin^2\theta}\frac{\partial^2\phi}{\partial\varphi^2}\right) \tag{1.25}$$

The volume element $d\tau = r^2 dr \sin\theta d\theta d\varphi$.

1.5.2 *Cylindrical Coordinates*

Many problems in electromagnetism have cylindrical symmetry, such as the field generated by a long wire carrying current. Cylindrical coordinates are helpful in such cases. These coordinates are generated by the transformation equations:

$\rho = \sqrt{x^2 + y^2};\ \ \varphi = \tan^{-1}\frac{y}{x}; z = z$

and their inverses

$x = \rho\cos\varphi;\ \ y = \rho\sin\varphi;\ \ z = z$

Volume element $= \rho d\rho d\varphi dz$

$$\nabla f = \hat{e}_\rho\frac{\partial f}{\partial\rho} + \hat{e}_\varphi\frac{1}{\rho}\frac{\partial f}{\partial\varphi} + \hat{e}_z\frac{\partial f}{\partial z} \tag{1.26}$$

$$\nabla\cdot\mathbf{A} = \frac{1}{\rho}\frac{\partial}{\partial\rho}(\rho A_\rho) + \frac{1}{\rho}\frac{\partial A_\varphi}{\partial\varphi} + \frac{\partial A_z}{\partial z} \tag{1.27}$$

$$\nabla\times\mathbf{A} = \hat{e}_\rho\left(\frac{1}{\rho}\frac{\partial A_z}{\partial\varphi} - \frac{\partial A_\varphi}{\partial z}\right) + \hat{e}_\varphi\left(\frac{\partial A_\rho}{\partial z} - \frac{\partial A_z}{\partial\rho}\right) + \hat{e}_z\frac{1}{\rho}\left(\frac{\partial}{\partial\rho}(\rho A_\varphi) - \frac{\partial A_\rho}{\partial\varphi}\right) \tag{1.28}$$

$$\nabla^2 f = \left(\underbrace{\frac{\partial^2 f}{\partial\rho^2} + \frac{1}{\rho}\frac{\partial f}{\partial\rho}}_{\frac{1}{\rho}\frac{\partial f}{\partial\rho}\left(\rho\frac{\partial f}{\partial\rho}\right)} + \frac{1}{\rho^2}\frac{\partial^2 f}{\partial\varphi^2} + \frac{\partial^2 f}{\partial z^2}\right) \tag{1.29}$$

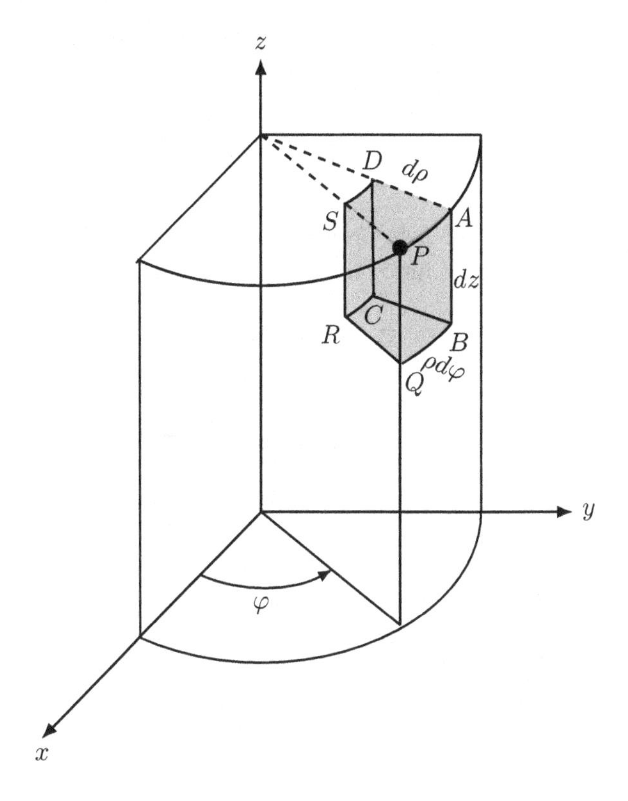

1.6 Vector Integral Calculus

1.6.1 *Line Integrals*

Suppose a particle experiences a force which is expressible as a vector field $\mathbf{F}$. A simple example is a gravitational field. The work done by such a force field upon the particle which undergoes an infinitesimal displacement $d\mathbf{s}$ is given by $dW = \mathbf{F} \cdot d\mathbf{s}$.

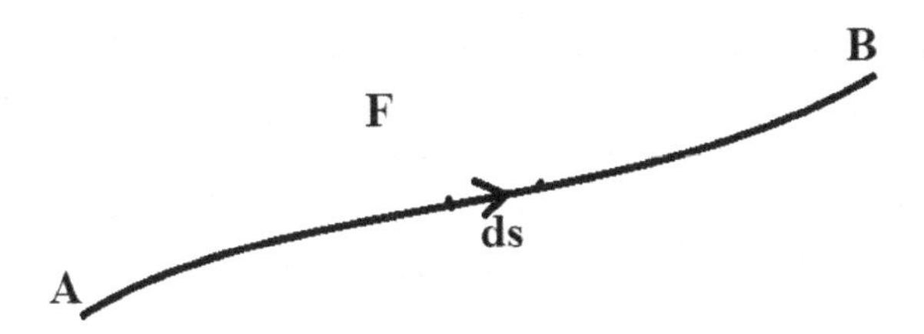

The total work done by the force field **F** on the particle which is displaced from A to B along the path shown above is

$$W = \int_A^B \mathbf{F} \cdot d\mathbf{s} \tag{1.30}$$

The work done by a force field on a particle which travels along a closed loop is given by the closed integral

$$W = \oint \mathbf{F} \cdot d\mathbf{s} \tag{1.31}$$

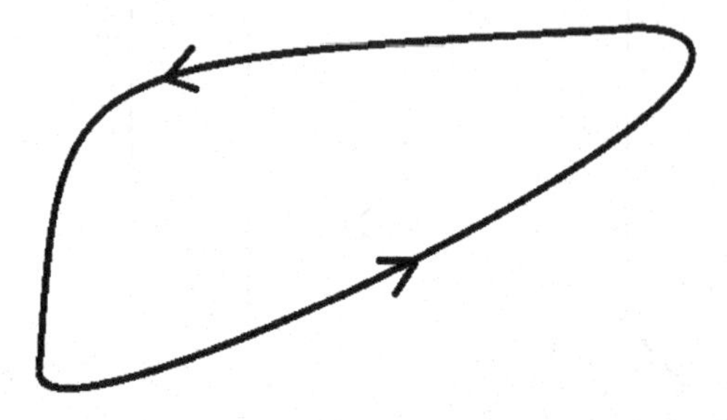

It is conventional to take a counterclockwise line integral as positive.

Suppose the particle in question is a vehicle that travels along the closed loop shown above, and let the force field be the friction between the car and the road. Let us assume that the surface is uniform throughout the path, and therefore the magnitude of the force of friction is constant. Since the force of friction is always directed opposite to the displacement, **F** and **ds** are opposite to each other. Therefore $\mathbf{F} \cdot d\mathbf{s} = -Fds$ at every point on the closed path. Now, F is constant, and therefore the total work done by friction on the car during its journey from start to finish is $-FS$ where S is the total length of the curved path. Negative work means mechanical energy is being converted to heat energy.

Next, we consider a cart traveling along a roller coaster. There are three forces acting on the cart at any time: the force of gravity acting downward, the normal force perpendicular to the surface of the roller coaster, and the force of friction opposite to the direction of motion. Since the normal force is always perpendicular to the displacement, the work done by the normal force is zero.

Next, we consider the gravitational force $\mathbf{F}$ acting on a planet orbiting the sun. The force of gravity acting on the planet (mass m) is everywhere directed towards the sun (mass M), and is inversely proportional to the square of the distance r of the planet from the sun, with magnitude

$$F = \frac{GmM}{r^2} \tag{1.32}$$

Planetary orbits are ellipses with the sun at one focus. Suppose the planet undergoes a small displacement along its path during which its distance from the sun increases by dr.

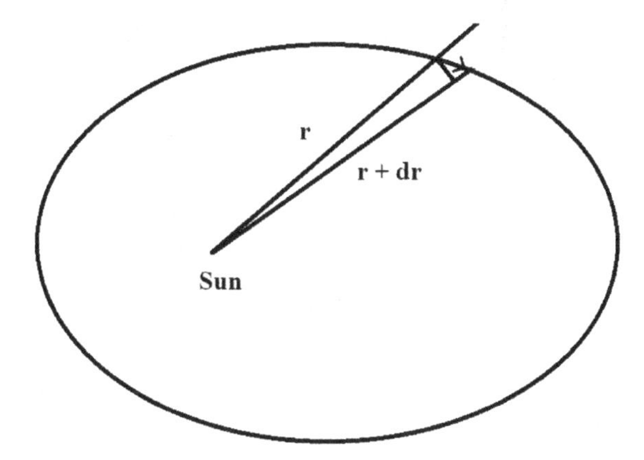

The work done by gravity on the planet during this small displacement is

$$dW = \mathbf{F} \cdot d\mathbf{s} \tag{1.33}$$

A simple geometrical calculation based on the angles and sides of the small triangle in the figure above yields the equation

$$dW = -Fdr = -\frac{GmM}{r^2}dr \tag{1.34}$$

It is evident that the work done depends only on the radial distance from the planet to the sun. The total work done by gravity on the planet as it moves from a point at a distance r_1 to a point at a distance r_2 is therefore

$$W = GmM\left(\frac{1}{r_2} - \frac{1}{r_1}\right) \tag{1.35}$$

The work done is independent of the path, which is the property of a conservative force field such as gravity, as distinct from a non-conservative force such as friction.

1.6.2 *Stokes' Theorem*

Consider a two-dimensional rectangle centered at the point (x, y). Let the dimensions of the rectangle be Δx and Δy, so that the vertices of this rectangle are at $A(x - \Delta x/2, y - \Delta y/2), B(x + \Delta x/2, y - \Delta y/2), C(x + \Delta x/2, y + \Delta y/2), D(x, y + \Delta y/2)$.

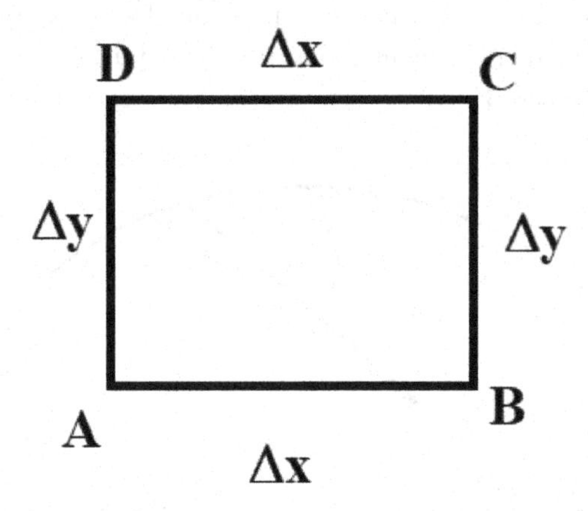

Let us take the line integral of a vector field **F** along the perimeter of this rectangle in the positive sense, i.e. counterclockwise.

Let us call the average value of F_x along AB as F_{x1}, and along CD as F_{x2}. And let us take the average value of F_y along AD as F_{y1} and along BC as F_{y2}. Let F_x, F_y denote the components at the point (x, y) at the center of the rectangle.

Since Δx and Δy are very small, we may write

$$F_{x1} = F_x - \frac{\Delta y}{2} \frac{\partial F_x}{\partial y} \tag{1.36}$$

and

$$F_{x2} = F_x + \frac{\Delta y}{2} \frac{\partial F_x}{\partial y} \tag{1.37}$$

We obtain similar expressions for F_{y1} and F_{y_2}.

So, the line integral amounts to a sum of four terms:

$$\oint \mathbf{F} \cdot d\mathbf{r} = F_{x1}\Delta x + F_{y2}\Delta y - F_{x2}\Delta x - F_{y1}\Delta y \tag{1.38}$$

Substituting and adding the four terms we obtain

$$\oint \mathbf{F} \cdot d\mathbf{r} = \Delta x \Delta y \left(\frac{\partial F_y}{\partial x} - \frac{\partial F_x}{\partial y} \right) \tag{1.39}$$

We define an area element vector $d\mathbf{S} \equiv \hat{n} dS$ as a vector of magnitude dS perpendicular to an infinitesimal area of magnitude dS. There is an ambiguity in the direction of the vector, which therefore is defined arbitrarily as the direction a right handed screw would advance if it were rotated in the direction taken by a line integral along the perimeter of the area element. So the quantity $\Delta x \Delta y$ can be taken as the magnitude of an area vector $\Delta\mathbf{S} \equiv \hat{k} \Delta S$ in the z direction. The term inside the brackets is simply the z component of $\nabla \times \mathbf{F}$. Hence

$$\oint \mathbf{F} \cdot d\mathbf{r} = \Delta\mathbf{S} \cdot \nabla \times \mathbf{F} \tag{1.40}$$

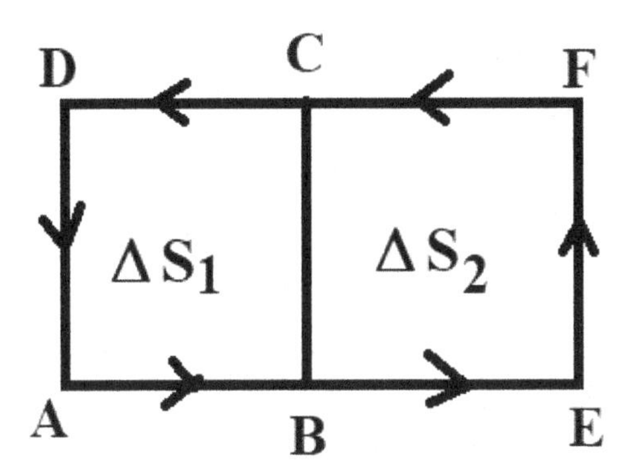

From the figure above, it is evident that we can add line integrals around each of the rectangles ABCDA and BEFCB to yield a line integral over the larger rectangle AEFDA. Hence

$$\oint \mathbf{F} \cdot d\mathbf{r} = \Delta\mathbf{S}_1 \cdot \nabla \times \mathbf{F}_1 + \Delta\mathbf{S}_2 \cdot \nabla \times \mathbf{F}_2 \tag{1.41}$$

Now, the quantity $\Delta\mathbf{S} \cdot \nabla \times \mathbf{F}$ is a scalar, and therefore independent of the coordinate system. So, we could generalize the figure above so that the two adjacent rectangles continue to share a common side but are in different planes, at a small angle to each other. And Eq. (1.41) would remain valid even though $\Delta\mathbf{S}_1$ is not in the same direction as $\Delta\mathbf{S}_2$.

If now we have a surface that is no longer a plane but is curved like a potato chip, we could analyze the curved surface into a large number of tiny rectangles, and perform the line integral around the perimeter of this curved surface to yield

$$\oint \mathbf{F} \cdot d\mathbf{r} = \Sigma_i \Delta \mathbf{S}_i \cdot \nabla \times \mathbf{F}_i \tag{1.42}$$

where the index i covers the very large number of area elements.

We now perform a Riemann sum, i.e., make each area element arbitrarily small while increasing the number of these elements, so that we obtain Stokes' theorem:

$$\oint \mathbf{F} \cdot d\mathbf{r} = \iint \nabla \times \mathbf{F} \cdot d\mathbf{S} \tag{1.43}$$

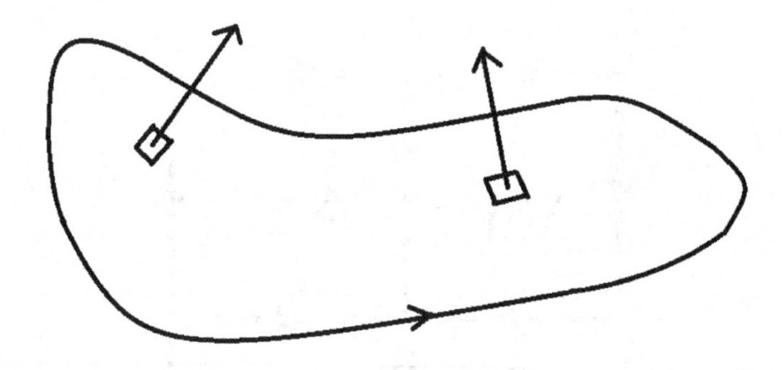

It could be objected that we have sort of "cut corners" in our derivation of Stokes' theorem. We divided up the curved surface into a large number of tiny rectangles. But the boundary is a curved loop, not a set of straight lines, so how do we justify our procedure? Let us consider a small surface $d\mathbf{S}$ in the xy plane. Then we can write $\mathbf{F} \cdot d\mathbf{r} = \mathbf{F} \cdot \hat{\imath} dx + \mathbf{F} \cdot \hat{\jmath} dy$. For an infinitesimal right triangle over which the value of $\mathbf{F}$ remains sensibly constant throughout, the line integral along the hypotenuse is equal to the sum of the line integrals along the other two sides. Thus the line integral along the boundary becomes equal to the line integral along the perpendicular lines in the limit as the rectangles become infinitesimally small. And the area of the surface enclosed by the curved boundary becomes equal to the sum of the areas of the rectangles, as shown in the following figure.

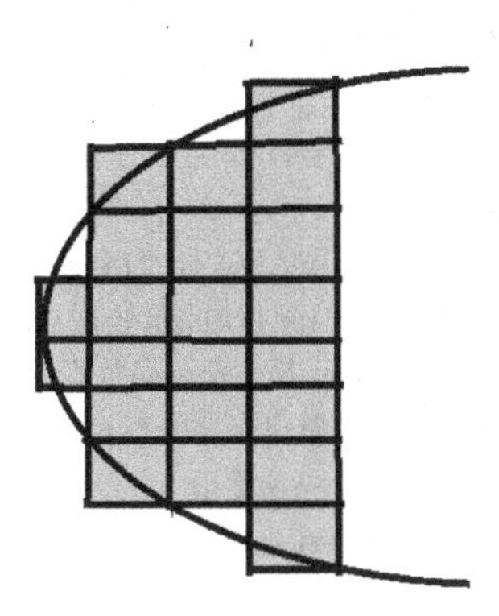

The work done by gravity on a planet that moves from a distance r_1 to a distance r_2 from the sun was obtained earlier (Eq. (1.35))to be

$$GmM\left(\frac{1}{r_2}-\frac{1}{r_1}\right)$$

It is evident that the total work done on the planet as it makes a complete orbit must be zero. This is consistent with Stokes' theorem.

If we place the sun at the origin of a coordinate system, and the position vector of the planet is $\mathbf{r}$, the force applied on the planet due to the sun's gravity becomes

$$\mathbf{F}=-\frac{GmM}{r^3}\mathbf{r} \tag{1.44}$$

The total work done on the planet by gravity in one complete orbit is $\oint \mathbf{F}\cdot d\mathbf{r}$. By Stokes' theorem this should equal

$$\iint \nabla\times\mathbf{F}\cdot d\mathbf{S}$$

Now, the curl of any radial vector function, i.e. of the form $f(r)\mathbf{r}$, is zero, as can be easily verified. Hence $\nabla\times\mathbf{F}=0$ and so $\oint \mathbf{F}\cdot d\mathbf{r}=0$. So no net work is done by gravity as the planet makes a complete orbit. So there is no net change in kinetic energy, and since there is no friction in the path of the planet through space, such a process can in principle be repeated forever, or at least as long as the mass of the sun does not change appreciably.

Exercise:
Show that $\nabla\times f(r)\mathbf{r}=0$ where $f(r)$ is a regular function of r.

1.6.3 *Divergence Theorem*

The flux of a vector field $\mathbf{A}$ across a small surface vector $d\mathbf{S} \equiv \hat{n} dS$ is defined as $\mathbf{A} \cdot d\mathbf{S}$. If the unit vector $\hat{n}$ is directed along the x axis, i.e. that the area element is perpendicular to the x axis, and $\hat{n} = \hat{\imath}$, the flux becomes $\mathbf{A} \cdot d\mathbf{S} = A_x dydz$. Let us consider a small rectangular block (a parallelepiped having all rectangular faces) of sides $\Delta x, \Delta y, \Delta z$. Let us call the point inside this block at its center as (x, y, z). The two faces perpendicular to the x axis are located at $x - \frac{\Delta x}{2}$ and $x + \frac{\Delta x}{2}$. The outward normal unit vectors $\hat{n}$ at these two faces are $-\hat{\imath}$ and $+\hat{\imath}$ respectively.

The total flux out of this block is the sum of the fluxes out of all six faces. Through a process analogous to the one we employed in deriving Stokes' theorem the total flux out of the faces perpendicular to the x axis can be evaluated as

$$\left(A_x + \frac{1}{2}\frac{\partial A_x}{\partial x}\Delta x \right) \Delta y \Delta z - \left(A_x - \frac{1}{2}\frac{\partial A_x}{\partial x}\Delta x \right) \Delta y \Delta z = \frac{\partial A_x}{\partial x} \Delta x \Delta y \Delta z \tag{1.45}$$

Thus the total flux out of the block is

$$\Delta \Phi = \left(\frac{\partial A_x}{\partial x} + \frac{\partial A_y}{\partial y} + \frac{\partial A_z}{\partial z} \right) \Delta x \Delta y \Delta z = \nabla \cdot \mathbf{A} \Delta \tau \tag{1.46}$$

Given a three-dimensional region of arbitrary shape and size, we can divide it into a very large number of very small blocks. Considering a larger block made of two adjacent blocks sharing a common face, it is evident that the total flux of out of the larger block is the sum of the fluxes out of each of the two smaller boxes. Continuing this way, it follows that the total flux out of a macroscopic region is simply the sum of the fluxes out of each of the microscopic blocks. Hence it follows that the flux of a vector field out of the closed surface of a three-dimensional region is the volume integral of the divergence of the vector over the region:

$$\oint\!\!\!\oint \mathbf{A} \cdot d\mathbf{S} = \iiint \nabla \cdot \mathbf{A} d\tau \tag{1.47}$$

This is called Gauss's Divergence Theorem.

Consider a compressible fluid whose density ρ is a scalar field which is a function of the spatial coordinates as well as time: $\rho(x, y, z, t)$. The velocity $\mathbf{v}$ is a vector field that also varies with time. The current density $\mathbf{J}$ is defined as $\mathbf{J} = \rho\mathbf{v}$. The rate at which the fluid passes through a small area element

$d\mathbf{S} \equiv \hat{n}dS$ is $\mathbf{J} \cdot d\mathbf{S}$. So the rate at which the fluid exits a region enclosed by a surface is the surface integral of $\mathbf{J} \cdot d\mathbf{S}$ over the entire surface. This must equal the rate $-\frac{dM}{dt}$ at which the mass of the liquid inside the region decreases with time. This latter quantity is the volume integral of the rate of decrease of the density of the fluid:

$$-\frac{dM}{dt} = -\iiint \frac{\partial \rho}{\partial t} d\tau \tag{1.48}$$

Therefore

$$-\iiint \frac{\partial \rho}{\partial t} d\tau = \oint\!\!\oint \mathbf{J} \cdot d\mathbf{S} \tag{1.49}$$

Applying the divergence theorem, we obtain

$$-\iiint \frac{\partial \rho}{\partial t} d\tau = \iiint \nabla \cdot \mathbf{J} d\tau \tag{1.50}$$

This equation is valid no matter what be the shape or size of the region over which the volume integral is evaluated. Hence the integrands must be equal, and so we obtain the equation of continuity

$$\nabla \cdot \mathbf{J} + \frac{\partial \rho}{\partial t} = 0 \tag{1.51}$$

1.6.4 *Gradient Theorem*

An important corollary to the divergence theorem is the gradient theorem. For a scalar field Φ defined in some region

$$\iiint \nabla \Phi d\tau = \oint\!\!\oint \Phi d\mathbf{S} \tag{1.52}$$

The proof of the gradient theorem is straightforward. Let Φ be a scalar field defined in some region. Suppose we multiply Φ by a constant vector $\mathbf{p}$ that is independent of the coordinates. By the divergence theorem

$$\oint\!\!\oint \mathbf{p}\Phi \cdot d\mathbf{S} = \iiint \nabla \cdot (\mathbf{p}\Phi) d\tau \tag{1.53}$$

Since $\mathbf{p}$ is a constant, we can take it out of the integral on the left side. One the right side, we note that $\nabla \cdot (\mathbf{p}\Phi) = (\nabla \cdot \mathbf{p})\Phi + \mathbf{p} \cdot \nabla \Phi = 0 + \mathbf{p} \cdot \nabla \Phi$. And so, we obtain the equation

$$\mathbf{p} \cdot \left[\oint\!\!\oint \Phi d\mathbf{S} - \iiint \nabla \Phi d\tau \right] = 0 \tag{1.54}$$

For this dot product to be zero for an arbitrary constant vector $\mathbf{p}$, the second vector must be zero, and so we obtain the gradient theorem:

$$\oiint \Phi d\mathbf{S} = \iiint \nabla\Phi d\tau \tag{1.55}$$

Archimedes' Principle: When a solid is immersed in a fluid, it experiences an upward force of buoyancy equal and opposite to the weight of the fluid displaced by the solid.

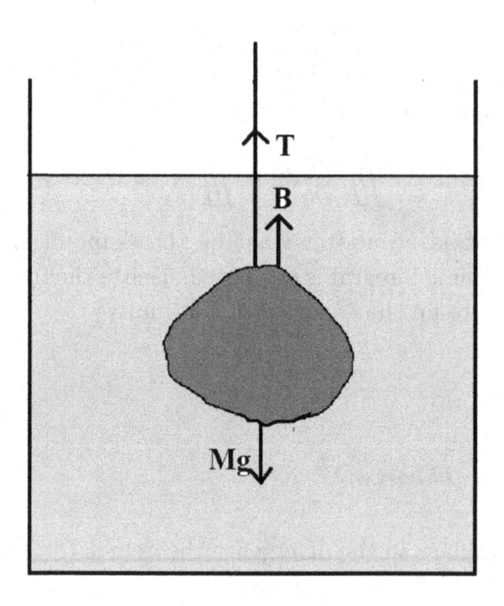

Consider a solid of mass M and density ρ_s fully immersed in a liquid of density ρ_ℓ. The downward gravitational force on the solid is Mg. The upward tension in the string by which the solid is suspended is T. The upward force of buoyancy is B. B is experimentally measurable as $T - Mg$. Archimedes' principle allows us to calculate B in terms of the density of the liquid and the volume of the solid. We will now prove Archimedes' principle by applying the gradient theorem.

The hydrostatic pressure at a depth y inside the liquid is given by $p = y\rho_\ell g$. This hydrostatic pressure will result in varying forces applied normally at every point on the immersed solid. The force applied on a surface of area dS is equal to $-p\hat{n}dS$ where the negative sign indicates that the force is directed onto the solid, whereas the unit vector is directed outwards.

So the net force applied by the liquid on the solid — which is the force of buoyancy — equals

$$\mathbf{B} = -\oiint y\rho_\ell g d\mathbf{S} = -\rho_\ell g \oiint y d\mathbf{S} \tag{1.56}$$

Applying the gradient theorem,

$$\mathbf{B} = -\rho_\ell g \iiint \nabla y d\tau \tag{1.57}$$

The coordinate y is directed downwards, and so $-\nabla y$ is a unit vector directed upwards, which we will call $\hat{u}$. Thus

$$\mathbf{B} = \rho_\ell g \hat{u} \iiint d\tau = \hat{u}\rho_\ell g V = \hat{u} W_\ell \tag{1.58}$$

Thus Archimedes' principle is proved.

1.7 General Orthogonal Coordinates

Cartesian coordinates are characterized by constant unit vectors $\hat{\imath}, \hat{\jmath}, \hat{k}$ which do not change direction from point to point. Coordinate systems that do not have this property — such as the spherical and cylindrical systems — are called curvilinear coordinates. Let us define an arbitrary curvilinear coordinate system (q_1, q_2, q_3). We label the unit vectors in this system as $\hat{e}_1, \hat{e}_2, \hat{e}_3$. An arbitrary vector $\mathbf{A}$ can be expanded in terms of its components as

$$\mathbf{A} = A_1\hat{e}_1 + A_2\hat{e}_2 + A_3\hat{e}_3 \tag{1.59}$$

and a displacement vector $d\mathbf{r}$ can be expanded as

$$d\mathbf{r} = \frac{\partial \mathbf{r}}{\partial q_1} dq_1 + \frac{\partial \mathbf{r}}{\partial q_2} dq_2 + \frac{\partial \mathbf{r}}{\partial q_3} dq_3 \tag{1.60}$$

Hence $\hat{e}_i = \frac{\partial \mathbf{r}}{\partial q_i}$. So the square of an interval can be expressed as

$$\begin{aligned} ds^2 &= d\mathbf{r} \cdot d\mathbf{r} \\ &= \left(\frac{\partial \mathbf{r}}{\partial q_1} dq_1 + \frac{\partial \mathbf{r}}{\partial q_2} dq_2 + \frac{\partial \mathbf{r}}{\partial q_3} dq_3 \right) \cdot \left(\frac{\partial \mathbf{r}}{\partial q_1} dq_1 + \frac{\partial \mathbf{r}}{\partial q_2} dq_2 + \frac{\partial \mathbf{r}}{\partial q_3} dq_3 \right) \end{aligned} \tag{1.61}$$

We will write this equation as

$$ds^2 = g_{ij} dq_i dq_j \tag{1.62}$$

where $g_{ij} = \frac{\partial \mathbf{r}}{\partial q_i} \cdot \frac{\partial \mathbf{r}}{\partial q_j}$ is called the *metric tensor* or simply the *metric* of the coordinate system.

We are particularly interested in orthogonal coordinate systems where the unit vectors are at every point mutually perpendicular to each other. Examples are the spherical and cylindrical systems. For such systems $\hat{e}_i \cdot \hat{e}_j = \delta_{ij}$ and the only non-vanishing elements of the metric are often expressed as $g_{11} = h_1^2, g_{22} = h_2^2, g_{33} = h_3^2 dq_3^2$, so that the displacement element is expressed as

$$ds^2 = h_1^2 dq_1^2 + h_2^2 dq_2^2 + h_3^2 dq_3^2 \tag{1.63}$$

Now, an infinitesimal area element $d\mathbf{S} \equiv \hat{n} dS$ takes the forms $\hat{i}dydz, \hat{j}dxdz, \hat{k}dxdy$ with areas perpendicular to the coordinate axes. It is evident that $\hat{k}dxdy = \hat{i}dx \times \hat{j}dy$, and so on cyclically. If we consider an arbitrary curvilinear coordinate system, if the sides of the elementary rectangle are bounded by the element vectors $d\mathbf{r}_1$, $d\mathbf{r}_2$ and $d\mathbf{r}_3$, where $d\mathbf{r}_i = \frac{\partial \mathbf{r}}{\partial q_i} dq_i$, then the volume of this elementary rectangle is the scalar triple product of the these three vectors:

$$d\tau = d\mathbf{r}_1 \cdot d\mathbf{r}_2 \times d\mathbf{r}_3 \tag{1.64}$$

Now, $d\mathbf{r}_i = \frac{\partial \mathbf{r}}{\partial q_i} dq_i = (\frac{\partial x}{\partial q_i}\hat{i} + \frac{\partial y}{\partial q_i}\hat{j} + \frac{\partial z}{\partial q_i}\hat{k})dq_i$, and so $d\mathbf{r}_2 \times d\mathbf{r}_3 =$

$$\begin{vmatrix} \hat{i} & \hat{j} & \hat{k} \\ \frac{\partial x}{\partial q_2} & \frac{\partial y}{\partial q_2} & \frac{\partial z}{\partial q_2} \\ \frac{\partial x}{\partial q_3} & \frac{\partial y}{\partial q_3} & \frac{\partial z}{\partial q_3} \end{vmatrix} dq_2 dq_3$$

and the volume element becomes, in curvilinear coordinates

$$d\tau = d\mathbf{r}_1 \cdot d\mathbf{r}_2 \times d\mathbf{r}_3 = \begin{vmatrix} \frac{\partial x}{\partial q_1} & \frac{\partial y}{\partial q_1} & \frac{\partial z}{\partial q_1} \\ \frac{\partial x}{\partial q_2} & \frac{\partial y}{\partial q_2} & \frac{\partial z}{\partial q_2} \\ \frac{\partial x}{\partial q_3} & \frac{\partial y}{\partial q_3} & \frac{\partial z}{\partial q_3} \end{vmatrix} dq_1 dq_2 dq_3 \tag{1.65}$$

The determinant in the equation is called the *Jacobian determinant*, or simply the *Jacobian*. Now, $\frac{\partial \mathbf{r}}{\partial q_i} = h_i \hat{e}_i$, and therefore the Jacobian determinant on the right side is simply $h_1 h_2 h_3 (\hat{e}_1 \cdot \hat{e}_2 \times \hat{e}_3) = h_1 h_2 h_3$ for orthogonal coordinates for which $\hat{e}_1 \cdot \hat{e}_2 \times \hat{e}_3 = 1$. Hence

$$d\tau = dxdydz = h_1 h_2 h_3 dq_1 dq_2 dq_3 \tag{1.66}$$

For spherical coordinates, $h_1 = 1, h_2 = r, h_3 = r \sin\theta$, and so we obtain $d\tau = r^2 \sin\theta dr d\theta d\varphi$.

For cylindrical coordinates, $h_1 = 1, h_2 = \rho, h_3 = 1$, and we get $d\tau = \rho d\rho d\varphi dz$.

1.7.1 *Differential Operations in Orthogonal Curvilinear Coordinates*

Gradient:

$$\nabla\Phi = \sum_i \hat{e}_i \frac{1}{h_i}\frac{\partial\Phi}{\partial q_i} \tag{1.67}$$

Divergence:

$$\nabla\cdot\mathbf{A} = \frac{1}{h_1h_2h_3}\left[\frac{\partial}{\partial q_1}(A_1h_2h_3) + \frac{\partial}{\partial q_2}(A_2h_1h_3) + \frac{\partial}{\partial q_3}(A_3h_1h_2)\right] \tag{1.68}$$

Curl:

$$\nabla\times\mathbf{A} = \frac{1}{h_1h_2h_3}\begin{vmatrix} h_1\hat{e}_1 & h_2\hat{e}_2 & h_3\hat{e}_3 \\ \frac{\partial}{\partial q_1} & \frac{\partial}{\partial q_2} & \frac{\partial}{\partial q_3} \\ h_1A_1 & h_2A_2 & h_3A_3 \end{vmatrix} \tag{1.69}$$

Laplacian:

$$\nabla^2\Phi = \frac{1}{h_1h_2h_3}\left[\frac{\partial}{\partial q_1}\left(\frac{h_2h_3}{h_1}\frac{\partial\Phi}{\partial q_1}\right) + \frac{\partial}{\partial q_2}\left(\frac{h_1h_3}{h_2}\frac{\partial\Phi}{\partial q_2}\right) + \frac{\partial}{\partial q_3}\left(\frac{h_2h_1}{h_3}\frac{\partial\Phi}{\partial q_3}\right)\right] \tag{1.70}$$

1.8 Imaginary Numbers

1.8.1 *The Argand Plane*

The electromagnetic field exists in a real three-dimensional space where each point is represented by a set of three real numbers. But real numbers are not the only numbers that can be generated by elementary algebra. So, the quadratic equation $x^2 + 1 = 0$ does not have any real solution. We therefore include a set of numbers called imaginary numbers. Since a quadratic equation has two solutions, we define the solutions of $x^2 + 1 = 0$ as the numbers i and $-i$, such that $i + (-i) = 0$ and, $i^2 = (-i)^2 = -1$. Imaginary numbers can be plotted along a line. If r is a real number (positive, negative, zero, rational or irrational), then ir is an imaginary number.

The imaginary numbers can be plotted along the imaginary number line which is perpendicular to the real number line. So, if the real number line is the x axis then the y axis is the imaginary number line. The plane so generated by these two axes is called the Argand plane.

Every point on the Argand plane represents a complex number. An arbitrary point in this plane represents an algebraic sum of a real and an imaginary number, called a complex number, usually written as $z = x + iy$, where x and y are both real numbers. x is called the real part of z and y is called the imaginary part of z. (Note that the imaginary part is also a real number.) If r is the distance of the point from the origin, and θ is the angle made by the radius vector of the point with the positive x axis, then $x = r\cos\theta$ and $y = r\sin\theta$. Thus $z = r(\cos\theta + i\sin\theta)$.

Now, $z^2 = r^2(\cos\theta + i\sin\theta)(\cos\theta + i\sin\theta) = r^2(\cos^2\theta - \sin^2\theta + i2\cos\theta\sin\theta) = r^2(\cos 2\theta + i\sin 2\theta)$.

Continuing this way, one can show that for any natural number n

$$(\cos\theta + i\sin\theta)^n = \cos n\theta + \sin n\theta \tag{1.71}$$

This is called De Moivre's theorem.

One can use Taylor's theorem to expand $\sin\theta$ in powers of θ to obtain

$$\sin\theta = \theta - \frac{\theta^3}{3!} + \frac{\theta^5}{5!}\cdots \tag{1.72}$$

Using the relation $\cos\theta = \frac{d\sin\theta}{d\theta}$ we can write the Taylor expansion of $\cos\theta$:

$$\cos\theta = 1 - \frac{\theta^2}{2!} + \frac{\theta^4}{4!}\cdots \tag{1.73}$$

Using the relation $\frac{de^\theta}{d\theta} = e^\theta$ we write the series expansion for e^θ as

$$e^\theta = 1 + \theta + \frac{\theta^2}{2!} + \frac{\theta^3}{3!} + \frac{\theta^4}{4!}\cdots \tag{1.74}$$

It follows that

$$e^{i\theta} = 1 + i\theta - \frac{\theta^2}{2!} - \frac{i\theta^3}{3!} + \frac{\theta^4}{4!}\cdots \tag{1.75}$$

Adding separately the real and the imaginary parts on the right side of this equation we obtain

$$e^{i\theta} = \cos\theta + i\sin\theta \tag{1.76}$$

De Moivre's theorem appears as a corollary to this important relation.

An arbitrary complex number is expressible as $z = re^{i\theta}$.

The absolute value of a complex number is defined as $|z| = \sqrt{x^2 + y^2}$.

$|z| = \sqrt{zz^*} = x^2 + y^2 = r^2$.

An algebraic function of z is in general a complex number.

1.8.2 *Gamma Function*

An important function called the gamma function is defined for a complex number z as

$$\Gamma(z) = \int_0^\infty e^{-t} t^{z-1} dt \tag{1.77}$$

It can be shown that $\Gamma(z+1) = z\Gamma(z)$.

From the definition of a gamma function it can be shown readily that $\Gamma(1) = 1$. It follows that for a natural number n, $\Gamma(n+1) = n!$.

A number that is important for physics is $\Gamma(\frac{1}{2})$. This can be calculated from the integral

$$\Gamma(1/2) = \int_0^\infty e^{-t} t^{-\frac{1}{2}} dt \tag{1.78}$$

We first make the substitution $t = x^2$, and the integral is converted to $2\int_0^\infty e^{-x^2} dx = \int_{-\infty}^\infty e^{-x^2} dx$.

Let us label this integral as I. So

$$I^2 = \int_{-\infty}^\infty e^{-x^2} dx \int_{-\infty}^\infty e^{-y^2} dy = \iint_{-\infty}^\infty e^{-x^2-y^2} dxdy \tag{1.79}$$

The term on the right is a surface integral over the entire xy plane. We can evaluate it by converting it to two-dimensional polar coordinates (r, φ), and replacing the area element $dxdy$ by $rdrd\varphi$ $(h_1 = 1, h_2 = r)$ we get

$$I^2 = \int_0^\infty re^{-r^2} dr \int_0^{2\pi} d\varphi = \pi \tag{1.80}$$

Thus

$$\Gamma\left(\frac{1}{2}\right) = \sqrt{\pi}$$

1.8.3 *Physical Reality of Imaginary Numbers*

A magnitude cannot be represented by an imaginary number. What is real to our senses is something that has magnitude — volume, mass, brightness, distance, etc, which are expressed as positive real numbers. But we are also accustomed to negative real numbers such as temperature, negative charge, etc. It is hard to think of imaginary numbers as having physical relevance. But we shall see in Chapter 7 that time can be thought of as imaginary space. Every point in space can be represented by a set of three real numbers. But to every point in space there is also a moment of time. The flow of time can be thought of as a motion in a different dimension, or as a motion in imaginary space. And further on, we will learn in Chapter 10 that at every point in real space there is a wave function which in general is a complex function of the coordinates of that point. Hence imaginary numbers are physically just as real as the mathematical real numbers.

Chapter 2

Fields Produced by Stationary Charges

2.1 Coulomb's Law

The earliest scientific study of electromagnetism began with the study of charged objects. When a piece of glass was rubbed with a silk cloth, the two bodies attracted each other. The glass and the cloth had become charged. When a charged glass rod was brought near another glass rod charged in the same manner, the two objects repelled each other. Likewise, two silk cloths charged in the same manner repelled each other.

So, there were two different kinds of charges. Like charges repelled each other, and unlike charges attracted each other.

Moreover, if, after rubbing the glass rod with the silk cloth, the silk cloth was draped around the glass rod, then after a short time both the cloth and the rod lost their charge.

This suggested that the charges could be described quantitatively. The unlike charges were opposite to each other — one was positive and the other was negative. These opposite charges were able to cancel each other, just as when a positive number is added to a negative number of the same magnitude the result is zero.

The charge on the glass rod was arbitrarily labeled as positive, and the charge on the silk cloth was therefore negative.

The force between two charges was found to be proportional to the product of the magnitudes of the charges. The force was also found to be inversely proportional to the square of the distance between the charges.

Force is a vector, and the direction of the force was found to be along the line joining the charges.

These experimental discoveries were put together in mathematical form to derive Coulomb's Law of force between charges. If a charge of magnitude q_1 is placed at a distance r from a charge of magnitude q_2, then it will experience a force $\mathbf{F}_{21}$ given by the following equation

$$\mathbf{F}_{21} = \frac{q_1 q_2}{4\pi\epsilon_0 r^2}\hat{e}_r \tag{2.1}$$

where ϵ_0 is a constant known as the permittivity of free space. If the two charges have the same sign, the unit vector $\hat{e}_r$ is directed away from the other charge q_2, and if the charges have opposite sign, then $\hat{e}_r$ is directed towards q_2. $\mathbf{F}_{21}$ is the force experienced by charge 1 due to charge 2. So, $\mathbf{F}_{12}$, the force experienced by charge 2 due to charge 1 must — by Newton's Third Law — be equal and opposite to $\mathbf{F}_{21}$:

$$\mathbf{F}_{12} = -\mathbf{F}_{21}$$

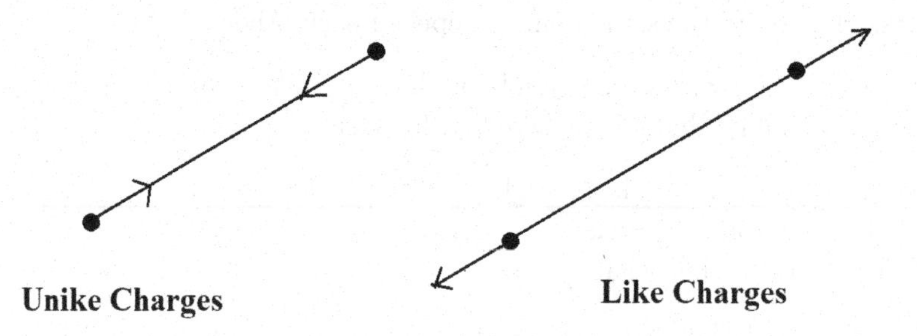

2.1.1 *An Inverse Square Law*

Coulomb's Law is mathematically similar to Newton's Law of Gravitation:

$$F_g = \frac{Gm_1 m_2}{r^2} \tag{2.2}$$

Both laws have forces proportional to the products of magnitudes and inversely proportional to the square of the distance between the objects.

A quantity that is inversely proportional to the square of a distance is said to obey an inverse square law. An example of an inverse square law is the law of sound intensity:

$$I = \frac{P}{4\pi r^2} \tag{2.3}$$

If a source emits sound at power P, the amount of energy passing unit area at a point a distance r from the source is given by the intensity I. Since the sound wave travels outwards equally in all directions, the energy per unit area drops off as the inverse of the area of the sphere centered at the source. The surface area of this sphere is $4\pi r^2$, which explains the inverse square law.

The inverse square law also applies to the light intensity generated by a light source. The equation is mathematically identical to that for sound intensity.

The inverse square law for both sound and light intensity is a consequence of the fact that space is three-dimensional, and also that our three-dimensional space is flat or Euclidean.

A flat space is said to be Euclidean, because it obeys the geometrical principles enunciated by Euclid. If we have a point P and a line L outside of P, it is possible to draw one and only one straight line through P that does not intersect L. If we draw a triangle in this space the sum of the three angles would total two right angles. The circumference of a circle would be $2\pi r$.

2.1.2 *Geometries of Space*

Now, suppose the universe is a two-dimensional surface curved in the third dimension. Consider two such possibilities:

1. The surface is curved like a saddle, called a hyperbolic surface. Such a curved surface obeys a geometry developed by Bolyai and Lobachevsky. We say that this surface has negative curvature. Since the surface is curved, we need to revise our concept of straight lines on this surface. One can generalize the concept of a straight line and define a straight line as the shortest distance between two points.

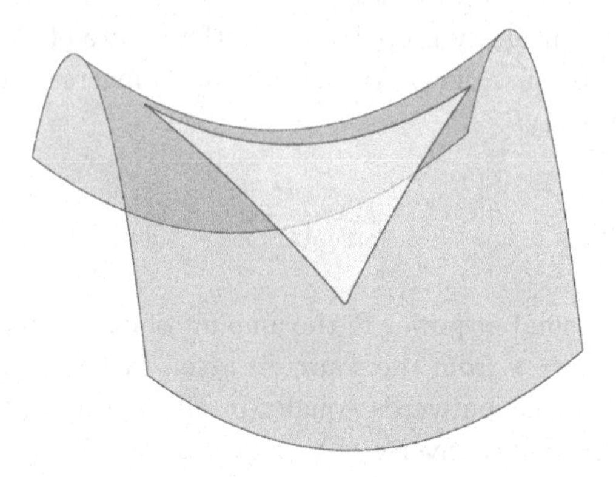

Suppose on this surface we have a point P and a "straight" line L not passing through P. Then it is possible to draw more than one "straight" line through P that does not intersect the line L. Now, if we were to draw a triangle with three vertices on this surface, the sum of the three angles would be less than two right angles. The circumference of a circle of radius r would be greater than $2\pi r$.

2. The surface is everywhere convex, like the surface of a sphere. This surface obeys a geometry developed by Riemann. We say that this surface has positive curvature. If we have a point P and a line L outside of P, it is impossible to draw a "straight" line through P that does not intersect the line L. The sum of the angles of a triangle would be greater than two right angles. The circumference of a circle would be less than $2\pi r$.

The surface of the earth is a good approximation to a surface of positive curvature. For small distances the surface of the earth is approximately flat, and the geometry is Euclidean. But for larger distances the geometry is Riemannian and therefore not Euclidean. Since the word *geometry* is Greek for "measurement of the earth", the notion of a Euclidean geometry is a sort of contradiction in terms.

2.1.3 *Higher Dimensions*

The surface of the earth is an example of a two-dimensional surface that is curved in the third dimension. Because gravity is a significant but not an overwhelming force on the surface of the earth, plant and animal life have been able to evolve, and birds are able to fly above the ground. So, humans became aware that they were living in a three-dimensional universe long before they discovered that the earth is round.

Intelligent two-dimensional entities could determine whether their world was flat, hyperbolic or Riemannian. They could draw triangles and measure the sum of the angles, or they could draw circles and measure the ratio of the circumference to the radius.

We live in a three-dimensional universe that appears to be flat or Euclidean. The inverse square law is an expression of the fact that the surface area of a sphere of radius r is $4\pi r^2$, which is a feature of flat three-dimensional space. However, just as early humans thought the surface of the earth was flat because it looked locally flat — the three angles of a small triangle drawn on the ground add up to two right angles — so, there is the possibility that the three dimensions of our space may be curved in the fourth dimension.

Naturally, the concept of a fourth space dimension is necessarily abstract. At any given point in our space we can draw three mutually perpendicular lines, but we cannot picture a fourth line perpendicular to the other three. The only way to conjure up a fourth dimension is through mathematics.

We will begin with a cube and go from there.

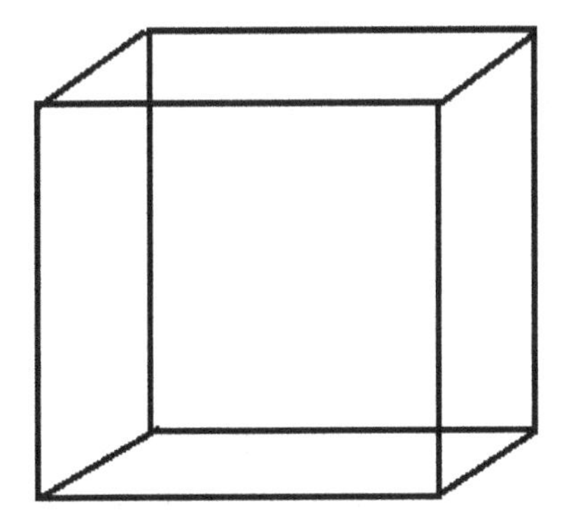

The number of vertices (points) = 8.

The number of squares (faces) = 6

(The figure shown above is a two-dimensional projection of a three-dimensional cube. The four parallelograms are projections of squares.

Adding these to the two actual squares seen in the figure, there is a total of 6 squares.)

The number of edges (lines) = 12.

The number of cubes (solids) = 1.

Next, consider the two-dimensional equivalent of a cube, which is a square.

The number of vertices (points) = 4.

The number of squares (faces) = 1.

The number of edges (lines) = 4.

Next, consider the one-dimensional equivalent of a square, which is a line segment bounded by two points, one at each extremity.

The number of vertices (points) = 2.

The number of edges = 1.

Finally, the zero-dimensional equivalent is a single point.

The number of vertices = 1.

We can find a simple algebraic formula that will generate all these numbers: $(2x + 1)^n$.

Dimension 0: $(2x + 1)^0 = 1$ (One vertex)

Dimension 1: $(2x + 1)^1 = 2x + 1$ (2 vertices + 1 line)

Dimension 2: $(2x + 1)^2 = 4x^2 + 4x + 1$ (4 vertices + 4 lines + 1 square)

Dimension 3: $(2x + 1)^3 = 8x^3 + 12x^2 + 6x + 1$ (8 vertices + 12 lines + 6 squares + 1 cube)

We could then extrapolate this formula to generate a four-dimensional equivalent of a cube, which is called a tesseract:

Dimension 4: $(2x + 1)^4 = 16x^4 + 32x^3 + 24x^2 + 8x + 1$ (16 vertices + 32 lines + 24 squares + 8 cubes + 1 tesseract)

The figure shown below is a two-dimensional projection of a four-dimensional tesseract:

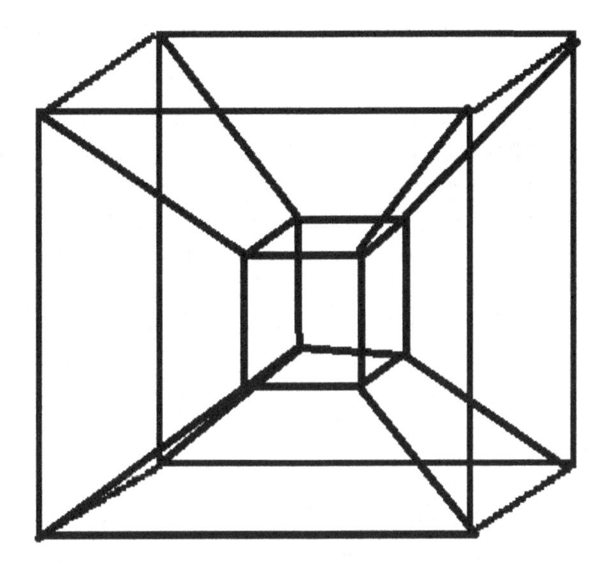

We see an inner cube, an outer cube, and six truncated pyramids that connect the inner to the outer cube. Each truncated pyramid is a projection of a cube. Thus there are 8 cubes in the tesseract.

2.1.4 *Round Objects*

The area of a circle of radius r is $A = \pi r^2$ and the circumference is $C = \frac{dA}{dr} = 2\pi r$.

The volume of a sphere of radius r is $V = \frac{4}{3}\pi r^3$ and its surface area is $S = \frac{dV}{dr} = 4\pi r^2$.

By an extension to n-dimensional geometry, it can be shown that the "volume" of an n-dimensional "hypersphere" is given by

$$V_n(r) = \frac{\pi^{n/2}}{\Gamma(\frac{n}{2}+1)} r^n \tag{2.4}$$

The "surface area" of such an n-dimensional "sphere" is therefore

$$S_n(r) = \frac{dV_n}{dr} = \frac{2\pi^{\frac{n}{2}}}{\Gamma(\frac{n}{2})} r^{n-1} \tag{2.5}$$

From these general formulas we obtain the four-dimensional "volume" of a hypersphere of radius r as $\frac{1}{2}\pi^2 r^4$ and the "area" of its surface is $2\pi^2 r^3$.

Of course, the "area" of a four-dimensional sphere is actually a three-dimensional volume.

Now, what if our three-dimensional universe is actually curved in the fourth dimension, somewhat like the apparently flat surface we live on is actually curved in the third dimension? We know that on the surface of the earth the circumference of a circle is less than $2\pi r$ where r is the radius measured along a geodesic. So, if our three-dimensional universe is curved with positive curvature, we would expect that the surface area of a sphere would be less than $4\pi r^2$ and if the curvature of our universe is negative, we would expect that the surface area of a sphere would be greater than $4\pi r^2$. In either case the inverse square law would not be obeyed.

Einstein's General Theory of Relativity states that three-dimensional space is curved by the presence of mass, and this curvature is positive.[1] One consequence is that straight lines in space are not Euclidean straight lines, but geodesics. And this affects not only the motion of matter, but also the motion of electromagnetic waves. Thus light is deflected by gravitational matter. It has been proved that light waves coming from a distant star undergo a very slight deflection when they pass close to the gravitational field of the sun. However, this deflection is very small, since the gravitational constant G is very small, much smaller than Coulomb's constant $\frac{1}{4\pi\epsilon_0}$.

To conclude, our universe is not exactly Euclidean, but the deviation from flatness is very small, and becomes significant in the presence of extremely powerful gravitational fields, such as close to a massive star or a black hole. But for all practical purposes we will assume that Coulomb's inverse square law is obeyed exactly. So we will assume the electromagnetic field exists in a flat, Euclidean three-dimensional space. And so we will treat Coulomb's Law as a universal law valid for large distances. We will have more to say about Coulomb's law at microscopic distances in a subsequent chapter, when we discuss the consequences of the quantum theory of matter and radiation. And we will examine the shape of the electromagnetic field in a strong gravitational field in a later chapter.

---

[1]We do not say that the universe is a three-dimensional space curved in the fourth dimension, but simply that it is curved. More will be said on this distinction in Chapter 12.

2.2 The Electric Field

Coulomb's Law suggests that when two charges are brought close together, one charge knows the magnitude, the sign and the exact relative position of the other, since the force experienced by one charge depends on all three factors. How did one charge obtain all this information about the other charge?

A similar question was raised a couple of centuries earlier when Newton developed his laws of motion and gravitation. Newton showed that the force between two objects of masses m_1 and m_2 is an attractive force of magnitude

$$F_g = \frac{Gm_1m_2}{r^2}$$

where r is the distance between the centers of gravity of the two masses and G is the Gravitational Constant. Newton's theory correctly explained the acceleration of an apple falling from a tree. But Newton could not explain how the apple knew that there was an earth towards which it should accelerate.

It was believed in ancient times that the effects of forces were transmitted instantaneously across distances. So, Archimedes claimed that he could lift an object as massive as the earth if he had a fulcrum and a sufficiently long lever. This implied that the force he applied at the end of the longer arm of the lever would be felt instantaneously at the opposite end of the lever. But we know that is not the case, since the force would have to travel at a finite speed from one end to the other, and in the process the lever would bend.

A cause that has an instant effect at a distance is called action at a distance. So, it was believed that a gravitational force was an action at a distance. No explanation could be provided for this action at a distance, but it was taken for granted.

However, when it came to the interaction between two electric charges, it was realized very soon that the interaction was not instant, but was communicated through a medium that is superimposed on empty space, a medium called the electric field. A charge creates an electric field which propagates outwards in all directions. A second charge placed some distance from the first picks up the field generated by the first charge. This second charge has no direct knowledge of the charge that generated the field, but experiences a force that is proportional to the strength of the field. Thus the field carries information about a charge in every direction.

How does the information travel through the field from one charge to another? We will suspend that question for the present, and limit our discussion to fields produced by static charges. We will assume that the charges in the systems we are studying have been stationary for a long time, and that the fields have had ample time to travel from charge to charge. So for now we will limit our attention to static fields.

Since the field carries quantitative information such as magnitude and direction, it is appropriate to define the field as a vector quantity that is a function of the coordinates. So, the electric field is a vector point function or a vector field, written as **E**. The force acting on a charge q at a point is given by the product of q and the field **E**:

$\mathbf{F} = q\mathbf{E}$, or more appropriately, $\mathbf{E} = \mathbf{F}/q$.

So the field at a point $P(\mathbf{r})$ generated by a positive charge q placed at the origin is given by

$$\mathbf{E}(\mathbf{r}) = \frac{q}{4\pi r^2}\hat{e}_r \tag{2.6}$$

where $\hat{e}_r$ is the unit vector directed from the origin to P.

We notice that the magnitude of the electric field is inversely proportional to the square of the distance from the charge: $E \propto \frac{1}{r^2}$. This leads to an interesting corollary. If the charge occupies zero volume — what we call a point charge — one can come arbitrarily close to the charge, and so r can become 0. But as $r \to 0$, we see that $E \to \infty$. So, the electric field diverges very close to a point charge. Since an electron is considered as a point charge, this becomes a real problem. In classical physics the problem is circumvented by assuming that the electron has a finite radius, and therefore the charge is spread out over a small but finite volume. The field generated by such a charge does not diverge. A different procedure is employed in quantum physics, which we will discuss in a subsequent chapter.

2.2.1 *Electric Flux*

Consider a region of space containing an electric field. Consider a small plane area dA whose normal $\hat{n}$ is at an angle to the electric field passing through dA. If dA is sufficiently small, we can treat the electric field as uniform across dA. We define the electric flux across dA as $d\Phi = \mathbf{E} \cdot \hat{n} dA$. If the angle between $\mathbf{E}$ and $\hat{n}$ is θ, then $d\Phi = EdA \cos\theta$. Now, $dA \cos\theta$ is equal to the area that is perpendicular to $\mathbf{E}$, which we shall call da. So $d\Phi = Eda$.

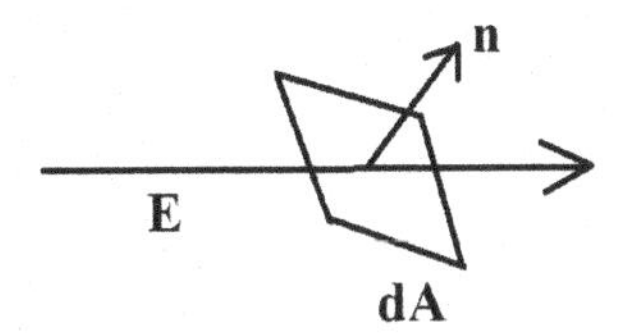

Suppose we have a charge q in a volume bounded by a closed surface. Let us calculate the total flux coming out of the closed surface from the charge inside.

Let us divide up the closed surface into a very large number of very small areas, and call one such area dA. The flux through this area is given by $d\Phi = \mathbf{E} \cdot \hat{n} dA = Eda$ where da is the area of the element that is perpendicular to $\mathbf{E}$.

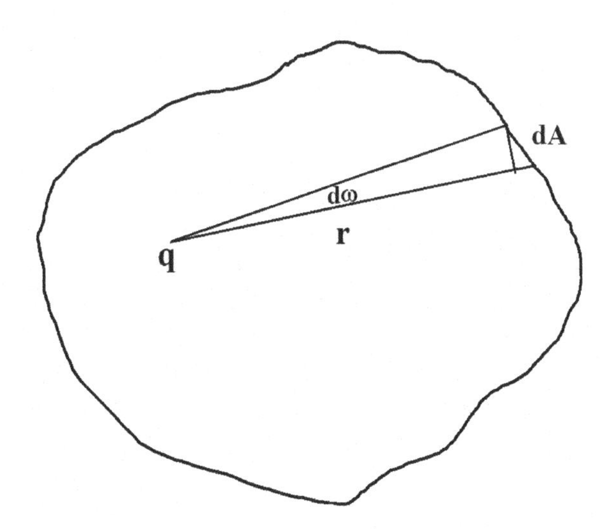

By Coulomb's law, the charge q will generate a field at the surface element dA equal in magnitude to $E = \frac{q}{4\pi\epsilon_0 r^2}$.

So, the flux out of $dA = d\Phi = \frac{qda}{4\pi\epsilon_0 r^2}$.

A solid angle is a three-dimensional angle, written as Ω. The base of a cone subtends a solid angle at the apex. Solid angles are measured in steradians (sr), the three-dimensional equivalent of the two-dimensional radians. So, if we have a sphere of radius r, 1 sr is the angle subtended by an area of r^2 on the surface of the sphere. So an entire sphere subtends a solid angle of 4π sr.

In the figure above, $d\Omega = \frac{da}{r^2}$, and so $da = r^2 d\Omega$.

Thus $d\Phi = \frac{qr^2 d\Omega}{4\pi\epsilon_0 r^2} = \frac{qd\Omega}{4\pi\epsilon_0}$.

So the total flux coming out of the enclosed surface $\Phi = \int \frac{q}{4\pi\epsilon_0} d\Omega = \frac{q}{\epsilon_0}$.

If there is more than one charge inside the closed surface, the total flux out of the surface $= \Sigma_i \frac{q_i}{\epsilon_0}$.

This is called Gauss' Law. The surface over which the flux was calculated is called a *Gaussian surface*, which may be physical or simply a construct of mathematical imagination.

Applying the divergence theorem, we obtain

$$\oiint_S \mathbf{E} \cdot \hat{n} dA = \iiint_V \nabla \cdot \mathbf{E}\, d\tau$$

So, if V is a volume containing charge Q,

$$\iiint_V \nabla \cdot \mathbf{E}\, d\tau = \frac{Q}{\epsilon_0}$$

If ρ is the charge density, then $Q = \iiint_V \rho d\tau$.

And so, we may write,

$$\iiint_V \nabla \cdot \mathbf{E}\, d\tau = \iiint_V \frac{\rho}{\epsilon_0} d\tau$$

This relationship is valid for regions of arbitrary shape and size. Thus the integrands on both sides must be equal. And so, we obtain the important equation

$$\nabla \cdot \mathbf{E} = \frac{\rho}{\epsilon_0} \tag{2.7}$$

This is one of the four Maxwell Equations. We shall call it Maxwell's First Equation.

Maxwell's first equation is a mathematical consequence of Coulomb's Law, which was obtained experimentally. Coulomb's law explains the force between two charges, whereas Maxwell's equation is an abstract statement that relates the divergence of the electric field at a point to the charge density at that point.

Maxwell's equation is very helpful for finding the electric fields due to distributions of charges, and the procedure becomes relatively easy when the system of charges has a simple geometric symmetry.

Suppose we have a spherical conductor of radius a which carries a charge Q, and we need to find the field at a distance from the conductor.

Let us draw a spherical Gaussian surface with radius r greater than a. The total electric flux out of this surface is given by $4\pi r^2 E$ where E is the electric field on this surface, and by symmetry has the same value at each point. By Gauss's law, $4\pi r^2 E = \frac{Q}{\epsilon_0}$. Hence,

$$E = \frac{Q}{4\pi\epsilon_0 r^2} \tag{2.8}$$

Thus, the electric field due a uniformly charged spherical shell at any point outside the shell is the same as if the radius of the shell were reduced to zero without changing the charge.

Of course, a charged spherical shell would have an outward pressure because the charges spread out over the shell would repel each other and attempt to move as far away from each other as possible. The smaller the radius of the shell becomes, the greater the pressure. If we were able to reduce the radius indefinitely, the charges on the surface would fly away into the surrounding atmosphere, or escape into the surrounding space if the conductor was surrounded by a vacuum. So, the conductor would lose its charge as it shrank.

2.2.2 *Field Near a Uniformly Charged Infinite Plane*

It is an interesting exercise to calculate the electric field due to an infinite uniformly charged plane at any point above the plane. We will first do it using Coulomb's law, and then we will derive the same result using Gauss's Law.

2.2.2.1 *Method A: Coulomb's Law*

First, we will consider a uniformly charged rod of infinite length, and a point at a distance a from this rod.

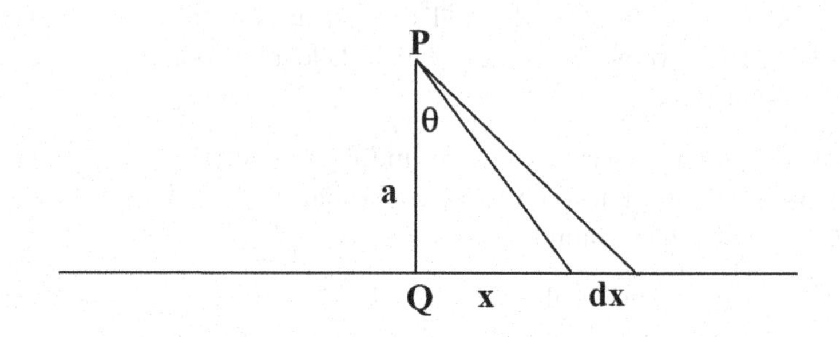

It is evident that the electric field at the point P will be directed away from the rod, along QP if the charge is positive, and along PQ if the charge is negative. We will assume the charge is positive. We will evaluate the magnitude of the field at P.

The charge is uniformly distributed on the rod. We can analyze the system by considering a small element of length dx on the rod. We will calculate the contribution of the charge on this element to the field at P:

The magnitude of the field at P due to the charge on dx =

$$\frac{\lambda dx}{4\pi\epsilon_0(a^2+x^2)}$$

Each such element on the rod will contribute a field given by the above formula. But only the component along QP will make a contribution, since components perpendicular to QP will cancel out when we consider contributions from the left and the right sides of Q. So, the actual contribution from the charge on dx to the field at P is given by

$$dE = \frac{\lambda dx}{4\pi\epsilon_0(a^2+x^2)}\cos\theta = \frac{\lambda dx}{4\pi\epsilon_0(a^2+x^2)}\frac{a}{\sqrt{a^2+x^2}} = \frac{\lambda a}{4\pi\epsilon_0}\frac{dx}{(a^2+x^2)^{\frac{3}{2}}} \tag{2.9}$$

By carrying out an integral over x from $-\infty$ to $+\infty$ we obtain the total electric field at P. This integral can be carried out by a suitable change of variables:

$x = a\tan\theta$. So $dx = a\sec^2\theta d\theta$, and $(a^2+x^2)^{\frac{3}{2}} = a^3\sec^3\theta$.

Hence

$$dE = \frac{\lambda a}{4\pi\epsilon_0}\frac{a\sec^2\theta}{a^3\sec^3\theta}d\theta = \frac{\lambda}{4\pi\epsilon_0 a}\cos\theta\,d\theta \tag{2.10}$$

Integrating, we obtain

$$E = \frac{\lambda}{4\pi\epsilon_0}\int_{-\pi/2}^{\pi/2}\cos\theta d\theta = \frac{\lambda}{2\pi\epsilon_0 a} \tag{2.11}$$

Next, we consider an infinite plane spanned by the xy coordinate system. We divide this plane into slices of width dy, each slice perpendicular to the y axis. Now, each slice can be considered an infinite rod. If P is a point at a distance b from the plane, then a typical slice or infinite rod is at a distance $\sqrt{b^2+y^2}$ from the point P.

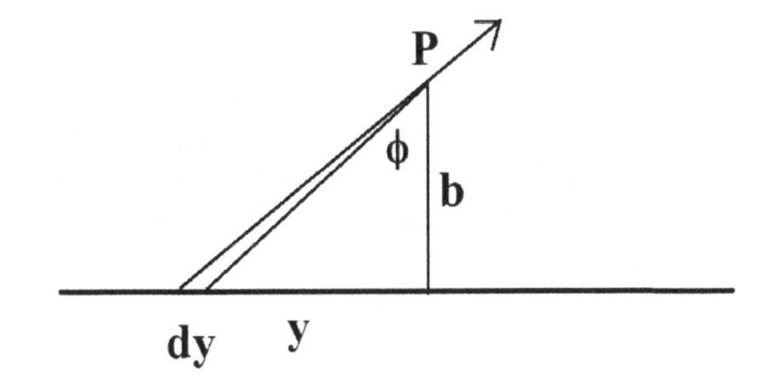

The figure above is a side view of the plane, which appears as the horizontal line. The slice with length along the x direction has width dy. The electric field at P due to the slice is directed along the arrow. The electric field due to this slice is obtained from Eq. (2.11), and we replace a by $\sqrt{b^2+y^2}$. Also, since we are considering areas and not just lengths, we replace the charge per unit length λ by the charge per unit area σ in the calculation. To obtain the total electric field at P due to the infinite plane, we add up the electric fields due to all the slices, adding up along the y axis from $-\infty$ to $+\infty$. Since the horizontal components cancel as we go from negative y to positive y, we need to add up only the vertical components of the field. The contribution due to each slice has the form

$$dE = \frac{\sigma dy}{2\pi\epsilon_0}\frac{\cos\phi}{\sqrt{b^2+y^2}} \tag{2.12}$$

Setting $y = b\tan\phi$, we get $dy = b\sec^2\phi$, and so the total field at P becomes

$$E = \frac{\sigma}{2\pi\epsilon_0}\int_{-\pi/2}^{\pi/2}d\phi = \frac{\sigma}{2\epsilon_0} \tag{2.13}$$

2.2.2.2 *Alternative method using Coulomb's Law*

An alternative method is to divide up the plane into concentric rings of small width dr:

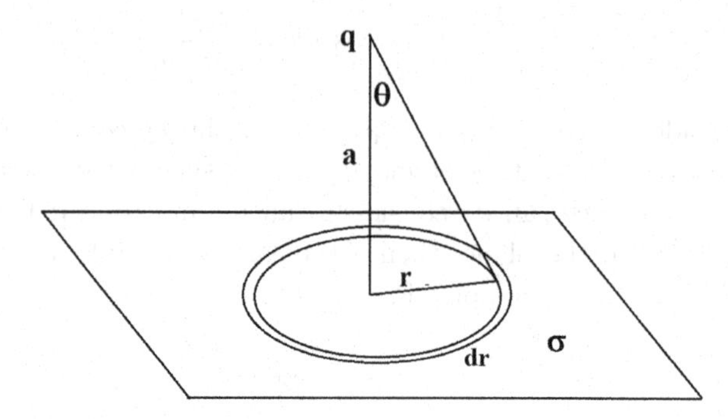

Consider a point P at a distance a from an infinite plane with a uniform charge density of σ. We will divide the plane into concentric rings of width dr. If the radius of the ring is r, the vertical component of the field at P due to the charge on this ring is

$$dE = \frac{2\pi r\sigma dr}{4\pi\epsilon_0(a^2 + r^2)}\cos\theta \tag{2.14}$$

Now, $r = a\tan\theta$, and so $dr = a\sec^2\theta d\theta$, and $a^2 + r^2 = a^2\sec^2\theta$.

$$E = \int_0^{\frac{\pi}{2}} \frac{2\pi\sigma a\tan\theta}{4\pi\epsilon_0}\frac{a\sec^2\theta d\theta}{a^2\sec^3\theta} = \frac{\sigma}{2\epsilon_0}\int_0^{\frac{\pi}{2}}\sin\theta d\theta = \frac{\sigma}{2\epsilon_0} \tag{2.15}$$

2.2.2.3 *Method B: Gauss's Law*

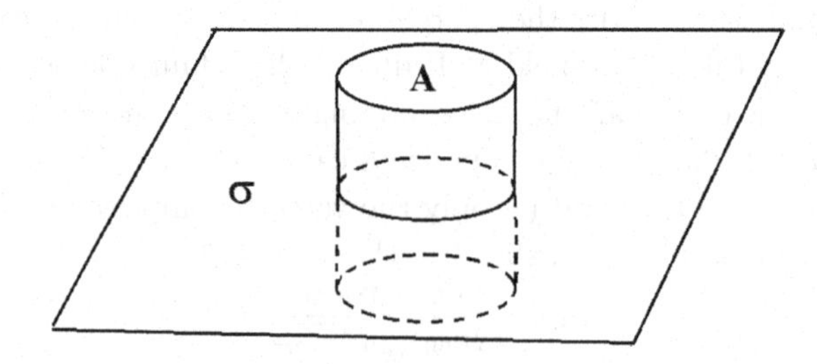

Consider an infinite plane with a uniform charge density σ, and a cylindrical box of cross-section A that is intersected by the plane as shown above.

Since the plane is infinite, by symmetry the field above and below the plane must be everywhere the same in magnitude, and, for positive σ, is directed upwards above the plane and directed downwards below the plane. Let the magnitude of the field be E.

The total flux out of the box $= EA + EA = 2EA$. By Gauss's law, this must equal $\frac{q}{\epsilon_0}$, where q is the total charge inside the box, which is σA. So $2EA = \frac{\sigma A}{\epsilon_0}$.

Therefore

$$E = \frac{\sigma}{2\epsilon_0} \tag{2.16}$$

2.3 Conservative Force

The electric field generated by a charge at the origin is $\mathbf{E} = \frac{q}{4\pi\epsilon_0}\hat{e}_r$. Let us find the curl of this vector field:

$$\nabla \times \mathbf{E} = \nabla\left(\frac{q}{4\pi\epsilon_0 r^2}\right) \times \hat{e}_r + \frac{q}{4\pi\epsilon_0 r^2}\nabla \times \hat{e}_r \tag{2.17}$$

It can be shown that $\nabla\left(\frac{1}{r^2}\right) = -\frac{2}{r^3}\hat{e}_r$, and so the first term on the right side of the equation is 0. Next, $\hat{e}_r = \frac{\mathbf{r}}{r}$. It is again easily demonstrated that $\nabla \times \hat{e}_r = 0$.

Hence, $\nabla \times \mathbf{E} = 0$ when the field is generated by a charge at the origin. If the charge is at a point other than the origin, that will not change the result. Suppose coordinate x is translated to $x' = x + a$. So, any function of x will become a function of $x' = x + a$. Now, $\frac{\partial f(x+a)}{\partial x} = \frac{\partial f(x+a)}{\partial (x+a)} = \frac{\partial f(x)}{\partial x}$ through a change of variables. So if $\nabla \times \mathbf{E} = 0$ for a charge at the origin, $\nabla \times \mathbf{E} = 0$ for fields generated by charges located at arbitrary positions other than the origin.

Hence we have an important rule concerning the electric field generated by stationary charges:

$$\boxed{\nabla \times \mathbf{E} = 0}$$

The work done by a force $\mathbf{F}$ upon an object that is displaced from point A to point B along a path C is defined as

$$W = \int_A^B \mathbf{F} \cdot d\mathbf{r} \tag{2.18}$$

Now, if a charge q is displaced along a closed path, the work done by the electric field on the charge is

$$W = \oint q\mathbf{E} \cdot d\mathbf{r} \tag{2.19}$$

By Stokes's theorem, this integral can be converted to a surface integral:

$$W = \oint q\mathbf{E} \cdot d\mathbf{r} = \iint q\nabla \times \mathbf{E} \cdot \hat{n} dA \tag{2.20}$$

Since $\nabla \times \mathbf{E} = 0$, it follows that for any arbitrary closed path, $\oint q\mathbf{E} \cdot d\mathbf{r} = 0$.

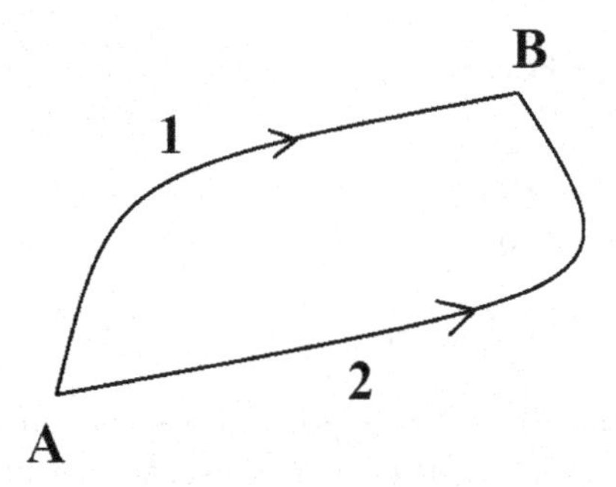

The line integral of $\mathbf{E} \cdot d\mathbf{r}$ along the closed path from A to B along path 2 and from B to A in the reverse direction along path 1 is zero for arbitrary shapes of the paths and arbitrary positions for A and B. This means that the path integral from A to B along path 2 must be equal to the path integral from A to B along path 1. So, the path integral from A to B is independent of the path taken.

A force field with the property that the work done by the field is independent of the path, and therefore depends only on the initial and the final positions, is called a conservative force field, and such a force is called a conservative force. The electric field is a conservative force field, and the electric force is a conservative force. Thus the field generated by static charges is similar to the gravitational field generated by static masses, which we discussed briefly in the previous chapter.

Now, if $\nabla \times \mathbf{E} = 0$, the inference is that $\mathbf{E}$ must be the gradient of a scalar field (which could also be a constant vector, having the same value everywhere).

So, if $\nabla \times \mathbf{E} = 0$, it follows that

$$\mathbf{E} = -\nabla\phi \tag{2.21}$$

where the negative sign has been chosen for later convenience. [This symbol ϕ is to be distinguished from Φ which was used for the electric flux, and from φ, which was used for the azimuthal angle.] ϕ is a scalar field, which is in general a function of the spatial coordinates. ϕ is called the *electric potential*, or simply the *scalar potential.* It is measured in volts (V).

So $\int_A^B \mathbf{E} \cdot d\mathbf{r} = -\int_A^B \nabla\phi \cdot d\mathbf{r} = -\int_A^B d\phi = \phi_A - \phi_B$ where ϕ_A and ϕ_B are the potentials at A and B respectively.

Maxwell's first equation $\nabla \cdot \mathbf{E} = \frac{\rho}{\epsilon_0}$ can be written in terms of the scalar potential as

$$\nabla^2\phi = -\frac{\rho}{\epsilon_0} \tag{2.22}$$

This is called *Poisson's equation.*

In empty space, where the charge density $\rho = 0$, this equation becomes

$$\nabla^2\phi = 0 \tag{2.23}$$

This is called *Laplace's equation.*

2.4 Potential and Field

2.4.1 *Equipotential Surfaces*

The relation between field and potential $\mathbf{E} = -\nabla\phi$ can be used to obtain the potentials in different situations. We know that the field due to a positive charge placed at the origin is given by

$$\mathbf{E} = \frac{Q}{4\pi\epsilon_0 r^2}\hat{e}_r$$

From this we can infer that the potential due to a charge Q at the origin is given by

$$\phi = \frac{Q}{4\pi\epsilon_0 r} \tag{2.24}$$

Since the charge Q can be positive or negative, the potential at a point can be positive or negative. The field due to a system of charges is simply the vector sum of the fields due to the charges in the system, and hence the potential due to a system of charges is the algebraic sum of the potentials due to these charges.

Fields are pictorially depicted by field lines, the direction of the lines showing the direction of the field, and the clustering density of the lines being proportional to the magnitude of the field.

There is also a pictorial depiction of electric potentials. The potential at a point does not have a direction, only a magnitude, which can be positive or negative. On a flat meteorological map points having the same temperature are connected by a line called an isotherm, and points having the same pressure by a line called an isobar. If a three-dimensional map could be generated either holographically or by constructing a solid model, the isotherms and isobars would not be lines but surfaces. In the same way, a two-dimensional diagram of an electric field would show points having the same potential by a line called an equipotential line, which is just a two-dimensional projection of a surface in three-dimensional space. Such a surface is called an equipotential surface or an equipotential. Analytically, the equation for an equipotential surface is given by

$$\phi(x, y, z) = c \tag{2.25}$$

Now, $\nabla\phi$ is a vector that is perpendicular to a surface along which ϕ is constant. Hence the electric field is perpendicular to the equipotential surface. So, the electric potential can be depicted pictorially as a Gaussian surface at every point on which the potential is constant, and at every point the field is perpendicular to the surface.

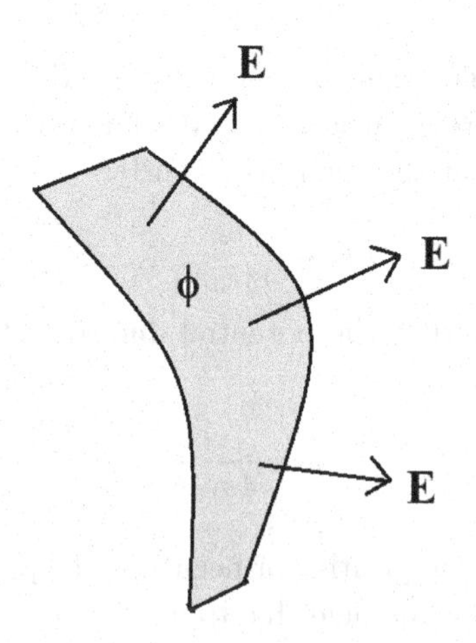

Thus we have two different ways of thinking of an electric field. We can either assign a vector with three components $\mathbf{E}$ or a scalar with one component ϕ. Often, it is easier to work with the electric potential than the electric field. So, if our objective is to find the force acting on a test charge q that is introduced at some point in the region, and if we find the electric potential at that point as a function of the coordinates, then we can take the gradient to calculate the force.

2.4.2 *Calculation of Potential at a Point*

If we have a distribution of charges with charge density as a function of the spatial coordinates $\rho(x, y, z)$, we can find the potential at the origin by evaluating the integral

$$\phi = \int_0^\infty \frac{\rho d\tau}{4\pi\epsilon_0 r} \tag{2.26}$$

where the symbol $d\tau$ is the element of volume (also written as dV, d^3x, or d^3r). For a point charge, this integral reverts to Eq. (2.24).

Now, if there are several charges in some region, and each charge generates a field at the origin, the net field at the origin is the vector sum of the fields due to the individual charges. This is a case of the *superposition principle.*

$$\mathbf{E} = \sum_i \mathbf{E}_i = -\sum_i \nabla\phi_i \tag{2.27}$$

So, if we call the potential at the origin as ϕ, so that $\mathbf{E} = -\nabla\phi$, then it follows that $\nabla\phi = \sum_i \nabla\phi_i$. Thus $\phi = \sum_i \phi_i$ plus an arbitrary constant, which we are free to set at 0. Thus the electric potential also obeys the principle of superposition. If there are several point charges q_1, q_2, q_3, etc. at different points at distances r_1, r_2, r_3 etc. respectively from the origin, the total potential at the origin is simply the algebraic sum of the potentials due to each individual charge.

$$\phi = \sum_i \frac{q_i}{4\pi\epsilon_0 r_i} \tag{2.28}$$

This equation holds good for arbitrary positions of the origin, and hence the potential at any point in a field is the algebraic sum of the potentials due to all the charges present.

Let us consider a system of three point charges, two of them along the x axis and one on the y axis, as shown:

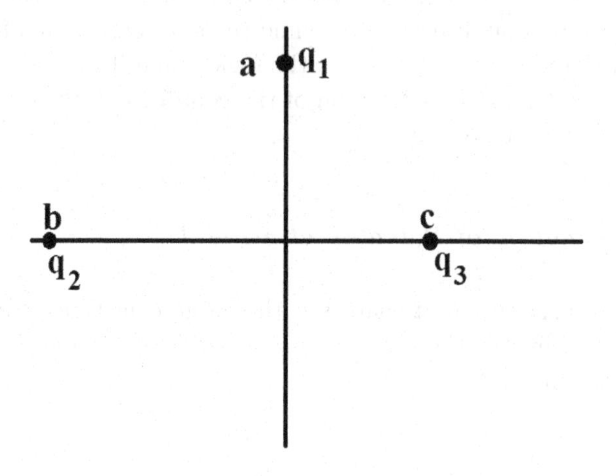

At every point in the plane there is a potential due to this system of charges. But we notice that the potential diverges to infinity at the points $(0, a)$, $(b, 0)$ and $(c, 0)$. This is a problem if we consider these charges as point objects having zero volume. We shall temporarily overrule that possibility and return to it in a later chapter when we examine the quantum theory of the electromagnetic field.

So we will assume that geometric point charges do not exist, and so we provisionally assign small finite diameters to these charges. But since these diameters are considered small, at distances close to a charge the magnitude of the potential would be large. So, if we were to bring a test charge q into the picture, and move it around without disturbing the other three charges, this charge would pick up very high potentials when it comes close to the fixed charges.

Suppose now we do not have a test charge, but only the three charges of the system. What is the potential picked up by, say charge q_1? The answer is the potential generated by the other two charges q_2 and q_3. So we have made q_1 the test charge in the field generated by charges q_2 and q_3. An important corollary is that each charge sees only the field generated by the other charges. A charge cannot experience the field or potential that it generates. In other words, a charge cannot detect itself. Putting it poetically, a charge cannot see itself.

Cassius: Tell me, good Brutus, can you see your face?
Brutus: No, Cassius, for the eye sees not itself. But by reflection, by some other things. (Shakespeare, *Julius Caesar*, Act 1, Scene 2)

As Brutus tells his friend Cassius, the eye can see itself only by reflection from some other objects. The ancient Romans saw their reflections in water or in polished metal surfaces. We shall see later that a charge cannot detect itself except by reflection, primarily from metallic surfaces.

Poisson's equation holds for the potential at any point:

$$\nabla^2 \phi = -\frac{\rho}{\epsilon_0} \tag{2.29}$$

And in any region where there is no charge, this equation is replaced by Laplace's equation

$$\nabla^2 \phi = 0 \tag{2.30}$$

Whereas this equation is physically equivalent to $\nabla \cdot \mathbf{E} = 0$, Laplace's equation is often mathematically easier to solve than the divergence equation for the same system.

Chapter 3

Electric Fields and Potentials

3.1 Solutions to Laplace's Equation

3.1.1 *Boundary Value Solutions*

In many real life situations we can measure the value of the electric potential ϕ on the boundary of some region, and we seek to find the potential inside the region — at least as a function of the coordinates. In such situations it is generally possible to solve Laplace's equation $\nabla^2\phi = 0$ to find $\phi(x, y, z)$ inside the region.

Example A: We are able to measure the potential at every point on the surface of a sphere, but not inside the sphere. We solve Laplace's equation for the inside, by using our knowledge of the potential at the boundary.

Example B: We know the potential at every point on the surface of a sphere. We seek to find the potential everywhere outside the sphere as a function of the coordinates. The potential at infinity is zero, far from all charges. Thus we know the potential at both the inner and the outer boundaries, and we can calculate the solution to Laplace's equation outside the sphere.

Example C: We want to find the potential everywhere inside a rectangular box. The sides of the box are kept at different potentials, which we are able to measure. Knowing these boundary conditions we can find the solution to Laplace's equation inside the box.

Such problems, where a scalar function — which in our case is the electric potential — is specified at every point of the boundary, are called Dirichlet problems, and the boundary condition is called a Dirichlet boundary condition. A different sort of boundary condition, called the Neumann boundary condition, specifies the gradient of the function — corresponding to the electric field — at every point on the boundary. For the present we deal with Dirichlet boundary conditions.

We will now prove that if ϕ is known at every point on the boundary, then there is a *unique* solution to Laplace's equation at every point within the boundary.

Proof:

Suppose there is more than one solution to Laplace's equation at any point within the boundary. Let us call two such distinct solutions ϕ_1 and ϕ_2. Let us define the function $\psi = \phi_1 - \phi_2$. We need to show that $\psi = 0$ everywhere within the boundary.

Now,

$$\nabla \cdot (\psi \nabla \psi) = \psi \nabla^2 \psi + \nabla \psi \cdot \nabla \psi \tag{3.1}$$

By the divergence theorem,

$$\iiint_V \nabla \cdot (\psi \nabla \psi) d\tau = \oiint_S \psi \nabla \psi \cdot \hat{n} dA \tag{3.2}$$

Therefore

$$\iiint_V (\psi \nabla^2 \psi + \nabla \psi \cdot \nabla \psi) d\tau = \oiint_S \psi \nabla \psi \cdot \hat{n} dA \tag{3.3}$$

(This equation is a particular case of a theorem called *Green's First Identity.*) Now, $\nabla^2 \phi_1 = \nabla^2 \phi_2 = 0$, and so $\nabla^2 \psi = 0$ within the boundary. This result holds even if there are charges in the space within the boundary, for then Poisson's equation applies to both potentials, $\nabla^2 \phi_1 = -\frac{\rho}{\epsilon_0}$ and $\nabla^2 \phi_2 = -\frac{\rho}{\epsilon_0}$. So, we obtain

$$\iiint_V |\nabla \psi|^2 d\tau = \oiint_S \psi \nabla \psi \cdot n dA \tag{3.4}$$

Since $\psi = \phi_1 - \phi_2 = 0$ everywhere on the boundary, the right side of this equation is zero. Since $|\nabla \psi|$ cannot be negative anywhere, it must be zero everywhere. Therefore $\nabla \psi = 0$ everywhere within the boundary. So ψ must be a constant, and since it is zero on the boundary, it must be zero everywhere inside as well. So $\phi_1 = \phi_2$.

3.1.2 *Rectangular Coordinates*

Laplace's equation in Cartesian coordinates is written as

$$\frac{\partial^2 \phi}{\partial x^2} + \frac{\partial^2 \phi}{\partial y^2} + \frac{\partial^2 \phi}{\partial z^2} = 0 \tag{3.5}$$

This form of Laplace's equation can be solved for geometries with right angles. Recall that a vector can be written in Cartesian coordinates as $\mathbf{A} = A_x\hat{\imath} + A_y\hat{\jmath} + A_z\hat{k}$. In this coordinate system the unit vectors are constant and do not vary from point to point, unlike in the spherical and cylindrical coordinate systems. So, the solution to Laplace's equation can be expressed as $\phi(x, y, z) = X(x)Y(y)Z(z)$ where X, Y, and Z are functions of the separate coordinates x, y and z respectively. Thus Eq. (3.5) becomes

$$\frac{1}{X}\frac{d^2X}{dx^2} + \frac{1}{Y}\frac{d^2Y}{dy^2} + \frac{1}{Z}\frac{d^2Z}{dz^2} = 0 \tag{3.6}$$

The first term contains no y or z, the second term contains no x or z, and the third term contains no x or y. Thus, each of the three terms must be independent of the coordinates, and therefore constant.

So, we can write $\frac{1}{X}\frac{d^2X}{dx^2} = \alpha$, $\frac{1}{Y}\frac{d^2Y}{dy^2} = \beta$, and $\frac{1}{Z}\frac{d^2Z}{dz^2} = \gamma$, where α, β, and γ are constant numbers, and $\alpha + \beta + \gamma = 0$. We shall assume these numbers are real. Since their sum is zero, at least one of them is negative, and at least one is positive. The physical geometry of the system, including the boundary conditions, will determine the signs of these numbers.

Consider the box shown below, with the bottom face in the xy plane, with one vertex at the origin, and the opposite vertex at the point (a, b, c), with a height of c along the z axis, where c is much greater than a and b. The bottom face of the box is maintained at a potential which is a function of x and y: $V(x, y, 0)$, and the other five faces are maintained at zero potential.

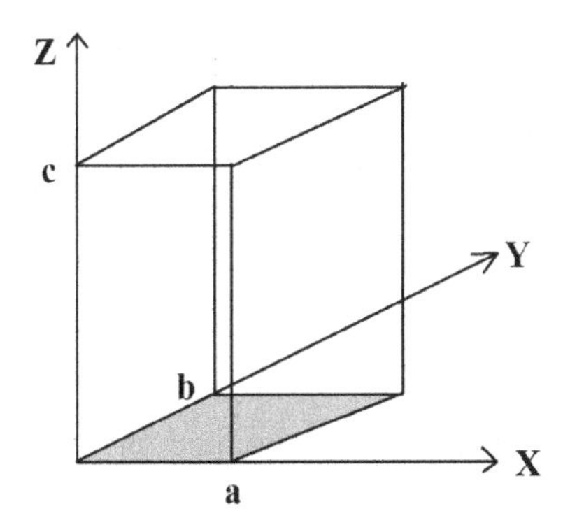

As we move upwards along z, we move further away from the bottom face having a non-zero potential to the top face which has zero potential. Hence we expect the magnitude of the potential inside the box to decrease all the way from bottom to top. We do not expect a similar variation along the x or y directions. These considerations imply, as will presently become clear, that γ should be positive, and α and β should be negative. Let $\alpha = -A^2$, $\beta = -B^2$ and $\gamma = C^2$, where A, B and C are real numbers.

$\frac{d^2X}{dx^2} = -A^2X;\ \frac{d^2Y}{dy^2} = -B^2Y;\ \frac{d^2Z}{dz^2} = C^2Z$ where $C^2 = A^2 + B^2$.

The general solutions of these equations are:

$X = A_1 \cos Ax + A_2 \sin Ax;\ \ Y = B_1 \cos By + B_2 \sin By;$
$Z = C_1 e^{Cz} + C_2 e^{-Cz}$

Since the potential is 0 for $x = 0, y = 0$, we can infer that $A_1 = B_1 = 0$. And since the potential is 0 for $x = a, y = b$, it follows that $A = n\pi/a$ and $B = m\pi/b$ where n and m are positive integers. Thus $C = \pi\sqrt{(n/a)^2 + (m/b)^2}$. We may then combine most of the remaining constants and write the solution as

$$\phi(x, y, z) = G\left(c_1 e^{\pi\sqrt{(n/a)^2+(m/b)^2}z} + e^{-\pi\sqrt{(n/a)^2+(m/b)^2}z}\right) \sin(n\pi x/a)\sin(m\pi y/b) \tag{3.7}$$

But this is just a special solution. The general solution is a sum of terms containing all possible values of n and m:

$$\phi(x, y, z) = \sum_{n=1}^{\infty}\sum_{m=1}^{\infty} G_{n,m}\left(c_1 e^{\pi\sqrt{(\frac{n}{a})^2+(\frac{m}{b})^2}z} + e^{-\pi\sqrt{(\frac{n}{a})^2+(\frac{m}{b})^2}z}\right) \sin\left(\frac{n\pi x}{a}\right)\sin\left(\frac{m\pi y}{b}\right) \tag{3.8}$$

This is the complete solution to Laplace's equation. But we can reduce the number of constants. Since the first term inside the brackets diverges for large z, this divergence can be eliminated by setting $c_1 = 0$, thereby yielding a general solution valid for arbitrary z. And the constants $G_{n,m}$ are not entirely arbitrary. We can apply the boundary condition at $z = 0$ to obtain

$$V(x, y, 0) = \sum_{n=1}^{\infty}\sum_{m=1}^{\infty} G_{n,m} \sin(n\pi x/a)\sin(m\pi y/b) \tag{3.9}$$

This equation also enables us to determine the coefficients $G_{n,m}$. The procedure is to multiply by the factor $\sin(n'\pi x/a)\sin(m'\pi y/b)$, where n'

and m' are arbitrary positive integers, and integrate over x and y

$$\sum_{n=1}^{\infty}\sum_{m=1}^{\infty} G_{n,m} \int_0^a \sin(n\pi x/a)\sin(n'\pi x/a)dx \int_0^b \sin(m\pi y/b)\sin(m'\pi y/b)dy \tag{3.10}$$

This is a double Fourier integral, and has the value $(ab/4)G_{n',m'}$.

Hence, we obtain (dropping the primes on n and m)

$$G_{n,m} = \frac{4}{ab}\int_0^a \int_0^b V(x,y,0)\sin(n\pi x/a)\sin(m\pi y/b)dxdy \tag{3.11}$$

This double integral can be carried out for each set of numbers (n,m) by using the explicit form of $V(x,y,0)$ and integrating between 0 and the known values of a and b.

3.1.3 *Spherical Polar Coordinates*

The Laplacian operator in spherical coordinates is as follows:

$$\nabla^2 = \frac{1}{r^2}\frac{\partial}{\partial r}\left(r^2\frac{\partial}{\partial r}\right) + \frac{1}{r^2\sin\theta}\frac{\partial}{\partial\theta}\left(\sin\theta\frac{\partial}{\partial\theta}\right) + \frac{1}{r^2\sin^2\theta}\frac{\partial^2}{\partial\varphi^2} \tag{3.12}$$

So, Laplace's equation in spherical coordinates may be written as

$$\frac{\partial}{\partial r}\left(r^2\frac{\partial\phi}{\partial r}\right) + \frac{1}{\sin\theta}\frac{\partial}{\partial\theta}\left(\sin\theta\frac{\partial\phi}{\partial\theta}\right) + \frac{1}{\sin^2\theta}\frac{\partial^2\phi}{\partial\varphi^2} = 0 \tag{3.13}$$

The first term on the left side contains no angular derivatives in θ or φ. We will consider a solution to this equation of the form $\phi = R(r)A(\theta,\varphi)$. Substituting into the equation and dividing by $R(r)A(\theta,\varphi)$, we obtain

$$\frac{1}{R}\frac{d}{dr}\left(r^2\frac{dR}{dr}\right) + \frac{1}{A}\frac{1}{\sin\theta}\frac{\partial}{\partial\theta}\left(\sin\theta\frac{\partial A}{\partial\theta}\right) + \frac{1}{A}\frac{1}{\sin^2\theta}\frac{\partial^2 A}{\partial\varphi^2} = 0 \tag{3.14}$$

We shall set the first term as a constant, which in light of later convenience we write as $\ell(\ell+1)$.

Hence we get a radial equation

$$\frac{d}{dr}\left(r^2\frac{dR}{dr}\right) = \ell(\ell+1)R \tag{3.15}$$

which is a second order differential equation with the solution

$$R(r) = Ar^{\ell} + Br^{-\ell-1} \tag{3.16}$$

(Note that here A is a constant number, distinct from the function $A(\theta, \varphi)$.) The angular equation takes the form

$$\frac{1}{A}\frac{1}{\sin\theta}\frac{\partial}{\partial\theta}\left(\sin\theta\frac{\partial A}{\partial\theta}\right) + \frac{1}{A}\frac{1}{\sin^2\theta}\frac{\partial^2 A}{\partial\varphi^2} = -\ell(\ell+1) \tag{3.17}$$

Next, writing the angular function as a product of functions of the polar angle θ and the azimuthal angle φ: $A(\theta, \varphi) = \Theta(\theta)\Phi(\varphi)$, we obtain the equation

$$\frac{\sin\theta}{\Theta}\frac{d}{d\theta}\left(\sin\theta\frac{d\Theta}{d\theta}\right) + \frac{1}{\Phi}\frac{d^2\Phi}{d\varphi^2} = -\ell(\ell+1)\sin^2\theta \tag{3.18}$$

Since the second term on the left side is independent of θ, and the remainder of the equation is independent of φ, we can treat this term as a constant, which for later convenience we write as $-m^2$ where m is an integer, which may be positive, negative or zero. The azimuthal equation now becomes

$$\frac{d^2\Phi}{d\varphi^2} = -m^2\Phi \tag{3.19}$$

The general solution is $\Phi = A\cos m\varphi + B\sin m\varphi$, or, in more common form using complex numbers, $\Phi = A_1 e^{im\varphi} + A_2 e^{-im\varphi}$. Recall that $e^{i\theta} = \cos\theta + i\sin\theta$.

That leaves us with the equation in the polar function $\Theta(\theta)$:

$$\frac{1}{\sin\theta}\frac{d}{d\theta}\left(\sin\theta\frac{d\Theta}{d\theta}\right) + \left[\ell(\ell+1) - \frac{m^2}{\sin^2\theta}\right]\Theta = 0 \tag{3.20}$$

It is easier to solve this equation if we make the substitution $x = \cos\theta$ and replace $\Theta(\theta)$ by $P(x)$. The polar equation becomes

$$\frac{d}{dx}\left[(1-x^2)\frac{dP}{dx}\right] + \left[\ell(\ell+1) - \frac{m^2}{1-x^2}\right]P = 0 \tag{3.21}$$

This is called the associated Legendre equation. In order to find the solutions to this equation, we first consider the solutions to the simpler equation obtained by putting $m = 0$:

$$\frac{d}{dx}\left[(1-x^2)\frac{dP}{dx}\right] + \ell(\ell+1)P = 0 \tag{3.22}$$

A solution to this differential equation can be obtained using Rodrigues' Formula

$$P_\ell(x) = \frac{1}{2^\ell \ell!}\frac{d^\ell}{dx^\ell}(x^2-1)^\ell \tag{3.23}$$

The P_ℓ are polynomials in x known as the Legendre polynomials:

$P_0 = 1$

$P_1 = x$

$P_2 = \frac{1}{2}(3x^2 - 1)$

$P_3 = \frac{1}{2}(5x^3 - 3x)$

$P_4 = \frac{1}{8}(35x^4 - 30x^2 + 3)$

$P_5 = \frac{1}{8}(63x^5 - 70x^3 + 15x)$

We see that for $x = 1$ or $\theta = 0$ all the Legendre polynomials equal 1.

The Legendre polynomials obey an orthogonality condition, expressed as

$$\int_1^1 P_{\ell'}(x) P_\ell(x) dx = \frac{2}{2\ell + 1} \delta_{\ell' \ell} \tag{3.24}$$

From the Legendre polynomials we define the associated Legendre functions as

$$P_\ell^m(x) \equiv (1 - x^2)^{|m|/2} \left(\frac{d}{dx} \right)^{|m|} P_\ell(x) \tag{3.25}$$

The associated Legendre functions are solutions of the associated Legendre Eq. (3.21). We may now replace the variable x by $\cos\theta$ and write some of the associated Legendre functions:

$P_0^0 = 1; \quad P_1^0 = \cos\theta; \quad P_2^0 = \frac{1}{2}(3\cos^2\theta - 1)$

$P_1^1 = \sin\theta; \quad P_2^1 = 3\sin\theta\cos\theta; \quad P_3^1 = \frac{3}{2}\sin\theta(5\cos^2\theta - 1)$

3.1.4 *Spherical Harmonics*

The angular Eq. (3.17) has a solution that is a product of the polar and azimuthal functions. We define an important function that satisfies Eq. (3.17), called a spherical harmonic, as

$$Y_{\ell m}(\theta, \varphi) = (-1)^m \sqrt{\frac{2\ell + 1}{4\pi} \frac{(\ell - |m|)!}{(\ell + |m|)!}} P_\ell^m(\cos\theta) e^{im\varphi} \tag{3.26}$$

These spherical harmonics have some interesting properties.

They are orthogonal to each other in both ℓ and m:

$$\int_0^{2\pi} d\varphi \int_0^{\pi} \sin\theta d\theta \, Y^*_{\ell' m'}(\theta,\varphi) Y_{\ell m}(\theta,\varphi) = \delta_{\ell' \ell}\delta_{m' m} \tag{3.27}$$

$d\varphi \sin\theta d\theta$ is the solid angle $d\Omega$, and the integration is done over the entire spherical surface, i.e. over the solid angle $\Omega = 4\pi$ sr.

An arbitrary function in θ and φ can be written as a series of spherical harmonics:

$$f(\theta,\varphi) = \sum_{\ell=1}^{\infty} \sum_{m=-\ell}^{\ell} A_{\ell m} Y_{\ell m}(\theta,\varphi) \tag{3.28}$$

The coefficients of this series may be obtained by carrying out integrals, making use of the orthogonality of the spherical harmonics (Eq. (3.27)):

$$A_{\ell m} = \int d\Omega \, Y^*_{\ell m}(\theta,\varphi) f(\theta,\varphi) \tag{3.29}$$

To get the general solution of Laplace's equation where there is no charge density we integrate these results with Eq. (3.16) and obtain

$$\phi(r,\theta,\varphi) = \sum_{\ell=0}^{\infty} \sum_{m=-\ell}^{\ell} \left[A_{\ell m} r^{\ell} + B_{\ell m} r^{-(\ell+1)}\right] Y_{\ell m}(\theta,\varphi) \tag{3.30}$$

Suppose this potential is generated by a point charge at the point $\mathbf{r}'$, and so has the form $\frac{q}{4\pi\epsilon_0 |\mathbf{r}-\mathbf{r}'|}$. Incorporating the constant terms ($4\pi\epsilon_0$ and q) into the coefficients on the right side, we obtain

$$\frac{1}{|\mathbf{r}-\mathbf{r}'|} = \sum_{\ell=0}^{\infty} \sum_{m=-\ell}^{\ell} \left[A_{\ell m} r^{\ell} + B_{\ell m} r^{-(\ell+1)}\right] P_\ell^m(\cos\theta) e^{im\varphi} \tag{3.31}$$

If we choose the coordinate system with the z axis pointing along $\mathbf{r}'$, the potential will have a symmetry about the z axis, and will be independent of φ and hence $m = 0$.

$$\frac{1}{|\mathbf{r}-\mathbf{r}'|} = \sum_{\ell=0}^{\infty} \left[A_{\ell} r^{\ell} + B_{\ell} r^{-(\ell+1)}\right] P_\ell(\cos\theta) \tag{3.32}$$

Now,

$$\frac{1}{|\mathbf{r}-\mathbf{r}'|} = \frac{1}{(r^2 + r'^2 - 2rr'\cos\theta)^{1/2}} \tag{3.33}$$

For $r \gg r'$, the right-hand side can be expanded as

$$\frac{1}{(r^2 + r'^2 - 2rr'\cos\theta)^{1/2}} = \frac{1}{r}\left(1 + \frac{r'}{r}\cos\theta\right)$$
$$= \frac{1}{r}\left[\left(\frac{r'}{r}\right)^0 P_0(\cos\theta) + \left(\frac{r'}{r}\right)^1 P_1(\cos\theta)\right] \quad (3.34)$$

This equation suggests the first two terms of a Taylor series in $\frac{r'}{r}$. Moreover, if the solution is valid for large r, the coefficients A_ℓ must be set to zero, and thus we obtain, for $r' < r$,

$$\frac{1}{|\mathbf{r} - \mathbf{r}'|} = \frac{1}{r}\sum_{\ell=0}^{\infty}\left(\frac{r'}{r}\right)^\ell P_\ell(\cos\theta) \quad (3.35)$$

We can check this equation for $\theta = 0$ or $x = 1$. Each Legendre polynomial becomes unity. And the two sides balance.

3.1.5 *Cylindrical Coordinates*

In cylindrical coordinates (ρ, φ, z) Laplace's equation becomes

$$\frac{\partial^2\phi}{\partial\rho^2} + \frac{1}{\rho}\frac{\partial\phi}{\partial\rho} + \frac{1}{\rho^2}\frac{\partial^2\phi}{\partial\varphi^2} + \frac{\partial^2\phi}{\partial z^2} = 0 \quad (3.36)$$

As usual, we can seek a solution as a product of functions of ρ, φ and z:

$$\phi(\rho, \varphi, z) = R(\rho)F(\varphi)Z(z) \quad (3.37)$$

We obtain three separate differential equations:

$$\frac{d^2Z}{dz^2} = k^2 Z \quad (3.38)$$

$$\frac{d^2F}{d\varphi^2} = -\nu^2 F \quad (3.39)$$

$$\frac{d^2R}{d\rho^2} + \frac{1}{\rho}\frac{dR}{d\rho} + \left(k^2 - \frac{\nu^2}{\rho^2}\right)R = 0 \quad (3.40)$$

The solutions to the first two equations are straightforward:

$$Z(z) = a_1 e^{kz} + a_2 e^{-kz} \quad (3.41)$$

$$F(\varphi) = b_1 e^{i\nu\varphi} + b_2 e^{-i\nu\varphi}$$

It is customary to solve the third equation by first making the substitution $x = k\rho$, and writing the equation as

$$\frac{d^2R}{dx^2} + \frac{1}{x}\frac{dR}{dx} + \left(1 - \frac{\nu^2}{x^2}\right) R = 0 \tag{3.42}$$

This is called the Bessel equation. We seek a solution in the form of a power series in x:

$$R(x) = x^\alpha \sum_{j=0}^{\infty} a_j x^j \tag{3.43}$$

Substituting into the Bessel equation we obtain solutions that can be expressed as power series in x. Two linearly independent solutions are

$$J_\nu(x) = \left(\frac{x}{2}\right)^\nu \sum_{j=0}^{\infty} \frac{(-1)^j}{j!\Gamma(j+\nu+1)} \left(\frac{x}{2}\right)^{2j} \tag{3.44}$$

known as the Bessel function of the first kind, and

$$N_\nu(x) = \frac{J_\nu(x)\cos\nu\pi - J_{-\nu}(x)}{\sin\nu\pi} \tag{3.45}$$

known as the Bessel function of the second kind, also known as the Neumann function, which can be constructed from the Bessel functions of the first kind.

Bessel functions of the first kind satisfy the generating function equation

$$e^{z(t-1/t)/2} = \sum_{n=-\infty}^{\infty} t^n J_n(z) \tag{3.46}$$

Exercise:

The potential due to a charge distribution $\rho(r, \theta, \varphi)$ is

$$\phi = \frac{q}{4\pi\epsilon_0} \frac{e^{-r/a}}{r}$$

where q is a positive point charge located at the origin.

(a) Solve Poisson's equation $\nabla^2\phi = -\frac{\rho}{\epsilon_0}$ to find ρ as a function of the polar coordinates (r, θ, φ).

(b) Calculate the total charge by taking an integral of ρ over all space.

3.2 A Puzzle

An electric field can be described either by its field vector $\mathbf{E}$ or by its scalar potential ϕ. In order to describe the field by its field vector we need to specify three scalar functions: $E_x(x, y, z), E_y(x, y, z)$ and $E_z(x, y, z)$. So three independent real numbers are needed. On the other hand, the field is also specified by the electric potential, one scalar function: $\phi(x, y, z)$. Thus a single real number suffices. This raises the question: how many real numbers are actually needed to specify the field at a point?

We could tackle this puzzle by considering a conductor or a dielectric that is an object of non-zero dimensions, however small these dimensions may be. If the object is placed in a non-vanishing electric field, then, since $\nabla\phi$ is not zero across the dimensions of the object, there is a variation in the value of ϕ across the different points on the object. Thus the object measures the value of ϕ not at one point, but at several points. In order to determine E_x it is necessary to measure ϕ at a minimum of two distinct points in the x direction, and likewise for the other components of the field.

But what if our object is a point charge with zero dimension? Such an object could pick up the scalar potential at only one point. In order to undergo a change of momentum, a particle must know in which direction it should accelerate. And the field cannot communicate this information to the point charge if this charge can measure the potential at only a single point. So a point charge presents a real problem. And this problem has no solution in classical electromagnetism. The solution is to be found in quantum mechanics, and we will discuss this in a subsequent chapter.

3.3 Electric Field Lines

3.3.1 *Parallel Plate Capacitor*

Consider two identical conducting plates parallel to each other, kept at a distance apart that is small compared to the length and width of each plate. Suppose one plate carries a surface charge density $+\sigma$ and the other a surface charge density of $-\sigma$.

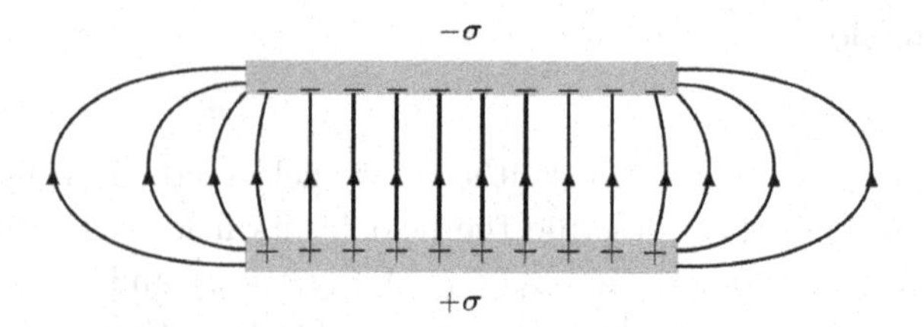

The lines with arrows indicate the direction of the electric field at points between the plates and at the sides of the plates. Note that within the space between the plates far from each end the electric field has the same direction at every point. Now, we will show that if the electric field has parallel vectors at two closely spaced points in space — with no charges in between them — then the field must have the same magnitude as well at these points.

To prove this, we will show that the converse cannot be true. Suppose we have two parallel but unequal electric fields close to each other. If such a thing is possible, then let us draw a rectangle such that one side is along one field and the other along the other parallel field. Let us assume that these fields are unequal in magnitude, and we shall call their magnitudes E_1 and E_2.

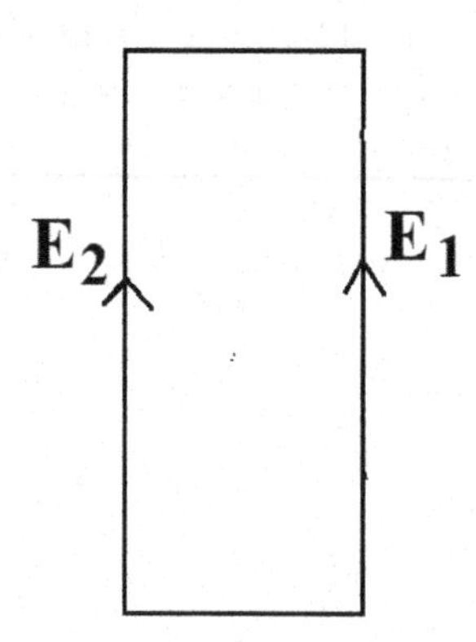

Now, let us carry out the closed line integral $\oint \mathbf{E} \cdot d\mathbf{r}$ in a positive (counter-clockwise) direction along the sides of this rectangle. Clearly, the value of this integral is $(E_1 - E_2)L$ where L is the side of the rectangle parallel to the electric field. By Stokes's theorem this integral equals $\iint \nabla \times \mathbf{E} \cdot \hat{n} dA$, and since $\nabla \times \mathbf{E} = 0$ for electrostatic fields, it follows that this integral is zero. Hence $E_1 = E_2$. Thus, if two electric fields in close proximity are parallel, they must have the same magnitude.

3.3.2 *Properties of Field Lines*

The foregoing diagram suggests a way of depicting electric fields pictorially, at least on a flat page. A line carrying an arrow indicates an electric field, the direction of the arrow showing the direction of the field. If the lines are parallel, the field is uniform. Since the electric field points away from a positive charge and towards a negative charge, the field lines originate in a positive charge and terminate in a negative charge. The diagram below shows the electric field lines due to a dipole, two equal and opposite charges situated at a short distance from each other. In this figure we have shown two charged spheres of equal radius:

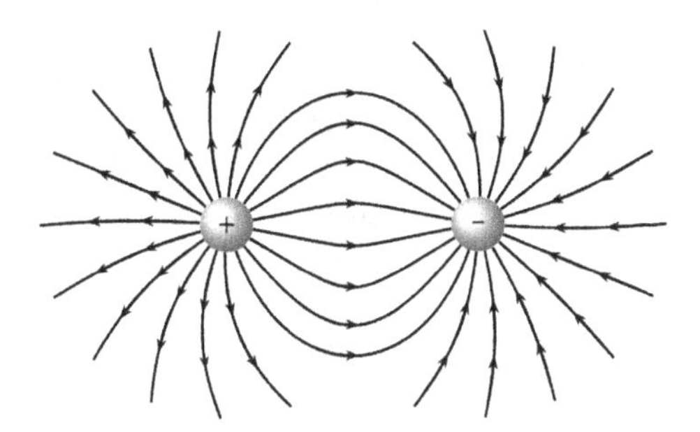

The field lines themselves are no more physical than the lines of latitude and longitude shown on maps. Electric field lines have some interesting properties, such as the following:

1. Field lines originate in a positive charge and terminate in a negative charge. (This rule will be modified when we bring in varying magnetic fields into the picture.)

2. Field lines never cross each other. Field lines appear to repel each other.

3. The stronger the field, the closer the field lines. The weaker the field, the more the lines are spread apart.

4. There is a tension along each field line. The lines tend to grow shorter in length, and to bring the opposite charges closer together.

5. In a uniform field, where the field has the same strength and direction at every point, the lines are parallel to each other.

3.3.3 *Conductors and Potential*

The degree to which charges can flow varies from material to material. The flow of charges through a substance is called conduction. Since charges flow relatively easily through metals, and with much difficulty through most non-metals, it appears that there is a natural distinction between substances that conduct and those that do not. Thus, we speak of conductors and non-conductors or insulators. Solid conductors are almost entirely metals and their alloys, with the notable exception of graphite, a crystalline form of carbon. Silver, gold and copper are excellent conductors. Non-metals are insulators, with the exception as noted of graphite, though diamond, another crystalline form of carbon, is a bad conductor.

Charge can be stored in a conductor if it is insulated from its surroundings. So, a metal sphere mounted on an insulating stand can be charged and will retain the charge. When a charge — in the form of billions of microscopic elementary charges — is given to such an insulated metal sphere, these microscopic charges will repel each other. This is because each charge will experience the field due to the other charges and be accelerated in the direction of the field (if the charge is positive) or in the opposite direction as the field (if the charge is negative), and the charges will tend to move far away from each other. But because they are confined to the conductor, they will occupy stable positions at which they will experience no net force. Each charge will experience an outward force of repulsion from the other charges, and an inward force compelling the charge to remain within the conductor. Thus, every charge will be in equilibrium, experiencing equal and opposite forces. We say that the conductor is in electrostatic equilibrium. The outward force felt by the charges per unit area will appear as an outward pressure acting on the conductor.

There are some important properties of such a charged conductor.

1. At any point close to the surface of a charged conductor in electrostatic equilibrium the field is perpendicular to the surface.

Suppose this were not true. Then there would be a component of the field parallel to the surface. Such a component would cause a charge on the surface to accelerate along the surface. But that is not possible, since the conductor is in equilibrium. Therefore, the field is perpendicular to the surface at all points close to the surface.

2. There is no field inside the conductor.

Suppose there is a field inside the conductor. Such a field must be representable by field lines that originate in a positive charge and terminate in a negative charge, or go to infinity. Within the conductor there cannot be charges of opposite signs, and the lines certainly do not go to infinity. Therefore there can be no fields inside a conductor.

3. All the charges reside on the surface of the conductor.

Suppose there is some charge inside the conductor. This charge would generate a field in its neighborhood. But since there can be no field inside the conductor, there cannot be a charge inside the conductor.

4. Charges tend to cluster at points of greater curvature on the surface.

The field lines everywhere are perpendicular to the surface. At a region of higher curvature the field lines diverge more than at a region of lower curvature.

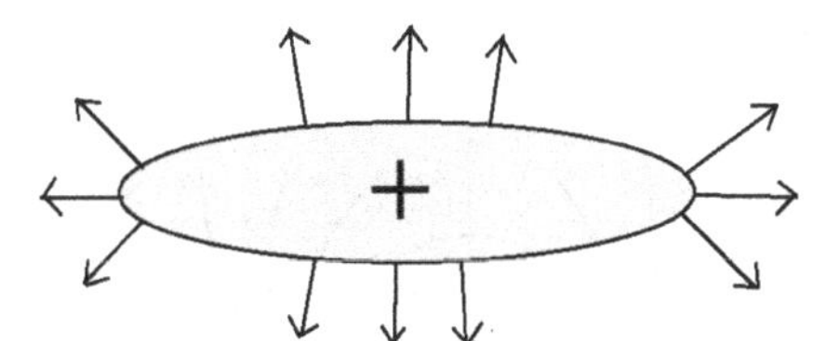

Field lines diverge more at surface regions of greater curvature

Therefore the field is stronger at regions of greater curvature (i.e. smaller radius of curvature) than at regions of smaller curvature. Now, the field close to a charged surface has magnitude $\frac{\sigma}{\epsilon_0}$, and so the surface charge density is greater at points of greater curvature. Consider a spherical conductor of radius r having surface charge density σ. The total charge on this conductor is $Q = 4\pi r^2\sigma$, and the potential at the surface of the conductor is $\phi = \frac{Q}{4\pi\epsilon_0 r} = \frac{r\sigma}{\epsilon_0}$. So, if the potential remains constant the charge density is inversely proportional to the radius: $\sigma \propto \frac{1}{r}$.

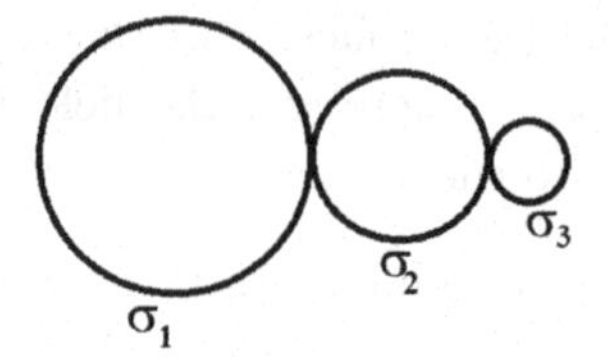

Suppose we have three conducting spheres of unequal radius in contact as shown above. At electrostatic equilibrium they will have the same potential. But they will not all have the same surface charge density.

5. Every point on and inside a conductor in electrostatic equilibrium is at the same electric potential.

If there were two neighboring points at different potentials, there would be a field between them, which is impossible.

3.3.4 *Image Charges*

Consider the field lines due to a dipole. Let us draw a plane perpendicular to the page intersecting the field lines halfway between the charges, as shown below:

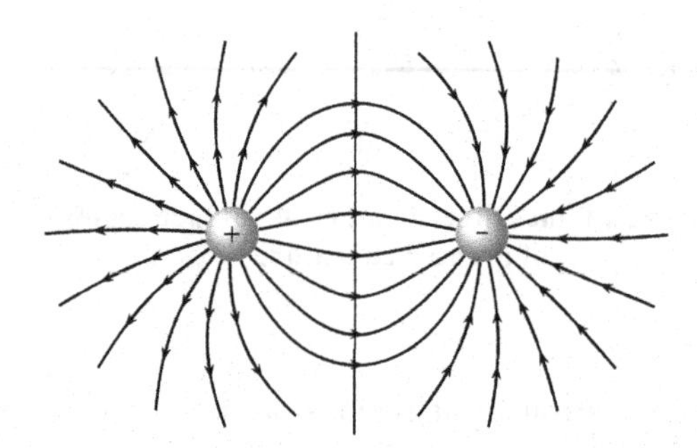

Because of the symmetry of the system consisting of two equal and opposite charges, the field lines must be perpendicular to the plane that divides the diagram at the center. At every point on the central plane, the potential must be zero, because every point is equidistant from the two equal and opposite charges.

Suppose now we were to remove the negative charge from the system, and place a grounded conductor with its plane edge along the central line:

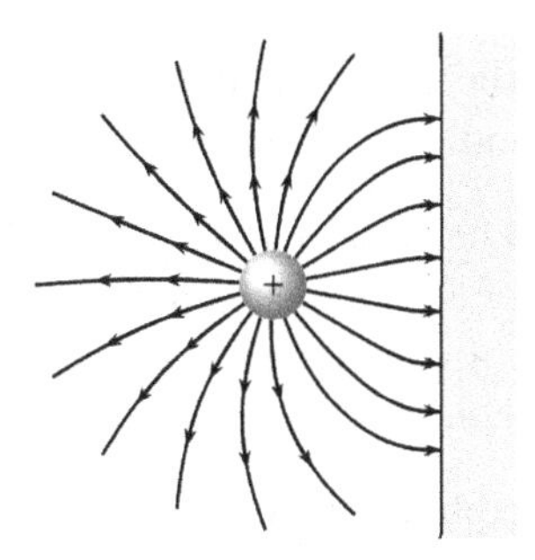

A grounded conductor is at zero potential. So every point on the conductor is at the same zero potential. So this grounded conductor has exactly the same potential as the plane intersecting the field lines midway between the charges.

We saw earlier that if the potential is specified at every point on the boundary, then the potential is determined at every point within the boundary, and this result applies even if there are charges within the boundary. Now, let us consider two different systems. System A consists of a positive charge $+Q$ placed at a distance d in front of an infinite grounded plane conductor. The grounded conductor is at zero potential. So, the region in front of the conductor is a semi-infinite space where the potential goes to zero at infinity, and is zero on the conducting surface. Next, consider System B which consists of two charges $+Q$ and $-Q$ placed at a distance of $2d$ from each other. Consider the semi-infinite space including the charge $+Q$ with the perpendicular midway plane as a boundary.

Both the spaces have the same boundaries with zero potential along the boundaries. Hence the potential at every point on one side of the mid-plane between the charges will be the same as if there was one charge and the midway plane was replaced by an infinite grounded conducting plane.

Thus, a charge placed in front of an infinite grounded conducting plane will generate a potential function in front of the plane which is identical to the potential due to a dipole.

So, as far as the field in front of the conducting plane is concerned, it is that due to the charge in front of the plane and an equal and opposite charge equidistant on the other side of the plane. This other charge is a fictitious charge, and is called the image of the real charge placed in front of the conductor.

This method of finding the potential due to a charge and a conducting surface is called the method of images.

The method of images can be used to find the potential due to a distribution of charges in the vicinity of a grounded conductor of any shape, though the method is useful only for a conductor of a regular shape such as a set of intersecting planes, a sphere, or a spheroid.

The method of images explains under what circumstances a charge can interact with itself. When placed near a grounded conductor, it generates an image of opposite sign and behind the surface of the conductor. The real charge interacts with its image. If the surface is a plane, the image has the same magnitude as the real charge. If the surface is convex, the image has a smaller magnitude. The rules for the image distance and image magnitude are essentially the same as those for virtual light images formed behind mirrors. Thus, with the help of a grounded conducting surface, a charge is able to "see itself".

3.4 Multipole Expansion of Potentials

3.4.1 *Dipoles*

A single charge of negligible dimensions can be treated as a point charge as long as we do not explore the regions microscopically close to the charge. Such a charge is called a *monopole.* A pair of equal and opposite point charges kept at a small distance from each other is called a *dipole.*

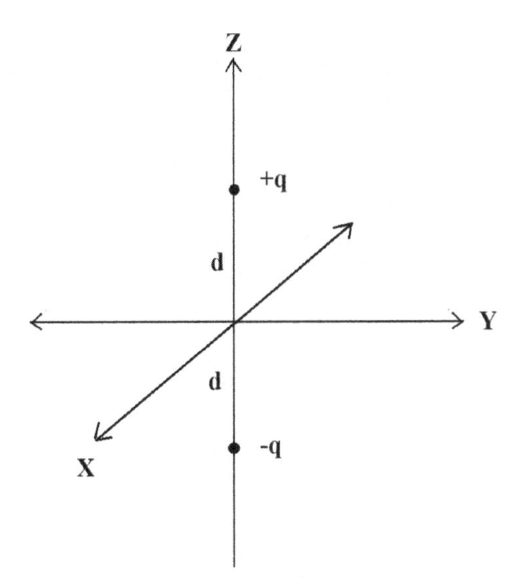

The *dipole moment* of a pair of charges q and $-q$ with the displacement vector $\mathbf{r}$ from $-q$ to $+q$ is defined as $\mathbf{p} = q\mathbf{r}$ and the SI unit of dipole moment is the C.m (Coulomb-meter). It is conventional to place the dipole centered at the origin pointing along the positive z axis. So $\mathbf{p} = 2qd\hat{k}$.

We can find the potential at a point $P(r, \theta, \varphi)$ by algebraically adding the potentials due to each charge:

$$\phi = \frac{q}{4\pi\epsilon_0}\left[\frac{1}{\sqrt{r^2 - 2rd\cos\theta + d^2}} - \frac{1}{\sqrt{r^2 + 2rd\cos\theta + d^2}}\right] \tag{3.47}$$

If $r \gg d$, we can write the potential as

$$\phi = \frac{p\cos\theta}{4\pi\epsilon_0 r^2} \tag{3.48}$$

Thus the potential of a dipole drops off as the inverse square of the distance. At a large distance along the positive z axis, $\phi = \frac{p}{4\pi\epsilon_0 r^2}$, and along the negative z axis, $\phi = -\frac{p}{4\pi\epsilon_0 r^2}$. At any point on the xy plane the potential is zero.

Using $\mathbf{E} = -\nabla\phi$, we can obtain the spherical polar components of the electric field as

$$E_r = \frac{2p\cos\theta}{4\pi\epsilon_0 r^3}$$

$$E_\theta = \frac{p\sin\theta}{4\pi\epsilon_0 r^3} \tag{3.49}$$
$$E_\varphi = 0$$

We may write the potential of a dipole in Cartesian coordinates as follows:

$$\phi = \frac{pz}{4\pi\epsilon_0(x^2+y^2+z^2)^{3/2}} \tag{3.50}$$

This will yield the Cartesian components of the electric field of a dipole:

$$E_x = \frac{3pzx}{4\pi\epsilon_0 r^5}$$
$$E_y = \frac{3pzy}{4\pi\epsilon_0 r^5} \tag{3.51}$$
$$E_z = -\frac{p}{4\pi\epsilon_0 r^3} + \frac{3pz^2}{4\pi\epsilon_0 r^5}$$

and these three equations can be combined into a single vector equation as

$$\mathbf{E} = \frac{1}{4\pi\epsilon_0 r^3}\left[\frac{3\mathbf{r}(\mathbf{p}\cdot\mathbf{r})}{r^2} - \mathbf{p}\right] \tag{3.52}$$

which is independent of the orientation of the dipole relative to the coordinates, though the dipole is still centered at the origin. For the more general case, where the dipole is at an arbitrary position at a distance r_0 from the origin,

$$\mathbf{E} = \frac{3\hat{n}(\mathbf{p}\cdot\hat{n}) - \mathbf{p}}{4\pi\epsilon_0|\mathbf{r}-\mathbf{r}_0|^3} \tag{3.53}$$

where $\hat{n}$ is a unit vector directed along $\mathbf{r} - \mathbf{r}_0$.

3.4.2 *Multipoles*

If we have four charges of equal magnitude, two positive and two negative, placed at the vertices of a rectangle, so that any two vertices at the ends of a side have opposite sign, then we have a quadrupole.

The potential of a quadrupole can be shown to drop off as the inverse cube of the distance: $\phi \propto \frac{1}{r^3}$.

An octopole has eight charges of equal magnitude with alternating signs along the adjacent vertices of a rectangular cuboid (also called a three-dimensional orthotope, rectangular prism, or rectangular parallelepiped).

The potential of an octopole can be shown to drop off as the inverse fourth power of the distance: $\phi \propto \frac{1}{r^4}$.

We will now draw an analogy with the study of sound waves. A pure note has a single frequency, but a note sounded on a musical instrument is a combination of harmonics, where each harmonic is a multiple of the fundamental mode. The potential generated by a distribution of charges, such as a continuous charge density ρ, can be analyzed as a combination of multipole potentials. We shall next set up the mathematical analysis of such potentials.

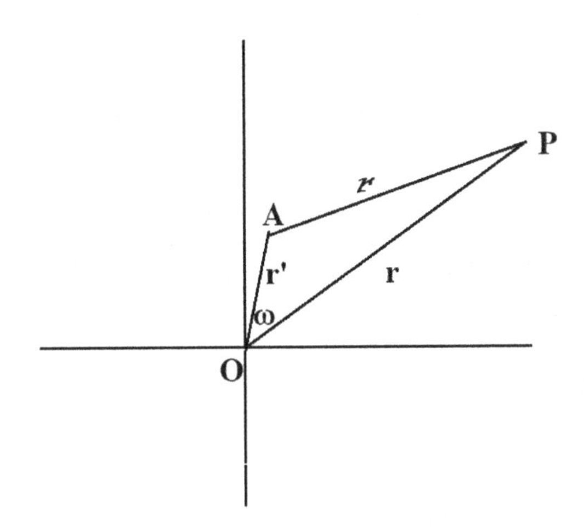

Since we are talking about an extensive distribution of charges, we can no longer limit ourselves to point charges placed at the origin, or a pair of charges placed on either side of the origin. Let us consider a point charge at the point A in the figure shown above, and let us calculate the potential at P due to this charge at A.

The potential at P due to a charge at A is proportional to $\frac{1}{|\mathbf{r}-\mathbf{r}'|}$.

Now,

$$\frac{1}{|\mathbf{r}-\mathbf{r}'|} = \frac{1}{\sqrt{r^2 + r'^2 - 2rr'\cos\omega}} \tag{3.54}$$

For $r' < r$, the right side can be expanded as a series of Legendre polynomials in $\cos\omega$:

$$\frac{1}{|\mathbf{r}-\mathbf{r}'|} = \frac{1}{r}\sum_{\ell=0}^{\infty}\left(\frac{r'}{r}\right)^{\ell} P_\ell(\cos\omega) \tag{3.55}$$

Now, as $\mathbf{r}' \to \mathbf{r}$, the left side diverges. The right side becomes $\frac{1}{r}\sum_{\ell=0}^{\infty} P_\ell(1)$. Since $P_\ell(1) = 1$ for all ℓ, the sum diverges. Thus this expansion is not helpful for evaluating the potential at positions close to the source.

A more complex expansion of $\frac{1}{|\mathbf{r}-\mathbf{r}'|}$ involves the spherical harmonics. We make use of the spherical harmonic addition theorem:

$$P_\ell(\cos\omega) = \frac{4\pi}{2\ell+1}\sum_{m=-\ell}^{\ell} Y_{\ell m}(\theta,\varphi)Y^*_{\ell m}(\theta'\varphi') \tag{3.56}$$

where (r',θ',φ') are the coordinates of A, and (r,θ,φ) are those of P. So

$$\frac{1}{|\mathbf{r}-\mathbf{r}'|} = 4\pi\sum_{\ell=0}^{\infty}\sum_{m=-\ell}^{m=\ell}\frac{1}{2\ell+1}\frac{r'^\ell}{r^{\ell+1}}Y^*_{\ell m}(\theta',\varphi')Y_{\ell m}(\theta,\varphi) \tag{3.57}$$

The potential at a point P $(\mathbf{r})$ due to a distribution of charges over a region described by the variable position vector $\mathbf{r}'$ is given by

$$\phi(\mathbf{r}) = \frac{1}{4\pi\epsilon_0}\int\frac{\rho(\mathbf{r}')}{|\mathbf{r}-\mathbf{r}'|}d^3\mathbf{r}' \tag{3.58}$$

The integral can be expanded using spherical harmonics as follows:

$$\phi(\mathbf{r}) = \frac{1}{\epsilon_0}\sum_{\ell,m}\frac{1}{2\ell+1}\left[\int Y^*_{\ell m}(\theta',\varphi')r'^\ell\rho(\mathbf{r}')d^3\mathbf{r}'\right]\frac{Y_{\ell m}(\theta,\varphi)}{r^{\ell+1}} \tag{3.59}$$

The integral inside the square brackets is called the *multipole moment* and is written as $q_{\ell m}$. And so

$$\phi(\mathbf{r}) = \frac{1}{4\pi\epsilon_0}\sum_{\ell=0}^{\infty}\sum_{m=-\ell}^{\ell}\frac{4\pi}{2\ell+1}q_{\ell m}\frac{Y_{\ell m}(\theta,\varphi)}{r^{\ell+1}} \tag{3.60}$$

where

$$q_{\ell m} = \int Y^*_{\ell m}(\theta',\varphi')r'^\ell\rho(\mathbf{r}')d^3\mathbf{r}' \tag{3.61}$$

The multipole moments have dimension L^ℓ.

Exercise:

Four charges of magnitude q are placed at the vertices of a square of side $2a$ in the xy plane. $P(r,\varphi)$ is a point in the plane. Positive charges $+q$ are placed at $(0,a)$ and $(0,-a)$ on the y axis. Negative charges $-q$ are placed at $(-a,0)$ and $(a,0)$ on the x axis. $r > a$. Evaluate expressions for
(a) the monopole moment of the system
(b) the dipole moment of the system
(c) the quadrupole moment of the system
(d) the potential at P due to the charge distribution
(e) the electric field at P due to the charge distribution.

3.5 Electrostatic Energy

3.5.1 *Potential and Potential Energy*

In order to bring two like charges closer, an external agent has to perform work on the charges. This work done by the agent is equal to the increase in potential energy of the system of the two charges. If the charges are unlike, the work done by the agent is negative, and the increase in potential energy is negative.

The magnitude of the force between two charges q_1 and q_2 at a distance r from each other is

$$F = \frac{q_1 q_2}{4\pi\epsilon_0 r^2} \tag{3.62}$$

The work done on the charges by an external force in bringing them from a distance r_1 to a distance r_2 from each other equals

$$W = \int_{r_1}^{r_2} -\frac{q_1 q_2}{4\pi\epsilon_0 r^2} dr = \frac{q_1 q_2}{4\pi\epsilon_0}\left(\frac{1}{r_2} - \frac{1}{r_1}\right) \tag{3.63}$$

The negative sign inside the integral indicates that the displacement dr is opposite to the force F. Suppose the charges are brought from infinity. We can do this in two steps. First, charge q_1 alone is brought from infinity. The work done is 0, because there are no other charges in the system, and so q_1 does not experience any force. Next, keeping charge q_1 fixed, we bring q_2 from infinity to a point at a distance r_2 from q_1. Since r_1 is infinity, the work done, which is the potential energy of the system of two charges, is

$$U = \frac{q_1 q_2}{4\pi\epsilon_0 r_2} \tag{3.64}$$

Now, we found earlier that the potential at a point at a distance r_2 from a charge q_1 is given by

$$\phi = \frac{q_1}{4\pi\epsilon_0 r_2} \tag{3.65}$$

So, the potential energy of a system of the two charges is given by $U = \phi q_2$ where ϕ is the potential due to the charge q_1 at the point occupied by charge q_2.

We will next calculate the potential energy of a charge and an infinite conducting plate at zero potential, and then show that this is equal to the potential energy of the charge and its image behind the plate.

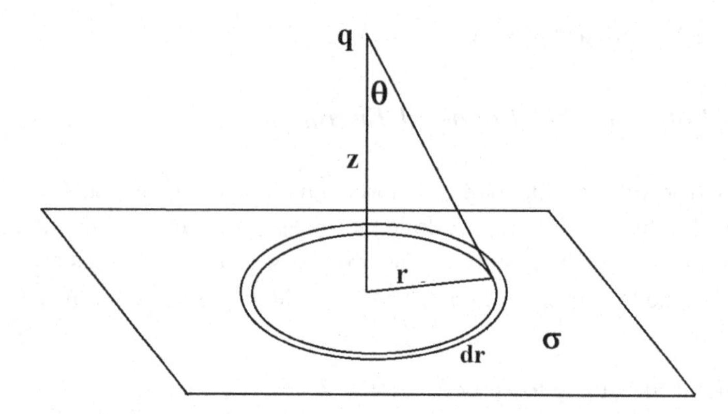

Consider a charge q at a distance z from a large grounded conducting plate at zero potential. The presence of the charge q will induce a surface charge on the plate. This charge will not be distributed uniformly on the plate, but its magnitude will diminish with the distance from the charge. So the plate will have an induced surface charge density that is a function of the distance r from the foot of the perpendicular drawn from the charge to the plate. Next, consider a ring of radius r and infinitesimal width dr centered at the foot of the perpendicular.

The amount of charge residing on this ring is equal to $2\pi r dr\sigma(r)$. σ is not constant, but is a function of the radial distance r. (σ is opposite in sign to q.)

So the potential energy of the charge q plus this ring equals

$$dU = \frac{q2\pi r dr\sigma(r)}{4\pi\epsilon_0\sqrt{r^2+z^2}} = \frac{q}{2\epsilon_0}\frac{r\sigma(r)dr}{\sqrt{r^2+z^2}} \tag{3.66}$$

This number is negative, because σ has the opposite sign as q. Now, the force component F_z on the charge q due the charge on the ring is obtained by taking the z derivative of the potential energy:

$$F_z = -\frac{\partial(dU)}{\partial z} = \frac{qr\sigma(r)drz}{2\epsilon_0(r^2+z^2)^{3/2}} \tag{3.67}$$

This force component is negative, because q and σ have opposite sign, and so the force is downward, attracting the charge q to the plate. The force acting on the ring itself is equal and opposite to this force, and so equal to $-F_z$, and is directed upward.

The upward force acting on the ring of charge is equal to the field at the ring divided by the charge on the ring. The electric field **E** immediately above the surface of the plate is perpendicular to the surface and so is in the z direction:

$$E = E_z = -F_z/(2\pi r dr \sigma) = -\frac{qz}{4\pi\epsilon_0(r^2+z^2)^{3/2}} \tag{3.68}$$

where $-F_z$ is the force acting on the ring.

And since $E = \frac{\sigma}{2\epsilon_0}$, this yields

$$\sigma(r) = \frac{-qz}{2\pi(r^2+z^2)^{3/2}} \tag{3.69}$$

Therefore

$$dU = \frac{-q^2 z}{4\pi\epsilon_0}\frac{rdr}{(r^2+z^2)^2} \tag{3.70}$$

So the potential energy of the entire system consisting of the charge q and the plate is

$$\begin{aligned} U &= \frac{-q^2 z}{4\pi\epsilon_0}\int_0^\infty \frac{rdr}{(r^2+z^2)^2} \\ &= \frac{-q^2 z}{4\pi\epsilon_0}\left[\frac{-1}{2(r^2+z^2)}\right]_0^\infty = -\frac{q^2}{4\pi\epsilon_0 2z} \end{aligned} \tag{3.71}$$

This is exactly equal to the potential energy of a charge q and its image $-q$ at a distance $2z$ from each other.

3.5.2 *Energy of an Electric Field*

In general, if the potential at a point is ϕ, the work done in bringing a charge q from infinity to that point is $q\phi$. If there is a system of charges already in place before charge q was brought in, then the work done in bringing this additional charge equals the increase in potential energy of the system. The total potential energy of the system is the total work done in bringing all the charges from infinity to the points occupied by the charges. If a set of n charges occupy positions $\mathbf{r}_i$, then the potential energy of a pair of such charges is $U_{ij} = \frac{q_i q_j}{4\pi\epsilon_0|\mathbf{r}_i - \mathbf{r}_j|}$. Thus the total potential energy of the system of n charges is given by

$$U = \sum_{i>j}^{n} \frac{q_i q_j}{4\pi\epsilon_0|\mathbf{r}_i - \mathbf{r}_j|} \tag{3.72}$$

If we write $|\mathbf{r}_i - \mathbf{r}_j| = r_{ij}$, then this equation can be written as

$$U = \frac{1}{2}\sum_{i \neq j}^{n} \frac{q_i q_j}{4\pi\epsilon_0 r_{ij}} = \frac{1}{2}\sum_{i}^{n} q_i \left(\sum_{j \neq i} \frac{q_j}{4\pi\epsilon_0 r_{ij}} \right) \tag{3.73}$$

The factor $\frac{1}{2}$ compensates for the double counting over pairs of charges. The expression within brackets is the potential generated by all the other charges q_j at the point occupied by the charge q_i. We shall call this quantity $\phi(\mathbf{r}_i)$, and so we can express the potential energy of the entire system of point charges as

$$U = \frac{1}{2}\sum_{i}^{n} q_i \phi(\mathbf{r}_i) \tag{3.74}$$

If we have a continuous distribution of charges within some region, we can calculate the potential energy of the distribution of charges using the formula

$$U = \frac{1}{2}\iiint_V \rho(\mathbf{r})\phi(\mathbf{r}) d\tau \tag{3.75}$$

Now, $\rho = -\epsilon_0 \nabla^2 \phi$. So we obtain for the potential energy

$$U = -\frac{\epsilon_0}{2}\iiint_V \phi \nabla^2 \phi d\tau \tag{3.76}$$

$\nabla \cdot (\phi \nabla \phi) = \nabla \phi \cdot \nabla \phi + \phi \nabla^2 \phi$. So

$$U = -\frac{\epsilon_0}{2}\iiint_V \nabla \cdot (\phi \nabla \phi) d\tau + \frac{\epsilon_0}{2}\iiint_V \nabla \phi \cdot \nabla \phi d\tau \tag{3.77}$$

The first integral can be converted into a surface integral over the closed surface enclosing the region containing the charges. And we shall use $\nabla \phi = -\mathbf{E}$ in the second integral.

$$U = -\frac{\epsilon_0}{2}\oint\!\!\oint_S \phi \nabla \phi \cdot \hat{n} dS + \frac{\epsilon_0}{2}\iiint_V \mathbf{E} \cdot \mathbf{E}\, d\tau \tag{3.78}$$

We will now let the closed surface recede to infinity, so that it is a sphere of very large radius R. At any point on this surface, the potential is of the order of $\frac{1}{R}$, and the gradient of the potential is of the order of $\frac{1}{R^2}$, and so the integrand drops off as $\frac{1}{R^3}$. This integrand is evaluated over a closed surface of area $4\pi R^2$. And so the integral itself is of the order of $\frac{1}{R}$, and vanishes as $R \to \infty$.

And so, the total potential energy of the charge distribution becomes

$$U = \frac{\epsilon_0}{2} \iiint_V E^2 \, d\tau \tag{3.79}$$

In order to find the potential energy of the charge distribution we evaluated the electric field throughout all space. This suggests that the repository of the potential energy of the charges is not within the charges themselves, but in the space that carries the field itself. This is a feature of every kind of potential energy. Whereas kinetic energy is the property of individual particles, potential energy is the property of systems. So the gravitational potential energy of the earth-sun system does not belong only to the earth or the sun, but is a property of the gravitational field between the two objects. So, we can drop the adjective potential and simply talk about the energy of a field. And since the space over which we have evaluated this electrical energy is infinite, it is more helpful to talk about energy per unit volume or energy density (u):

$$u = \frac{\epsilon_0}{2} E^2 \tag{3.80}$$

which is measured in $\mathrm{J/m^3}$. It is evident that the energy density can never be negative.

3.5.3 *Energy of a Charged Capacitor*

If a charge q is displaced by an external force from a point where the potential is ϕ_i to a point where the potential is ϕ_f, the work done by the external force, or the increase in potential energy of the system $= q(\phi_f - \phi_i)$. This increase could be positive or negative. In the event that the initial potential $\phi_i = 0$, the work done, or increase in potential energy $= q\phi_f$. Where would we find points having zero potential? A point very far from any charge, or at infinity, is at zero potential. Also, a point equidistant from two equal and opposite charges, or every point in the plane halfway between the plates of a capacitor, are all at zero potential.

Consider a parallel plate capacitor having charges $+Q$ and $-Q$ with corresponding charge densities $+\sigma$ and $-\sigma$ on each plate. Let the area of each plate be A and the distance between the plates be d. Now, let us picture the two charged plates brought really close together so they almost touch each other. At this stage the potential energy of the capacitor would be very small, practically zero. Now, suppose the positive plate is slowly moved away from the negative plate through a distance d. The potential energy of the capacitor now increases, and is equal to the work done by the external agent who is carrying out this displacement.

The force acting on the positive plate is the charge on the plate multiplied by the field experienced by the plate, which is the field due to the negative plate, with magnitude $E_1 = \frac{\sigma}{2\epsilon_0}$ where $\sigma = Q/A$. So $Q = 2\epsilon_0 E_1 A$. So, the potential energy of the capacitor U = force × displacement = $QE_1 d = 2A\epsilon_0 E_1^2 d$. Now, the electric field between the plates is twice the field generated by a single plate, $E = 2E_1$. And so we can write the potential energy of the capacitor as $U = \frac{1}{2}\epsilon_0 E^2 Ad$. Since $Ad = V$, the volume of the space between the plates, the energy per unit volume between the plates is $u = \frac{1}{2}\epsilon_0 E^2$.

3.5.4 *Potential Energy and Field Energy of a Pair of Charges*

The potential energy of a pair of equal and opposite charges q and $-q$ at a distance r from each other is

$$-\frac{q^2}{4\pi\epsilon_0 r}$$

This is clearly negative. However, we know that the energy of an electric field can never be negative. So this is a paradox. But only apparently. In physics what is real — i.e. measurable — is not the potential energy, but the change in potential energy. For a conservative force system — like electrical or gravitational forces, a change in potential energy is measurable as a corresponding change in kinetic energy. We saw earlier that a single charge q in an otherwise empty space has zero potential energy. But this is not quite true. In order to create that charge, parts of the charge have to come together, and that requires work done by the agent responsible for bringing the parts of the charge together, and so even an isolated charge has electrical energy.

Let us model a charge as a hollow sphere of radius r_0 with a total charge q. The electric field generated by this charge at any point outside the sphere is $\frac{q}{4\pi\epsilon_0 r^2}$. So the total electrical energy in all of space is given by

$$U = \frac{1}{2}\epsilon_0 \iiint_{r_0}^{\infty} \left(\frac{q}{4\pi\epsilon_0 r^2}\right)^2 4\pi r^2 dr = \frac{q^2}{8\pi\epsilon_0 r_0} \tag{3.81}$$

We would get an identical expression if the charge were negative. So, for two such charged spheres at great distance from each other, the total field energy is $\frac{q^2}{4\pi\epsilon_0 r_0}$. We may call this the initial potential energy of the two charges kept at great distance from each other. The work done by an external agent in bringing them closer to a distance r from each other is $-\frac{q^2}{4\pi\epsilon_0 r}$. Hence the total potential energy — or more accurately, the field energy — of the charges at a distance r from each other becomes

$$\frac{q^2}{4\pi\epsilon_0}\left(\frac{1}{r_0} - \frac{1}{r}\right)$$

Since $r_0 < r$, this field energy is always positive.

For a given pair of charges only the second term is measurable. But the first term cannot be neglected in quantum theory or relativity, as we shall see in a later chapter. So, if two equal and opposite charges are suddenly created at large distances from each other, the energy needed to create them is

$$\frac{q^2}{8\pi\epsilon_0 r_0}$$

But if they are created at a distance $r = r_0$ from each other, the energy required is 0. Thus, it is possible for equal and opposite charges to appear out of nowhere and then disappear as long as the distance between them is of the order of r_0, for then not only the total charge but also the total energy of the system will be conserved.

Chapter 4

Magnetostatics

4.1 The Phenomenon of Magnetism

4.1.1 *Properties of Magnets*

Magnetism as a natural phenomenon has been attested in antiquity. Certain pieces of metallic ores were found to have some peculiar properties: they attracted pieces of iron, and when suspended freely from its center of mass, such a piece of ore swung to a preferred orientation that was approximately north-south. The two ends of such a piece were dubbed the north-seeking or north pole, and the south-seeking or south pole. When two such pieces of ore were brought close, the ends of opposite polarity attracted, and the ends of like polarity repelled.

A piece of ore having these properties came to be called a magnet in English, because a large amount of such ores were mined in Magnesia in Greece. Soon artificial magnets came to be made through contact with natural magnets.

A tiny magnet in the shape of a needle pivoted at its center of mass — called a compass — points in a particular direction at any point on the surface of the earth. In analogy with the electric field, we say there is a magnetic field at every point close to the surface of the earth. A magnetic piece of iron such as a bar magnet also generates its own magnetic field. Again, in analogy to an electric field, a magnetic field can be represented by field lines. These field lines diverge close to a north or south pole, and are parallel at great distances from a pole:

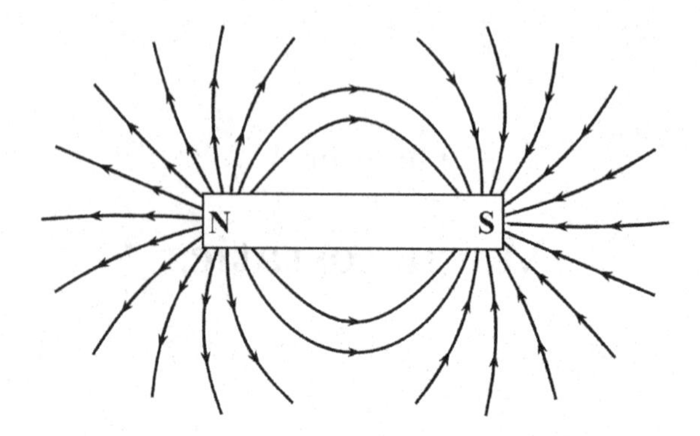

It is seen that the field lines are very similar to those of an electric dipole. But there is a difference. Whereas positive and negative charges can be isolated, a north pole can never be separated from a south pole. So, if we were to cut the bar magnet in half, we would obtain not a north pole and a south pole, but two shorter bar magnets, each with a north and a south pole:

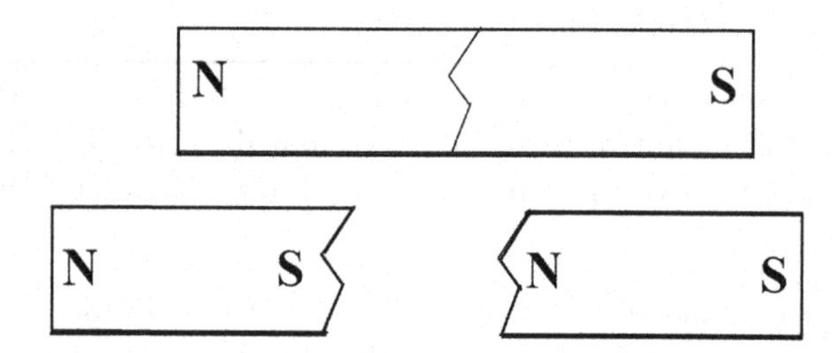

The inseparability of magnetic poles indicates that magnetic field lines may have some fundamental differences from the electric field lines.

Suppose we consider an imaginary Gaussian surface that intersects a bar magnet as shown below:

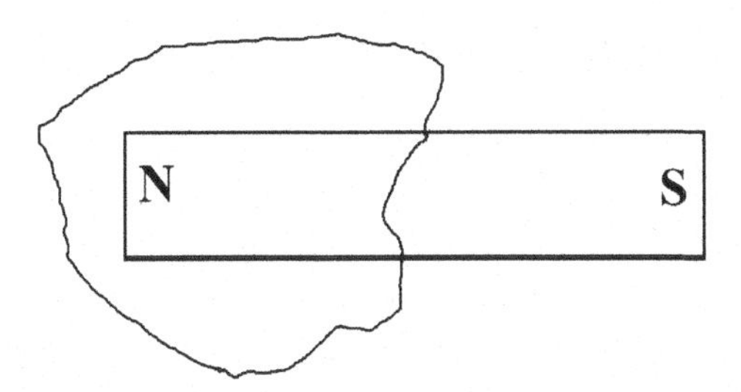

If a north pole could be isolated, then the total divergence of the magnetic field out of the Gaussian surface would be non-zero. But, if the north pole cannot be isolated from the south pole, then we should assume that a south pole is generated inside the Gaussian surface and a north pole outside the surface:

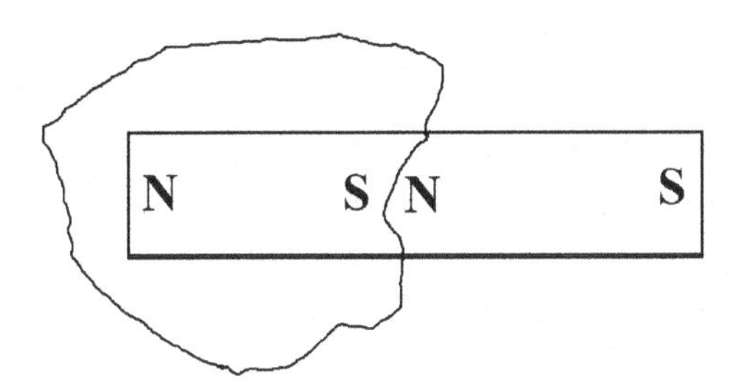

Thus the flux of the magnetic field through any closed surface must be zero.

Let us denote the magnetic field by the vector $\mathbf{B}$ leaving the precise definition of this field for later.

So, our hypothesis suggests that for any closed surface, $\oiint_S \mathbf{B} \cdot \hat{n} dS = 0$. So by the divergence theorem, the volume integral of the divergence of $\mathbf{B}$ over the volume enclosed by the surface is zero. Since this is true for any volume enclosed by any surface, it follows that $\nabla \cdot \mathbf{B} = 0$ at all times. This is the magnetic equivalent of Maxwell's first equation $\nabla \cdot \mathbf{E} = \frac{\rho}{\epsilon_0}$ and a comparison of the two suggests that there is no such thing as a magnetic charge.

Our inability to isolate a magnetic pole has therefore yielded a mathematical equation expressing a fundamental property of the magnetic field. A rigorous treatment will serve to confirm and corroborate our intuitive reasoning.

4.1.2 *Coulomb's Law for Magnets*

Magnetic poles cannot be isolated. But if we have two very long magnets, the forces between their poles can be investigated. Coulomb discovered the inverse square law for magnetic poles. He found that the force between two poles is directly proportional to the products of their pole strengths, inversely proportional to the square of the distance between the poles, and acts along the line joining the poles. In today's SI units the force in newtons (N) can be expressed as

$$\mathbf{F} = \mu_0 \frac{p_1 p_2}{4\pi r^2} \hat{e}_r \tag{4.1}$$

In this equation we have expressed the constants in the light of later knowledge regarding magnetism. And so we will not bother to assign units to the different quantities. We will measure the pole strengths p_1 and p_2 in unnamed SI units. r is the distance between the poles in meters, and $\hat{e}_r$ a unit vector directed along the line between the poles. μ_0 is called the *permeability of free space*, and in the SI system has the exact value $4\pi \times 10^{-7}$.

The strength of a magnet is generally expressed in terms of its magnetic moment. If $\mathbf{s}$ is the displacement vector directed from the south pole to the north pole, and the strength of each pole has magnitude p, then the magnetic moment is defined as

$$\mathfrak{m} = p\mathbf{s} \tag{4.2}$$

Unlike the pole strength, the magnetic moment — also called the magnetic dipole moment — is an important quantity, and we will express it in its modern units, which are ampère meters squared (A.m^2).

A magnetic pole experiences a force in a magnetic field. Such a field can be represented as a vector field with magnitude and direction at any point in space. In analogy to the definition of the electric field, we may define the magnetic field $\mathbf{B}$ by the force experienced by a pole in the magnetic field,

$$\mathbf{F} = p\mathbf{B} \tag{4.3}$$

Now, a pole cannot be isolated, and so a more useful formula would be the magnetic field due to a dipole at a large distance (large compared to the length of the magnet) from the center of the magnet. Since the same law — Coulomb's law — holds for both electric charges and magnetic poles, we seek an equation analogous to the equation for the electric field due to an electric dipole:

$$\mathbf{E}_{dip}(\mathbf{r}) = \frac{1}{4\pi\epsilon_0} \frac{1}{r^3} \left[3(\mathbf{p} \cdot \hat{r})\hat{r} - \mathbf{p}\right] \tag{4.4}$$

And so the magnetic field due a magnet having dipole moment $\mathfrak{m}$ at a long distance r from the magnet is given by

$$\mathbf{B}_{dip}(\mathbf{r}) = \frac{\mu_0}{4\pi}\frac{1}{r^3}\left[3(\mathfrak{m}\cdot\hat{r})\hat{r} - \mathfrak{m}\right] \tag{4.5}$$

4.2 Current and Magnetic Field

4.2.1 *Electric Current*

A charge q experiences a force in an electric field according to $\mathbf{F} = q\mathbf{E}$. In a vacuum a charged particle would accelerate according to Newton's Second Law. But in a conductor, free charges collide with the atoms and with other free charges and as a result they cannot accelerate freely. So, when an electric field is applied to a conductor, the free electrons would attempt to accelerate, but would end up moving with a constant velocity called the *drift velocity*. This flow of electrons leads to an electric current. Because of historical reasons, the direction of electric current was chosen so that it came out to be the opposite of the actual direction of electron flow. So it is conventional to speak of positive charges moving along a conductor. The magnitude of electric current I is defined as the amount of charge crossing any section of the conducting wire per second:

$$I = \frac{dQ}{dt} \tag{4.6}$$

The vector current density $\mathbf{J}$ at a point is defined as the product of the charge density and the velocity of charge at that point:

$$\mathbf{J} = \rho\mathbf{v} \tag{4.7}$$

Experiments have shown that the direction of flow of the current is the direction of the electric field, and the magnitude of the current is proportional to the strength of the field.

So, we can write: $\mathbf{E} = \varrho\mathbf{J}$. The proportionality constant ϱ is called the resistivity of the conducting material. This ϱ is to be distinguished from the charge density ρ. The greater the resistivity, the smaller the current, for a particular electric field.

Suppose the current flows along a cylindrical conductor of constant cross-section A. Let $\mathbf{L}$ be the displacement vector of a section of the conductor of length L. Then

$$\mathbf{E}\cdot\mathbf{L} = \varrho\mathbf{J}\cdot\mathbf{L} \tag{4.8}$$

Now, $\mathbf{E} \cdot \mathbf{L} = \Delta V$, the potential difference across the ends of the length L of the conductor, $\mathbf{J} \cdot \mathbf{L} = JL$, and $J = I/A$. So

$$\Delta V = \frac{\varrho L}{A} I \tag{4.9}$$

The ratio $\Delta V/I$ is called the resistance R (measured in ohms, Ω) of the section of the wire. So

$$R = \frac{\varrho L}{A} \tag{4.10}$$

4.2.2 *Oersted's Law*

Oersted discovered that when current flows through a wire, a curling magnetic field appears concentric with the wire in a plane perpendicular to the wire. The direction of the magnetic field has a particular relation to the direction of the current. The current flows in the direction that a right handed screw would advance if rotated along the direction of the magnetic field. This rule is also called the right-hand rule, as indicated in the figure below. If the thumb of the right hand indicates the direction of the current, the curved fingers indicate the direction of the field.

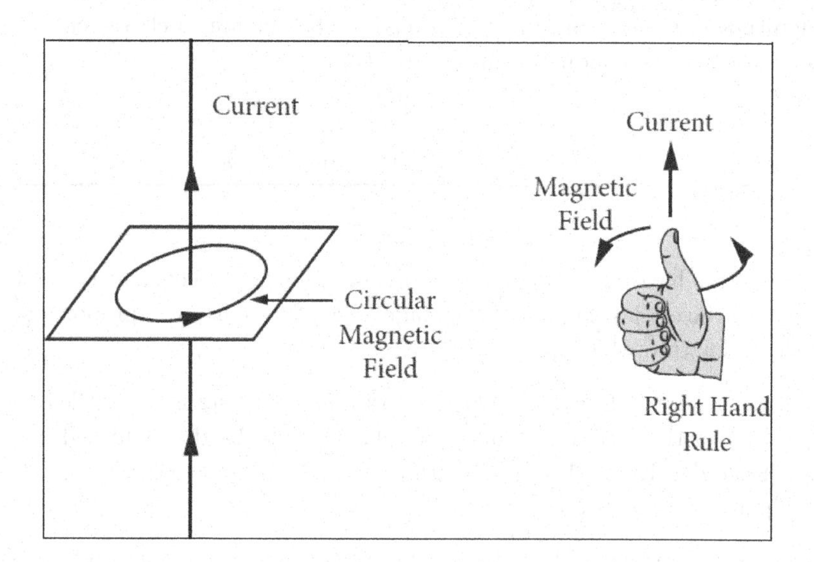

Oersted's law is stated in mathematical form as

$$\oint_C \mathbf{B} \cdot d\mathbf{r} = \mu_0 I \tag{4.11}$$

where μ_0 is the permeability of free space. If we consider a point at a distance r from the center of the wire, then the above equation yields the magnitude of the magnetic field at that point as

$$B = \frac{\mu_0 I}{2\pi r} \tag{4.12}$$

If we take the direction of the current to be the z axis, the magnetic field is perpendicular to the z axis, and can be expressed in vector form (in cylindrical coordinates) as

$$\mathbf{B} = \frac{\mu_0 I}{2\pi \rho}\hat{e}_\varphi \tag{4.13}$$

where $\hat{e}_\varphi$ is a unit vector tangential to the magnetic field at every point.

Oersted's law indicates that a time independent magnetic field is generated by a time independent electric current. Taking the divergence of Eq. (4.13) and expressing the result in cylindrical polar coordinates — whereby $B_\rho = 0, B_z = 0$, and $B_\varphi = \frac{\mu_0 I}{2\pi\rho}$ — we obtain

$$\nabla \cdot \mathbf{B} \equiv \frac{1}{\rho}\frac{\partial(\rho B_\rho)}{\partial \rho} + \frac{\partial B_z}{\partial z} + \frac{1}{\rho}\frac{\partial B_\varphi}{\partial \varphi} = 0 \tag{4.14}$$

Thus Oersted's law provides a mathematical proof for a divergence free magnetic field generated by a steady current. Later we will see that regardless of how a magnetic field is generated, it will always be divergence free: $\nabla \cdot \mathbf{B} = 0$.

Oersted's law introduces a concept of profound importance for the electromagnetic field — handedness. A current generates a field of a definite handedness. This phenomenon raised some questions of symmetry, and the physicist Ernst Mach expressed his shock on seeing that nature was apparently not ambidextrous. The relationship between the current and the field is illustrated by the right-hand rule, and thus lacks a bilateral symmetry.

We have thus far considered a current traveling through a wire, which has a relatively small and uniform radius. In the more general case, we could consider current moving through a medium, and the rate of flow of charge may not be uniform throughout the medium. Suppose at a given point within the medium the charge density is ρ and the velocity of the charge is $\mathbf{v}$. We define the current density

$$\mathbf{J} = \rho \mathbf{v} \tag{4.15}$$

The rate of flow of charge across a small area element dA is given by $\mathbf{J}\cdot\hat{n}dA$ where $\hat{n}$ is a unit vector perpendicular the area element. The rate of flow of charge out of a volume V contained by a closed surface S is

$$\oiint_S \mathbf{J}\cdot\hat{n}dA = \iiint_V \nabla\cdot\mathbf{J}d\tau \tag{4.16}$$

Conservation of electric charge requires that as charge flows out of a volume, the amount of charge within the volume should decrease, and therefore the charge density within should also decrease. So

$$\iint_V \nabla\cdot\mathbf{J}d\tau = -\iiint \frac{\partial\rho}{\partial t}d\tau \tag{4.17}$$

which leads to the *equation of continuity* for charge

$$\nabla\cdot\mathbf{J} = -\frac{\partial\rho}{\partial t} \tag{4.18}$$

So, for a steady flow of charge within a conductor, where the amount of charge within any volume does not change with time,

$$\nabla\cdot\mathbf{J} = 0.$$

Now, consider a conductor carrying a total steady current I. This is the rate at which charge crosses any section of the wire. So, if $\mathbf{J}$ is the current density at any point within the conductor, then

$$\iint \mathbf{J}\cdot\hat{n}dA = I = \frac{1}{\mu_o}\oint_C \mathbf{B}\cdot d\mathbf{r} = \frac{1}{\mu_o}\iint \nabla\times\mathbf{B}\cdot\hat{n}dA \tag{4.19}$$

and hence we obtain the equation

$$\nabla\times\mathbf{B} = \mu_0\mathbf{J} \tag{4.20}$$

This is true only for steady currents, and fields that do not change with time. Equation (4.20) is called Ampère's Law. It is the differential form of Oersted's Law.

4.3 Force on a Charge in a Magnetic Field

4.3.1 *Forces between Wires carrying Current*

Consider two wires carrying current in the same direction placed close to each other in parallel directions:

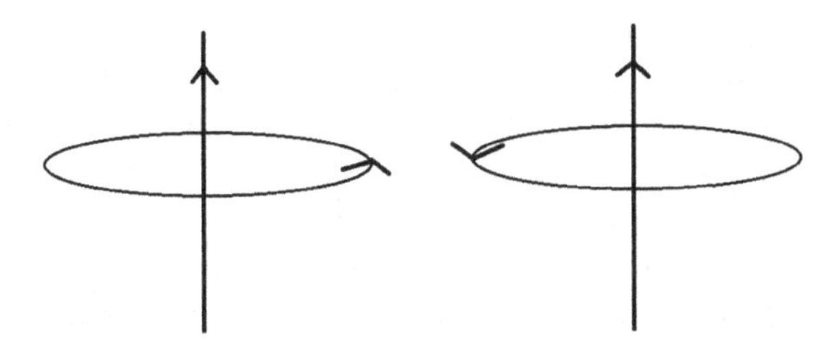

The magnetic lines are in opposite directions in the space between the wires. Since these lines represent the magnetic field vector, they would cancel each other at all points along the line joining the wires. A view of the wires in which the currents flow into the page would show a typical magnetic field line as a loop with an approximate elliptical shape:

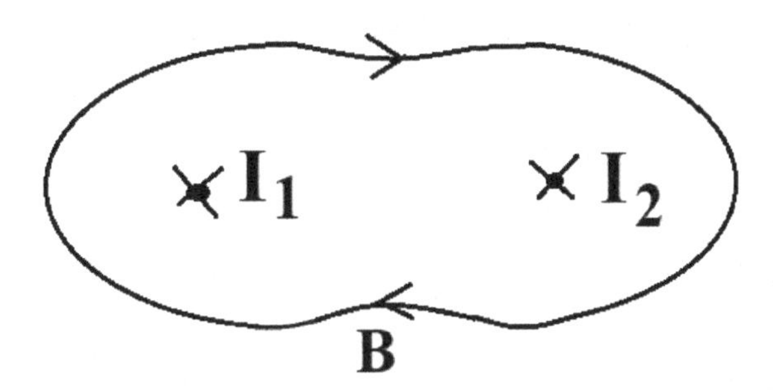

When we studied electrostatics, we noted that the electric field lines behave as though there is a tension along each line, so that the line tries to become as short as possible. Thus, a line flowing from a positive to a negative charge has a tendency to shorten its own length, thereby drawing the opposite charges towards each other. A similar rule applies to magnetic field lines. In the diagram above, the field line shown curving around the two wires (with parallel currents flowing into the page) tends to minimize its length or perimeter, thereby drawing the wires closer together. This, then, is our theoretical guess. It turns out that this guess is substantiated by experiment.

When two long wires carrying parallel currents I_1 and I_2 are placed alongside each other at a distance a from one another, they are found to experience a force of attraction per unit length given by

$$f = \frac{\mu_0 I_1 I_2}{2\pi a} \tag{4.21}$$

This is the magnetostatic equivalent to Coulomb's law in electrostatics. The correspondence is not obvious at first sight, since Coulomb's law is an inverse square law, and above we have an inverse law without the square. But that is because when we consider two infinitely long wires parallel to each other, we have reduced the three-dimensional system to a two-dimensional problem. In order to compare apples with apples, we need to set up an analogous two-dimensional problem in electrostatics. Consider two infinitely long uniformly charged straight wires placed at a distance a parallel to each other. It is not hard to show that the force per unit length between these two wires is

$$f = \frac{\lambda_1 \lambda_2}{2\pi\epsilon_0 a} \tag{4.22}$$

The lambdas are the linear charge densities on the wires. Equation (4.22) is the two-dimensional version of Coulomb's law, and it is a simple inverse law.

Exercise:
Prove Eq. (4.22).

Equation (4.21) gives the force per unit length between two wires. Now, a force is a vector quantity that is applied to one object by another. So wire 1 applies this force to wire 2, and wire 2 applies an equal and opposite force to wire 1, by Newton's third law.

If the two parallel wires have current flowing in opposite directions, so that the currents are antiparallel in the neighboring wires, these wires would repel each other, and the magnitude of the force would remain the same. So the rule is: parallel currents attract, antiparallel currents repel. This rule might seem in some ways the opposite of the corresponding rule in electrostatics. But when we examine the situation more closely, as the ensuing discussions will clarify, the rule regarding forces between current carrying conductors actually confirms the rule: likes repel, unlikes attract.

4.3.2 *The Lorentz Force*

Consider a line segment of length d on wire 2. The force experienced by this segment of wire 2 is

$$F = fd = \frac{\mu_0 I_1 I_2 d}{2\pi a} \tag{4.23}$$

Suppose the charges flow at speed v along wire 2, and that the time taken for a charge q to travel the distance d along wire 2 is t seconds. Since $v = d/t$ and $I_2 = q/t$, it follows that $qv = I_2 d$ and so the force can be written as

$$F = \frac{qv\mu_0 I_1}{2\pi a} \tag{4.24}$$

The magnetic field experienced by wire 2 due to the current in wire 1 has magnitude $B = \frac{\mu_0 I_1}{2\pi a}$, and therefore

$$F = qvB \tag{4.25}$$

Now, the charge velocity $\mathbf{v}$, the magnetic field $\mathbf{B}$ and the force $\mathbf{F}$ are all vectors. Examining the directions of these vectors, for this particular configuration we can write

$$\mathbf{F} = q\mathbf{v} \times \mathbf{B} \tag{4.26}$$

This suggests that a moving charge experiences a force that is perpendicular to the magnetic field. A magnetic field has no effect on a stationary charge. This formula has been confirmed experimentally by sending charges at different velocities through magnetic fields of varying strengths. If a region of space contains both an electric field $\mathbf{E}$ and a magnetic field $\mathbf{B}$, a charge q traveling at velocity $\mathbf{v}$ through that region will experience a force $\mathbf{F}$ given by

$$\mathbf{F} = q(\mathbf{E} + \mathbf{v} \times \mathbf{B}) \tag{4.27}$$

The total force experienced by a charge in an electromagnetic field is called the *Lorentz force.*

4.4 The Biot-Savart Law

Let us consider a coordinate system in which the origin passes through a wire carrying current I. Consider an element of length ds represented by the vector $\mathbf{ds}$ along the wire at the origin. According to Biot and Savart, the magnetic field $d\mathbf{B}$ at a point at a displacement $\mathbf{r}$ relative to the origin is given by

$$d\mathbf{B} = \frac{\mu_0 I}{4\pi} \frac{\mathbf{ds} \times \mathbf{r}}{r^3} \tag{4.28}$$

The Biot-Savart law is the three-dimensional equivalent of Oersted's law. Whereas Oersted's law is an inverse distance law, the Biot-Savart law is an inverse square distance law.

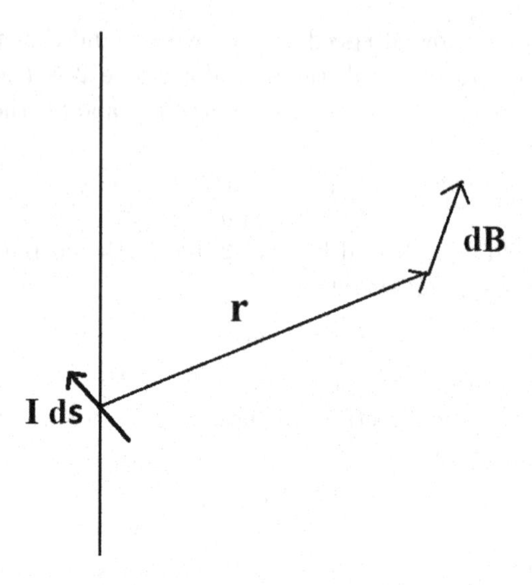

The following figure denotes a section of wire of arbitrary shape carrying constant current:

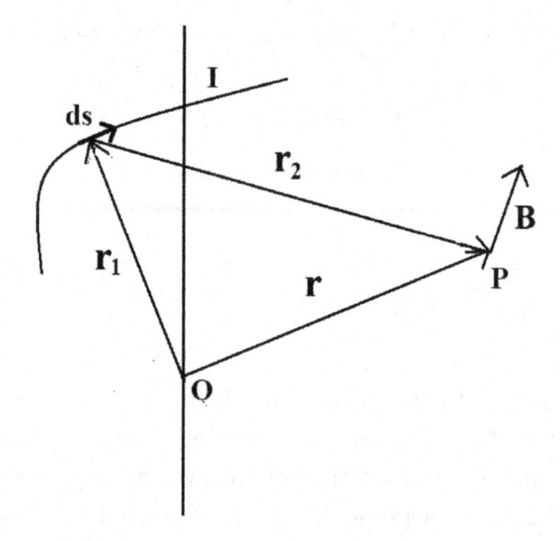

The resultant magnetic field at a point $P(\mathbf{r})$ generated by a current flowing through a wire of arbitrary shape is

$$\mathbf{B} = \frac{\mu_0 I}{4\pi} \int \frac{\mathbf{ds} \times \mathbf{r}_2}{|\mathbf{r}_2|^3} = \frac{\mu_0 I}{4\pi} \int \frac{\mathbf{ds} \times (\mathbf{r} - \mathbf{r}_1)}{|\mathbf{r} - \mathbf{r}_1|^3} \tag{4.29}$$

The variable in the integral is the coordinate $\mathbf{r}_1$. This variable gets integrated out, and the result is a function of $\mathbf{r}$.

This equation is the Biot-Savart law in integral form. It provides the most general expression for a magnetic field due to a steady current. A steady current generally flows through a closed circuit, and so the integral in the above equation is a line integral taken around the loop of the circuit. In any case, the result of the integration is a function of the position $\mathbf{r}$ of the point at which the magnetic field is evaluated.

It is not hard to show that $\nabla \cdot \mathbf{B} = 0$ from the above integral. We need to use the identity $\nabla \cdot (\mathbf{a} \times \mathbf{b}) = \mathbf{b} \cdot (\nabla \times \mathbf{a}) - \mathbf{a} \cdot (\nabla \times \mathbf{b})$. Since the integral is a function of $\mathbf{r}$, the divergence acts only on functions of $\mathbf{r}$, and so $d\mathbf{s}$ and $\mathbf{r}_1$ must be treated as constants. $\nabla \times d\mathbf{s}$ is obviously 0, and it is easy to show that $\nabla \times \frac{(\mathbf{r}-\mathbf{r}_1)}{|\mathbf{r}-\mathbf{r}_1|^3} = 0$. The Biot-Savart law of Eq. (4.29) is the most general formula for the generation of magnetic fields by steady currents.

Exercise:
Using Eq. (4.29) prove that $\nabla \cdot \mathbf{B} = 0$.

4.5 Magnetic Field of a Circular Circuit

The Biot-Savart law enables us to find the magnetic field at the center of a circular circuit carrying a constant current. If a circular circuit of radius a carries a steady current I then the magnetic field at the center is axial, perpendicular to the plane of the circuit. Performing the integral over the entire circumference, we obtain the magnitude of the magnetic field at the center as

$$\frac{\mu_0 I}{4\pi} \frac{2\pi a^2}{a^3} = \frac{\mu_o I}{2a} \tag{4.30}$$

Exercise:
Derive Eq. (4.30) using the Biot-Savart law of Eq. (4.29).

Equation (4.29) also enables us to find the direction of this field. The field has the direction in which a right handed screw would advance if rotated along the direction of the current.

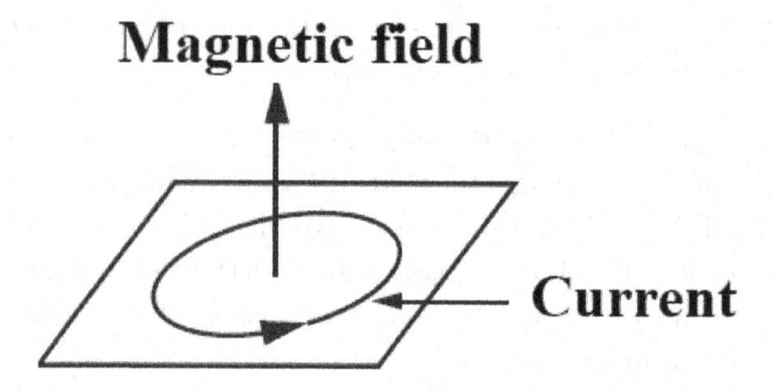

The field has the same direction at every point inside the circuit in the plane of the circuit. The following figure shows the lines of the magnetic field due to a circuit in the shape of a ring or torus seen edge on:

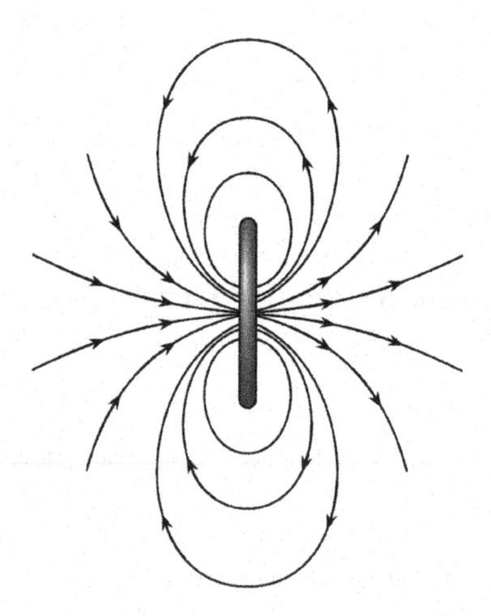

In the figure above, the current in the ring — when viewed from the left — flows clockwise. The magnetic field lines are very similar to those of a bar magnet. Thus the current carrying ring is a magnet, with a north pole on the right face and a south pole on the left face.

The above figure also provides a schematic model of the magnetic field of the earth. The earth contains a circulating current below its surface, which causes the entire planet to behave like a large magnet. The axis of rotation of this subterraneous current is not exactly aligned with the axis of rotation of the earth itself, and so the magnetic poles are not exactly at the geographical poles. So there is a small angle called the *declination* between the magnetic north-south and the geographical north-south at any point on the surface of the earth.

If we place two identical rings carrying current coaxially close to each other, so that the direction of current will be the same in both, then they will attract each other. We can also examine this phenomenon in terms of the magnetic fields generated by the rings:

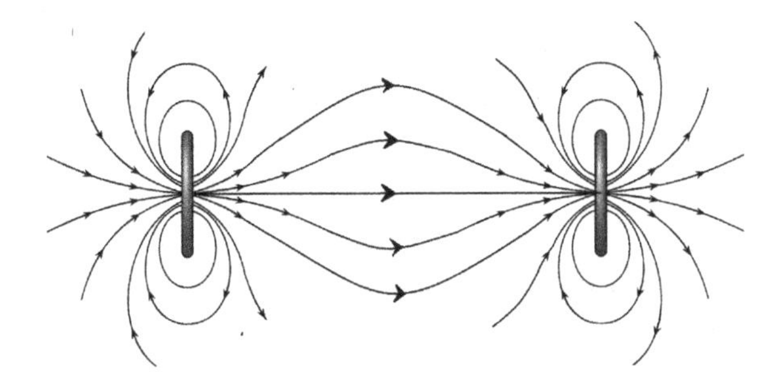

Between the two rings the field lines flow as though from a north pole to a south pole. There is attraction between the rings because the currents are parallel. The attraction can also be described as an attraction between opposite magnetic poles. In terms of the field lines, one could describe the attraction as a tension along the lines, tending to minimize the lengths of the lines between the rings. Thus each ring is a tiny magnet. One face is a north pole, and the other face is a south pole. When the rings are lined up with currents in parallel, a south pole faces the north pole of the other ring, and there is attraction. So, in this sense, it is the opposites that attract.

Now, if one of the rings is flipped, the rings would repel, because the currents would be flowing in opposite directions. The rings are now arranged with like poles facing each other, and there is repulsion. So, likes do repel.

A wire in the shape of a coil — called a *solenoid* — can be thought of as a large number of rings arranged parallel to each other. When current flows through such a solenoid there is a magnetic field generated inside and outside the solenoid. The magnetic field lines outside the solenoid mimic those of a bar magnet:

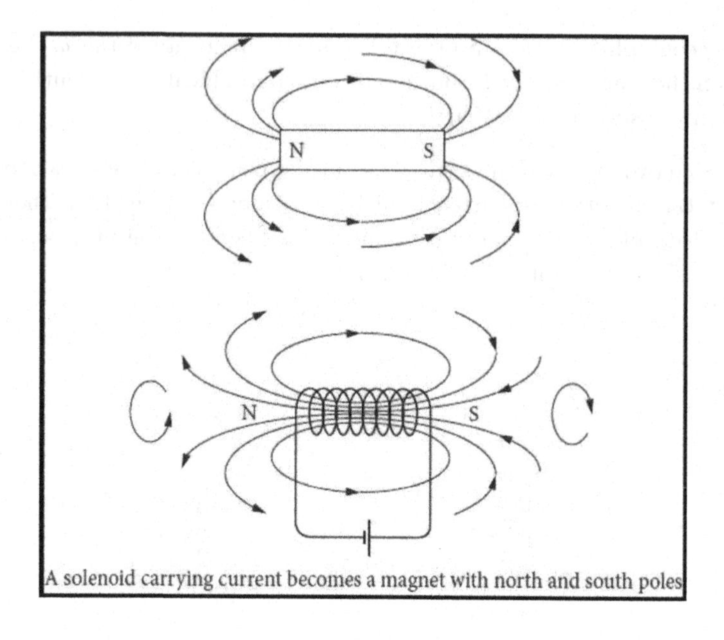

A solenoid carrying current becomes a magnet with north and south poles

Thus, a bar magnet — or any other magnet — can be thought of as containing circling currents that generate the magnetism. This would also explain why it is impossible to separate the poles of a magnet — it would be like attempting to separate the two faces of a coin. In the case of a magnetizable metal like iron, the circling current is due to the atomic electrons which have orbital and spin angular momentum. While it is incorrect to say an individual electron spins around its axis, or even that it revolves around the nucleus, when a large number of atoms are brought together it is possible for the spins to be aligned in the atoms that constitute a neighborhood within the metal. When such an alignment takes place, the effect is like a circulating current within the neighborhood. So such a cluster of atoms becomes a magnet with a north and a south pole. Each such cluster — also called a *domain* — is quite small. In an unmagnetized piece of iron these domains have random orientations and so their magnetic fields cancel each other. But when such a piece of iron is magnetized the domains align themselves and their magnetic moments add up so that the iron piece has become a magnet.

Because the phenomenon of magnetism arises from the flow of electric charges, it is better to define the magnetic field **B** in terms of the electric charge and its velocity, and not in terms of a force acting on a magnetic pole. And so, we define the magnetic field via the force experienced by a charge moving through the field:

$$\mathbf{F} = q\mathbf{v} \times \mathbf{B} \tag{4.31}$$

So a magnetic field of 1 Tesla (T) imparts a force of 1 N to a charge of 1 C traveling at 1 m/s perpendicular to the field.

Unlike the electric field, the magnetic field applies a force perpendicular to the field. One consequence is that magnetic forces do no work.

The work done by a force on an object is defined as $dW = \mathbf{F} \cdot d\mathbf{r}$ and so the work done on a charge q in an electromagnetic field is

$$\mathbf{F} \cdot d\mathbf{r} = q(\mathbf{E} + \mathbf{v} \times \mathbf{B}) \cdot d\mathbf{r} = q\mathbf{E} \cdot d\mathbf{r} + q(\mathbf{v} \times \mathbf{B}) \cdot d\mathbf{r} \tag{4.32}$$

Now, $d\mathbf{r} = \frac{d\mathbf{r}}{dt}dt = \mathbf{v}dt$, and so $(\mathbf{v} \times \mathbf{B}) \cdot d\mathbf{r} = (\mathbf{v} \times \mathbf{B}) \cdot \mathbf{v}dt = 0$. Thus the work done by a magnetic field on a charge is zero. Hence $dW = q\mathbf{E} \cdot d\mathbf{r}$. The work done by an electromagnetic field on a charge is purely electrical.

4.5.1 *Magnetic Field inside a Long Solenoid*

According to Ampère's law, $\nabla \times \mathbf{B} = \mu_0 \mathbf{J}$. By Stokes's theorem, $\oint_S \mathbf{B} \cdot d\mathbf{r} = \iint \nabla \times \mathbf{B} \cdot \hat{n} dA$, and so the line integral of the magnetic field along any closed loop equals the integral $\mu_0 \iint \mathbf{J} \cdot \hat{n} dA$ which is equal to $\mu_0 I$ where I is the total current flowing through the loop. Such a loop is called an *Ampèrian loop*.

The following illustrates a long solenoid carrying a steady current I:

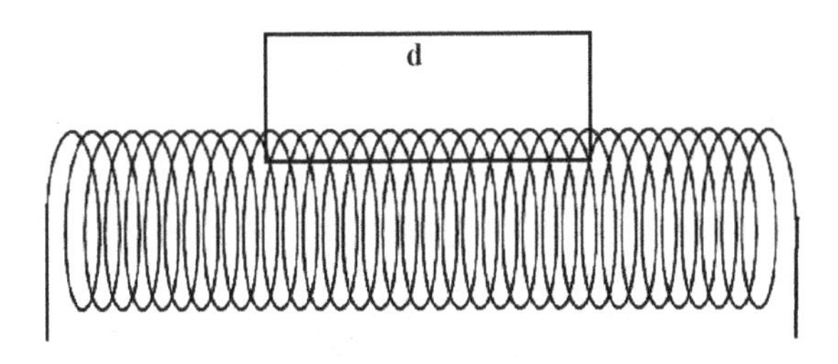

For a long solenoid, the field outside and close to the solenoid is zero. Let the field inside the solenoid have magnitude B. Consider an Ampèrian loop in the shape of a rectangle with the long side equal to d. Let the number of turns of the solenoid per meter be n. So the total current flowing through the Ampèrian loop is nId. This is proportional to the line integral of the magnetic field around the loop. And so we obtain $Bd = \mu_0 nId$ and thus

$$B = \mu_0 nI \tag{4.33}$$

We notice that — as long as the field outside the solenoid is zero — the field inside the solenoid is uniform, since a vertical shift in the Ampèrian loop will not affect the result of the integral, as long as we keep one of the horizontal sides fully inside and the other fully outside the solenoid.

4.6 The Vector Potential A

In a time independent system where the charges and the currents do not change with time, we found that $\nabla \cdot \mathbf{B} = 0$. But we may reasonably assume that this is valid even for time-dependent systems, as otherwise we would be able to separate the poles of a magnet, which, as we saw, is impossible as long as we understand magnetism as caused by circulating currents even at the microscopic level. And so we can reasonably assume that $\nabla \cdot \mathbf{B} = 0$ is a universal law of physics. We defer the rigorous mathematical proof to a later chapter.

Since the divergence of the curl of any vector is identically zero, we can write the magnetic field as the curl of some other vector field:

$$\mathbf{B} = \nabla \times \mathbf{A} \tag{4.34}$$

The vector $\mathbf{A}$ is called the vector potential, and is associated with the scalar potential ϕ that we learned in an earlier chapter.

One can find the vector potential quite easily for a uniform magnetic field, for which $\mathbf{B}$ is a constant, independent of the coordinates. This is the case inside a long solenoid.

Let us consider a circle C of radius r inside the solenoid, with its center on the axis of the solenoid. So

$$\oint_C \mathbf{A} \cdot d\mathbf{r} = \iint \nabla \times \mathbf{A} \cdot \hat{n} dA = \iint \mathbf{B} \cdot \hat{n} dA \tag{4.35}$$

Thus, we obtain $2\pi r A = B\pi r^2$ and hence

$$A = \frac{Br}{2} = \frac{1}{2}\mu_0 nIr \tag{4.36}$$

The vector **A** curls along a circle perpendicular to the axis of the solenoid. Hence **A** is parallel to the current flowing along the turns of the solenoid. The fact that **A** follows the actual motion of the charges suggests that **A** may be more real than **B** in relation to the charges themselves.

When we defined the scalar potential or electric potential ϕ by $\mathbf{E} = -\nabla\phi$, we could have added any constant number to ϕ without affecting the definition. Physically, this means that it is not the actual potential, but the difference of potential, that can be physically measured in any situation. Now, as we define the vector potential (which we do not call the magnetic potential) by $\mathbf{B} = \nabla \times \mathbf{A}$, we see that we can add any function of the form $\nabla\chi$ to **A** without affecting the definition, because $\nabla \times \nabla\chi = 0$. So, the vector potential is undefined to within an additive function $\nabla\chi$. This property of the vector potential is called *gauge invariance.*

Now, suppose we have two different vector potentials, $\mathbf{A}^{'}$ and **A**, which are related by the gradient of some scalar function χ such that $\mathbf{A} = \mathbf{A}^{'} + \nabla\chi$. Now, both **A** and $\mathbf{A}^{'}$ are equally valid vector potentials for some magnetic field **B**. Let us choose the function χ such that $\nabla^2\chi = -\nabla \cdot \mathbf{A}^{'}$. Clearly, therefore, for this choice of gauge, $\nabla \cdot \mathbf{A} = 0$. We will work in this gauge

According to Ampère's theorem $\nabla \times \mathbf{B} = \mu_0 \mathbf{J}$. Putting in $\mathbf{B} = \nabla \times \mathbf{A}$, we obtain

$$\nabla \times (\nabla \times \mathbf{A}) = \mu_0 \mathbf{J} \tag{4.37}$$

Now, $\nabla \times (\nabla \times \mathbf{A}) = \nabla(\nabla \cdot \mathbf{A}) - \nabla^2 \mathbf{A}$. Within our choice of gauge, $\nabla \cdot \mathbf{A} = 0$, and so we obtain

$$\nabla^2 \mathbf{A} = -\mu_0 \mathbf{J} \tag{4.38}$$

This is the magnetic equivalent of the electrical Poisson's equation $\nabla^2\phi = -\frac{\rho}{\epsilon_0}$. Whereas the electrostatic potential is a scalar with a single component, the magnetostatic potential is a vector with three components.

When we study time varying systems we will see that we can no longer make a clean distinction between electric and magnetic phenomena. And so the potential ϕ is simply called the scalar potential, and **A** is called the vector potential. In analogy with the solution to Poisson's equation in electrostatics, we can find a solution to Eq. (4.38) as follows:

$$\mathbf{A}(\mathbf{r}) = \frac{\mu_0}{4\pi} \int \frac{\mathbf{J}(\mathbf{r}^{'})}{|\mathbf{r} - \mathbf{r}^{'}|} d^3\mathbf{r}^{'} \tag{4.39}$$

4.7 Magnetic Moment and Rotating Charges

4.7.1 *Magnetic Dipole Moment of a Current Loop*

We will now evaluate the vector potential at a point P some distance away from the currents, i.e. if r is the distance from the origin to P, and r' the distance from the origin to a point through which a current flows, then we assume that $r' \ll r$. In such a situation the following approximation is valid:

$$\frac{1}{|\mathbf{r}-\mathbf{r}'|} = \frac{1}{r} + \frac{\mathbf{r}\cdot\mathbf{r}'}{r^3} \tag{4.40}$$

Exercise:
Prove Eq. (4.40).

Within this approximation we can obtain an expansion for the vector potential at some point. The expression for the ith component of the vector potential becomes, within this approximation:

$$A_i(\mathbf{r}) = \frac{\mu_0}{4\pi}\left[\frac{1}{r}\int J_i(\mathbf{r}')d\tau' + \frac{\mathbf{r}}{r^3}\cdot\int J_i(\mathbf{r}')\mathbf{r}'d\tau'\right] \tag{4.41}$$

In analogy with electric multipoles, the first integral is a magnetic monopole term, and the second is a dipole term. If all the currents are contained within a certain region, we can choose the volume of integration to include this region. Clearly, the net current within this volume is zero, since no current flows in or flows out of the region. Thus the monopole term vanishes. This reflects the physical fact that magnetic poles cannot be isolated, unlike electric charges.

Now, if u and v are scalar fields and $\mathbf{J}$ the current density, then

$$\iiint_V \nabla\cdot(uv\mathbf{J})d\tau = \oint\!\!\oint_S uv\mathbf{J}\cdot\hat{n}dA \tag{4.42}$$

If the closed surface S is extended to a sphere of sufficiently large radius, the right side will vanish, since we assume the currents are limited to a small region close to the origin. The integral on the left side can be expanded as

$$\iiint_V (v\nabla u\cdot\mathbf{J} + u\nabla v\cdot\mathbf{J} + uv\nabla\cdot\mathbf{J})d\tau = 0 \tag{4.43}$$

We now assume the currents are steady, and so $\nabla \cdot \mathbf{J} = -\frac{\partial \rho}{\partial t} = 0$. Next, let us set $u = r_i^{'}$, the ith component of the vector $\mathbf{r}^{'}$. We shall set $v = r_j^{'}$. So the above integral becomes

$$\iiint (r_j^{'} J_i + r_i^{'} J_j) d\tau^{'} = 0 \tag{4.44}$$

We will express the integral in the second term within the square brackets in Eq. (4.41) as

$$\mathbf{r} \cdot \int \mathbf{r}^{'} J_i d\tau^{'} = \sum_j r_j \int r_j^{'} J_i d\tau^{'} \tag{4.45}$$

Making use of the result of Eq. (4.44), we can write

$$\mathbf{r} \cdot \int \mathbf{r}^{'} J_i d\tau^{'} = -\frac{1}{2} \sum_j r_j \int (r_i^{'} J_j - r_j^{'} J_i) d\tau^{'} \tag{4.46}$$

An inspection of the right side of the equation shows that it is the ith component of a cross product. The vector equation can then be written as

$$\mathbf{r} \cdot \int \mathbf{r}^{'} \mathbf{J} d\tau^{'} = -\frac{1}{2} \mathbf{r} \times \int (\mathbf{r}^{'} \times \mathbf{J}) d\tau^{'} \tag{4.47}$$

If we consider a current I flowing through a wire in the shape of a loop, then $\mathbf{J}(\mathbf{r}^{'}) d\tau^{'} = I d\mathbf{r}^{'}$, where $d\mathbf{r}^{'}$ is the displacement vector of an element of length dr' along the wire. Thus,

$$\mathbf{r} \cdot \int \mathbf{r}^{'} \mathbf{J} d\tau^{'} = -\mathbf{r} \times I \oint \frac{1}{2} \mathbf{r}^{'} \times d\mathbf{r}^{'} \tag{4.48}$$

Now, $\frac{1}{2}\mathbf{r}^{'} \times d\mathbf{r}^{'}$ is a vector perpendicular to the plane of $\mathbf{r}^{'}$ and $d\mathbf{r}^{'}$ and having magnitude equal to the area of the triangle formed by $\mathbf{r}^{'}, \mathbf{r}^{'} + d\mathbf{r}^{'}, d\mathbf{r}^{'}$. If the loop carrying current is very small, the integral is a vector of magnitude A (= area of the loop) and perpendicular to the plane of the loop. When multiplied by the current I one obtains a quantity defined as the magnetic dipole moment $\mathbf{m}$ for the small loop. This definition can be applied also to an extended current density. So, the magnetic dipole moment — or simply magnetic moment — of a current distribution is

$$\mathbf{m} = \frac{1}{2} \int \mathbf{r}^{'} \times \mathbf{J}(\mathbf{r}^{'}) d\tau^{'} \tag{4.49}$$

This is a very different definition of magnetic dipole moment than the one we introduced earlier in this chapter, where we defined the magnetic moment of a magnet as $\mathfrak{m} = p\mathbf{s}$ in terms of the pole strength p and the length vector of the magnet $\mathbf{s}$. (Note the difference between the symbols. $\mathbf{m}$ is not the same as $\mathfrak{m}$.) But we shall see that both provide the same results for the magnetic field due to a dipole, whether the dipole is due to a magnet or due to a circulating current. In other words, there is no difference between $\mathfrak{m}$ and $\mathbf{m}$ as far as the magnetic field is concerned.

So, using Eq. (4.49), we obtain for the vector potential:

$$\mathbf{A}(\mathbf{r}) = \frac{\mu_0}{4\pi}\frac{\mathbf{m} \times \mathbf{r}}{r^3} \tag{4.50}$$

The magnetic field due to this vector potential can be obtained from $\mathbf{B} = \nabla \times \mathbf{A}$, and so we get

$$\mathbf{B}_{dip}(\mathbf{r}) = \frac{\mu_0}{4\pi}\frac{1}{r^3}\left[3(\mathbf{m} \cdot \hat{r})\hat{r} - \mathbf{m}\right] \tag{4.51}$$

This equation is identical to Eq. (4.5). The bottom line therefore is that a magnet can be thought of either as a pair of poles separated by a distance or a circling current with an enclosed area. The magnetic moment can be defined as either the product of a pole strength and the separation between the poles or the current multiplied by the area of the loop. So $ps = IA$.

Exercise:
Prove Eq. (4.51).

4.7.2 *Gyromagnetic Ratio*

The gyromagnetic ratio or the magnetogyric ratio of a particle is the ratio of its magnetic moment to its angular momentum. Suppose there are n different charges q_i contributing to the current density $\mathbf{J}$. If the velocity of such a charge occupying a position $\mathbf{r}_i$ is $\mathbf{v}_i$, then the current density $\mathbf{J} = \sum_i q_i \mathbf{v}_i \delta(\mathbf{r} - \mathbf{r}_i)$. The magnetic moment of such a system of particles is

$$\mathbf{m} = \frac{1}{2}\int \mathbf{r} \times \sum_i q_i \mathbf{v}_i \delta(\mathbf{r} - \mathbf{r}_i) d\tau = \frac{1}{2}\sum_i q_i(\mathbf{r}_i \times \mathbf{v}_i) \tag{4.52}$$

Suppose the ith particle has mass M_i. Its angular momentum about the origin is given by $\mathbf{L}_i = M_i(\mathbf{r}_i \times \mathbf{v}_i)$. Then

$$\mathbf{m} = \sum_i \frac{q_i}{2M_i}\mathbf{L}_i \tag{4.53}$$

So, if we are considering a collection of particles of the same mass and charge, $\frac{q_i}{M_i}$ is the same for all of them, and we shall call this ratio $\frac{q}{M}$. Let $\mathbf{L} = \sum_i \mathbf{L}_i$. The magnetic moment of this system of particles is

$$\mathbf{m} = \frac{q}{2M}\mathbf{L} \tag{4.54}$$

The gyromagnetic ratio g of such a system of charges is therefore

$$g = \frac{q}{2M} \tag{4.55}$$

If now we have a uniformly charged sphere carrying a total charge q and having mass m, its gyromagnetic ratio is

$$\gamma = \frac{q}{2m} \tag{4.56}$$

Later we will learn that this formula does not work for an electron's magnetic moment in terms of its intrinsic angular momentum. This is a consequence of the quantum theory. For elementary particles such as the electron, we introduce what is called the g factor, a dimensionless number. The electron g factor is sometimes written as g_e, and is slightly greater than 2. So the gyromagnetic ratio of an electron is about twice the value of a classical rotating charged sphere.

$$\gamma_{electron} = \frac{e}{2m} g_e \tag{4.57}$$

g_e has been measured with great accuracy. Expressed to 8 significant figures, $g_e = 2.0023193$.

4.8 Magnetic Monopoles

We have shown that classical macroscopic laws forbid the separation of magnetic poles. However, at the subatomic level, certain phenomena have been predicted and detected that seem to suggest that it may be possible to separate the north and the south poles in this sort of magnetism. Researchers at the London Centre for Nanotechnology have coined the term *magnetricity* for this new phenomenon whereby magnetic north and south poles could travel independently.[1] If this line of investigation is correct, then the ideas presented in this chapter would require some readjustments.

[1] Bramwell, S., Giblin, S., Calder, S. *et al.* "Measurement of the charge and current of magnetic monopoles in spin ice," *Nature* **461**, 956–959 (2009). https://doi.org/10.1038/nature08500.

Chapter 5

Fields Produced by Time Varying Sources

5.1 Equation of Continuity and Applications

Thus far we have studied electric fields that are generated by static charge configurations that do not vary with time, and magnetic fields that are generated by steady time-independent currents. So time as a variable did not enter any of the equations for the electric or the magnetic fields, and while current was defined as the time rate of flow of charge, this rate of flow was constant, without any changes.

When an object is being charged or discharged the charge on the object varies with time. In order to charge up an object, charges need to flow into it, i.e. there must be a flow of current. A charge cannot appear out of nowhere, and also cannot disappear at any point. The equation of continuity expresses the relationship between varying charge and flowing current in any region of space:

$$\nabla \cdot \mathbf{J} = -\frac{\partial \rho}{\partial t} \tag{5.1}$$

It is evident that $\mathbf{J}$ and ρ are functions of position and time.

According to Maxwell's differential equation the charge density is the source of the electric field:

$$\nabla \cdot \mathbf{E} = \frac{\rho}{\epsilon_0} \tag{5.2}$$

Let us now examine the insights this equation can provide if the charge density — and hence the electric field — are no longer time independent.

First, we take the time derivative on both sides:

$$\frac{\partial}{\partial t}(\nabla \cdot \mathbf{E}) = \frac{1}{\epsilon_0}\frac{\partial \rho}{\partial t} \tag{5.3}$$

Substituting Eq. (5.1), we obtain

$$\nabla \cdot \frac{\partial \mathbf{E}}{\partial t} = -\frac{1}{\epsilon_0}\nabla \cdot \mathbf{J} \tag{5.4}$$

which we can rewrite as

$$\nabla \cdot \left(\frac{\partial \mathbf{E}}{\partial t} + \frac{\mathbf{J}}{\epsilon_0}\right) = 0 \tag{5.5}$$

If the divergence of a vector is zero, then the vector can be written as the curl of some vector field. And so, we can write

$$\frac{\partial \mathbf{E}}{\partial t} + \frac{\mathbf{J}}{\epsilon_0} = \nabla \times \mathbf{G} \tag{5.6}$$

where $\mathbf{G}$ is a yet undetermined vector field. For this equation to be valid at all times, it must be valid for time invariant fields as well. So, setting the first term on the left side to zero, we get

$$\nabla \times \mathbf{G} = \frac{\mathbf{J}}{\epsilon_0} \tag{5.7}$$

We know that static currents obey Ampère's Law

$$\nabla \times \mathbf{B} = \mu_0 \mathbf{J} \tag{5.8}$$

This correspondence enables us to identify the vector field $\mathbf{G}$ as

$$\mathbf{G} = \frac{1}{\mu_0 \epsilon_0}\mathbf{B} \tag{5.9}$$

Substituting into Eq. (5.6), we get the equation

$$\nabla \times \mathbf{B} = \epsilon_0 \mu_0 \frac{\partial \mathbf{E}}{\partial t} + \mu_0 \mathbf{J} \tag{5.10}$$

This equation shows that a curling magnetic field is produced not only by a current, but also by a time varying electric field.

The physical meaning of this equation can be illustrated by a simple example. Consider a circuit with a capacitor formed of two identical parallel disks that is being discharged at a steady rate, thereby providing a steady current in the wires of the circuit. We seek to calculate the magnetic field at two different positions: first, at some distance perpendicular from the wire, and second, at the same distance but measured from the central point midway between the plates of the capacitor, as shown in the figure.

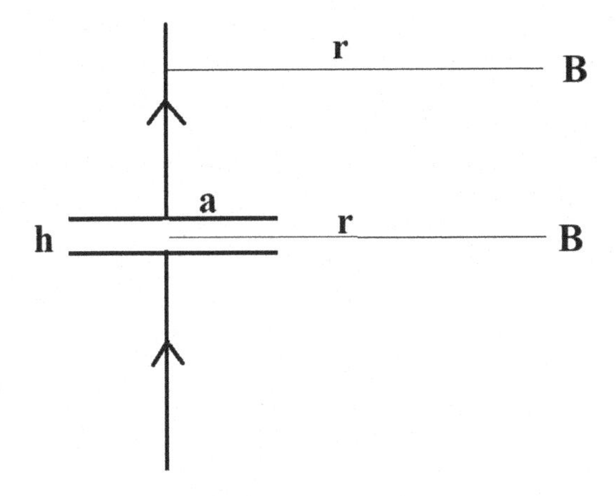

Let the radius of each plate of the capacitor be a, and let the separation between the plates be h. Let the charge on the capacitor at any time be Q. Let the current in the wires be I.

By the continuity of charge and current, $I = -\frac{dQ}{dt}$.

The magnetic field at a point at a distance r from the wire, from Oersted's law, is a circular vector field of magnitude $B = \frac{\mu_0 I}{2\pi r}$.

Let us now evaluate the magnetic field at the same distance from the center of the capacitor.

We will use Eq. (5.10) and apply Stokes's theorem in a plane perpendicular to the wire and passing between the plates of the capacitor. Here there is no current, but a varying electric field. There is no electric field outside the plates of the capacitor. The magnetic field at a distance r from the center of the capacitor has the magnitude

$$B = \frac{1}{2\pi r}\epsilon_0\mu_0\frac{\partial E}{\partial t}(\pi a^2) = \frac{\epsilon_0\mu_0 a^2}{2r}\frac{\partial E}{\partial t} \tag{5.11}$$

Between the capacitor plates,

$$E = \frac{\sigma}{\epsilon_0} = \frac{Q}{\pi a^2\epsilon_0} \tag{5.12}$$

Hence, we obtain the magnitude of the curling magnetic field at a point at the same distance r from the center of the capacitor

$$B = \frac{\mu_0 I}{2\pi r} \tag{5.13}$$

As this calculation shows, a curling magnetic field cannot tell if the source is a steady current or a steadily changing electric field.

Mathematically, the left side of Eq. (5.10) vanishes if the two terms on the right side cancel each other. We will next examine a physical realization of this mathematical result.

Suppose a capacitor is discharged directly across its plates as shown:

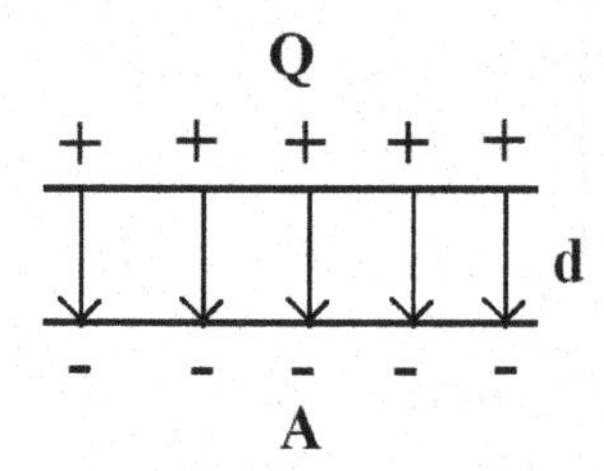

Let the magnitude of the charge on one plate at any time be Q. Let the area of each plate be A and let the distance between the plates be d. Suppose now the charges are discharged across the plates. So, the current density $\mathbf{J}$ has the opposite sign as $\frac{\partial \mathbf{E}}{\partial t}$. Now $JA = \frac{dQ}{dt}$. So $\frac{J}{\epsilon_0} = \frac{dQ}{dt}\frac{1}{A\epsilon_0}$. The field across the plate at any time is $\frac{\sigma}{\epsilon_0} = \frac{Q}{A\epsilon_0}$. So the rate of decrease of the field has magnitude $\frac{dQ}{dt}\frac{1}{A\epsilon_0}$. Thus,

$$\frac{\mathbf{J}}{\epsilon_0} = -\frac{\partial \mathbf{E}}{\partial t} \tag{5.14}$$

So the changing electric field cancels the current density between the plates, resulting in a net zero magnetic field.

Equation (5.10) is one of the Maxwell equations of the electromagnetic field. This equation implies that a source of a magnetic field can be either a current or a varying electric field. And either of these sources generates a curling magnetic field with zero divergence, and hence magnetic monopoles cannot be generated by currents or by electric fields. This result enhances a claim for the universal validity of the equation $\nabla \cdot \mathbf{B} = 0$. We are now ready to elevate this equation to the status of a law. But it will not hurt to examine an important phenomenon that will finally put to rest any residual doubts.

5.2 Faraday's Law

5.2.1 *Magnetic Flux Increase and EMF in a Circuit*

The figure below shows a wire in the shape of a rectangular C (⊏) with a rod placed perpendicularly across the arms of the wire. This rod is capable of sliding to the left or right along the arms of the wire. A uniform magnetic field **B** perpendicular to the plane of the figure is directed into the page.

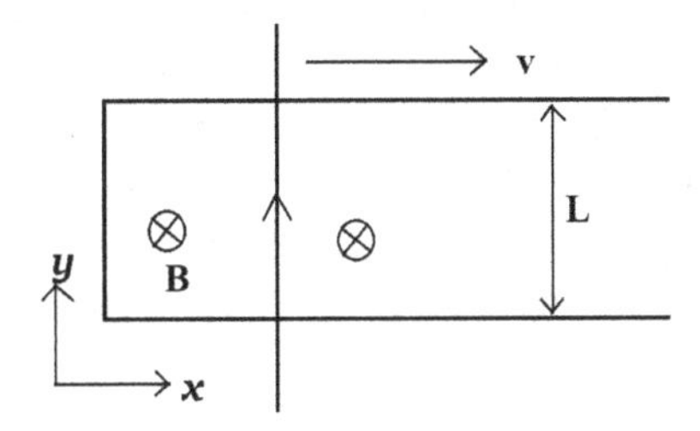

Let the distance between the arms of the wire be L. Suppose the rod is moved to the right at a constant speed $v = \frac{dx}{dt}$. Now, the conducting rod has charges that are free to move along the rod. So, if we consider a charge q sitting on the rod, this charge is being carried to the right at speed v. So, by the Lorentz force law, it would experience a force in the positive y direction. The magnitude of this force would be

$$F = qvB \tag{5.15}$$

The work done by this force in displacing the charge along the rod between the arms of the wire is $FL = qvBL = qBL\frac{dx}{dt} = qB\frac{dA}{dt}$ where A is the area of the rectangle constituted by the wire and the rod, i.e. the area formed by the completed circuit. Now, $B\frac{dA}{dt} = \frac{d\Phi}{dt}$ where Φ is the total magnetic flux through the circuit. FL is the work done in taking a charge q across the rod, and since there is no force acting on the charge in the remaining portion of the circuit, FL is the work done in taking a charge q around the entire circuit. Therefore

$$FL = q\oint \mathbf{E} \cdot d\mathbf{r} \tag{5.16}$$

The integral is carried out in the positive sense — i.e. counterclockwise — in the xy plane. A counterclockwise rotation in the xy plane is represented as a rotation vector along the positive z direction. But the magnetic field is directed along the negative z axis, and so we can write

$$\oint \mathbf{E} \cdot d\mathbf{r} = -\frac{d\Phi}{dt} = -\iint \frac{d\mathbf{B}}{dt} \cdot \hat{n} dA \tag{5.17}$$

By Stokes's theorem this yields the relation

$$\nabla \times \mathbf{E} = -\frac{d\mathbf{B}}{dt} \tag{5.18}$$

In this final equation we have abstracted away from the physical situation that led to the creation of the equation. In the actual derivation of this equation we considered the magnetic field as a constant vector field that does not change with time, but the equation we came up with is a relationship between a curling electric field and a magnetic field that varies with time. We need to test the general validity of this equation, and examine whether it is possible to generate a curling electric field from a time varying magnetic field.

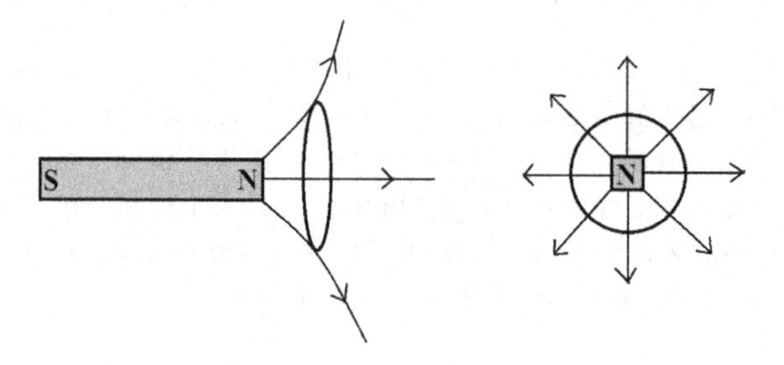

Such a process is illustrated in this figure, which shows two views of a circular loop of wire in motion relative to a bar magnet. The field lines due to the north pole of the magnet are shown in both views.

Case 1. The magnet is kept stationary and the loop is moved towards the magnet.

A charge sitting on the loop will experience a force perpendicular to the magnetic field and perpendicular to the direction of motion of the loop, and so a clockwise current is induced in the loop.

Case 2. The loop is stationary and the magnet is brought towards the loop. The magnetic flux will increase and so a curling electric field is generated along the loop which induces a current in the loop. In accordance with Eq. (5.18) a clockwise current is induced in the loop.

Experimental evidence shows that in both cases, as long as the relative velocity is the same, the induced emf in the loop is the same. Thus Case 2 establishes the validity of Eq. (5.18) as a standalone equation expressing a fundamental fact of physics.

5.2.2 *Magnetic Flux Inertia*

As the magnetic field through the loop is changed, a current is induced in the loop. This induced current will have a magnetic field of its own, and it is not hard to see that this induced field will have the opposite direction as the external field due to the bar magnet. This has two implications.

First, the current induced in the loop will cause a repulsion between the magnet and the loop, since the induced current will generate a north pole that will be facing the north pole of the approaching bar magnet. And so, the external agent that is bringing the two objects closer together must exert a force to overcome this repulsion.

Second, since the induced magnetic field inside the loop has the opposite direction as the external magnetic flux, it has the effect of at least partly neutralizing the increasing magnetic flux within the loop. Suppose now the loop was initially close to the magnet, with some flux passing through it. If now the loop and the magnet are pulled away from each other, the flux within the loop decreases, but the loop will generate a current with a flux that is in the same direction as the original flux, thus attempting to neutralize the change in flux. So, the current induced in the loop always attempts to neutralize any change in the flux through the loop. This is a sort of inertia of magnetic flux. This phenomenon is called Lenz's Law.

5.3 Sources and Fields

5.3.1 *Maxwell's Equations*

When Eq. (5.18) is abstracted from specific experimental situations it is a statement that a curling electric field is equal to a time varying magnetic field. To express this relation between a spatial variation of one field and a time variation of another, the partial time derivative is employed:

$$\nabla \times \mathbf{E} = -\frac{\partial \mathbf{B}}{\partial t} \tag{5.19}$$

This equation has the status of a universal law.

Let us now take the divergence of both sides:

$$\nabla \cdot (\nabla \times \mathbf{E}) = -\frac{\partial}{\partial t}(\nabla \cdot \mathbf{B})$$

Since the left side is identically zero, it follows that $\nabla \cdot \mathbf{B}$ must always be 0. Hence, we can state as a universal law that the divergence of a magnetic field is always zero, regardless of how the field is generated — whether by a static current or by a varying electric field.

So the three equations that we have established in this chapter are the following:

$$\nabla \cdot \mathbf{B} = 0 \tag{5.20}$$

$$\nabla \times \mathbf{B} = \epsilon_0 \mu_0 \frac{\partial \mathbf{E}}{\partial t} + \mu_0 \mathbf{J} \tag{5.21}$$

$$\nabla \times \mathbf{E} = -\frac{\partial \mathbf{B}}{\partial t} \tag{5.22}$$

To these we must add the equation established in an earlier chapter:

$$\nabla \cdot \mathbf{E} = \frac{\rho}{\epsilon_0} \tag{5.23}$$

And these are all the laws we need for the complete description of classical (i.e. pre-quantum) electromagnetism. What is remarkable is that all four equations contain the del operator. Electromagnetism is a thoroughly spatial theory. It is a field theory.

5.3.2 *Fields and Potentials*

For a system consisting only of static charges, the electric field is entirely expressible in terms of the scalar potential: $\mathbf{E} = -\nabla\phi$, and the magnetic field does not exist. If some of these charges are set in arbitrary motion, a magnetic field also arises, which can be fully expressed in terms of the vector potential: $\mathbf{B} = \nabla \times \mathbf{A}$.

The electric field is independent of the vector potential as long as the system is time independent, i.e. when the currents are stationary. But if the currents vary with time, the magnetic field also varies with time, and a time varying magnetic field induces a curling electric field according to the equation

$$\nabla \times \mathbf{E} = -\frac{\partial \mathbf{B}}{\partial t} \tag{5.24}$$

Using the equation $\mathbf{B} = \nabla \times \mathbf{A}$ we obtain

$$\nabla \times \left(\mathbf{E} + \frac{\partial \mathbf{A}}{\partial t} \right) = 0 \tag{5.25}$$

Thus the vector $\mathbf{E} + \frac{\partial \mathbf{A}}{\partial t}$ must be the gradient of some scalar. Since $\mathbf{E} = -\nabla\phi$ for static fields, i.e. those for which $\mathbf{A}$ is independent of time, we may write the electric field in terms of both the scalar and the vector potentials:

$$\mathbf{E} = -\nabla\phi - \frac{\partial \mathbf{A}}{\partial t} \tag{5.26}$$

The first term on the right is the contribution to the electric field from charges, and the second term from changing currents. Steady currents do not generate electric fields.

A magnetic field is expressible entirely by the curl of the vector potential.

$$\mathbf{B} = \nabla \times \mathbf{A} \tag{5.27}$$

A time varying vector potential will generate a time varying magnetic field. Since $\frac{\partial \mathbf{A}}{\partial t} = -\nabla\phi - \mathbf{E}$, taking the time derivative of Eq. (5.27) will yield the Maxwell equation expressing Faraday's law Eq. (5.24).

5.3.3 *Causality*

Suppose we have a system consisting of charges in arbitrary motion. The fields generated by them would in general be quite complicated. Time varying electric fields generate magnetic fields and time varying magnetic fields generate electric fields. These fields have an important property that we shall now examine.

Two philosophers named Publius and Vitellius from ancient Rome fell into a spacetime warp and found themselves on a beach in 21st century USA. While taking in their new surroundings, they observed a water skier being pulled along by a speeding boat. The following is a literal translation of their conversation. (Latin has no articles.)
Publius: Why boat go so fast?
Vitellius: Man on string chase him!

For these philosophers motion was explicable in terms of causality. But it is not just ancient philosophers who worried about causes and effects. Causality is an important issue in modern thought, including modern science. Just as in algebra there are independent and dependent variables, so too in physics there are causes and effects. The cause is created in the setup of the experimental apparatus and the effects are measured. But not all equations are causal statements. The popular statement of Newton's Third Law — every action has an equal and opposite reaction — is misleading, because the word *reaction* has the connotation of a response to an action, and therefore temporally posterior to the action. But the correct understanding of the Third Law is that when an object A applies a force on an object B, the object B *simultaneously* applies an equal and opposite force on A. It is misleading to think of these two opposite forces as a cause followed by an effect. However, in the equation for Newton's Second Law $\mathbf{F} = m\mathbf{a}$ the force is understood as the cause and the acceleration as the effect. Inside a turning vehicle the effect of inertia is experienced by the passengers as the pseudoforce called centrifugal force. We do not say that the turning — which is a centripetal acceleration — is the cause and the centrifugal force the effect. Rather, we say that the cause is the centripetal force of friction applied by the road to the turning wheels of the vehicle. So equations in physics can be both causal and acausal or reversible. We now raise the question whether (5.18) is causal or reversible. We have already shown that a changing magnetic flux can produce a curling electric field. So the right side is the cause and the left side is the effect. But can the process be reversed?

For this process to be reversed it would be necessary to create a curling electric field in a closed loop so that $\oint \mathbf{E} \cdot d\mathbf{r}$ has a non-zero value along the loop. But this cannot be done by any assembly of static electric charges, since for fields generated by static charges $\nabla \times \mathbf{E} = 0$. Having a source of potential difference like a battery or a capacitor will not help, since the net potential difference across a closed circuit included these sources will be zero. The only way to generate such a curling electric field along the loop is to generate a varying magnetic flux within the loop. Thus Eq. (5.18) is irreversible. The changing magnetic flux is the source, and the curling electric field is the effect. And since a magnetic field is generated by an electric current, electric charges and electric currents are the sources that generate electric and magnetic fields. Causality is important in the study of electromagnetic fields.

5.3.4 *Speed of Electromagnetic Fields*

Causality implies an earlier cause and a later effect. We shall see in this section that electric and magnetic fields are not created instantaneously at arbitrary distances from the sources, but they propagate through space at a finite speed.

In this following discussion we are interested only in the fields in spatial regions where there are no charges — stationary or in motion — and hence where $\rho = 0$ and $\mathbf{J} = 0$ everywhere in the region of interest to us. Such fields are called fields in empty space.

Taking the curl of Eq. (5.21) after setting $\mathbf{J} = 0$,

$$\nabla \times (\nabla \times \mathbf{B}) = \epsilon_0 \mu_0 \frac{\partial}{\partial t}(\nabla \times \mathbf{E})$$

$$\nabla(\nabla \cdot \mathbf{B}) - \nabla^2 \mathbf{B} = -\epsilon_0 \mu_0 \frac{\partial^2 \mathbf{B}}{\partial t^2}$$

which yields

$$\nabla^2 \mathbf{B} = \epsilon_0 \mu_0 \frac{\partial^2 \mathbf{B}}{\partial t^2} \tag{5.28}$$

Following an analogous procedure with Eq. (5.22) we obtain

$$\nabla^2 \mathbf{E} = \epsilon_0 \mu_0 \frac{\partial^2 \mathbf{E}}{\partial t^2} \tag{5.29}$$

These equations are of the form $\nabla^2 u = \frac{1}{v^2} \frac{\partial^2 u}{\partial t^2}$ which represents a plane wave traveling at speed v through a medium. Thus, in empty space an electric field $\mathbf{E}$ and a magnetic field $\mathbf{B}$ propagate as waves at the finite speed

$$v = \frac{1}{\sqrt{\epsilon_0 \mu_0}} \tag{5.30}$$

The experimentally determined values of these constants are, to three significant figures, $\epsilon_0 = 8.85 \times 10^{-12}$ C^2/N.m^2, and $\mu_0 = 1.26 \times 10^{-6}$ T.m/A ($4\pi \times 10^{-7}$ T.m/A). This yields the velocity of the wave $v = 3.00 \times 10^8$ m/s. Maxwell realized that this is so close to the known value of the speed of light c that he concluded that light is a form of electromagnetic radiation consisting of oscillating electric and magnetic fields.

Apart from identifying the nature of light as an electromagnetic wave, the electromagnetic wave equations provide proof that any influence generated by a moving charge cannot travel instantaneously through space, and although c is a very large number by ordinary standards, it is not infinity.

Einstein's theory of Special Relativity is based partly on the premise that nothing, not even a signal, can travel faster than c. Thus c is a fundamental constant of nature. And since $c = \frac{1}{\sqrt{\epsilon_0 \mu_0}}$, we can eliminate one of the constants in this equation. It is common to eliminate μ_0 and replace it with $\frac{1}{\epsilon_0 c^2}$. With this substitution, we may write the four Maxwell equations as follows:

Maxwell's Equations

$$\nabla \cdot \mathbf{E} = \frac{\rho}{\epsilon_0} \tag{5.31}$$

$$\nabla \cdot \mathbf{B} = 0 \tag{5.32}$$

$$\nabla \times \mathbf{E} = -\frac{\partial \mathbf{B}}{\partial t} \tag{5.33}$$

$$\nabla \times \mathbf{B} = \frac{\mathbf{J}}{\epsilon_0 c^2} + \frac{1}{c^2}\frac{\partial \mathbf{E}}{\partial t} \tag{5.34}$$

and the electric and magnetic fields are defined via the Lorentz force

$$\mathbf{F} = q(\mathbf{v} \times \mathbf{B}) \tag{5.35}$$

The origin of the magnetic field is contained in the fourth equation. Let us play a game in which the constant c becomes a parameter that we can vary at will. Suppose c becomes arbitrarily large, i.e. $c \to \infty$. The right side of the equation then vanishes, and so the magnetic field $\mathbf{B}$ reduces to zero. Thus, if c is infinity, then there is no magnetism. The magnetic field $\mathbf{B}$ exists because the effect of disturbing an electric charge travels at a finite speed through space.

5.3.5 *Plane Wave Electric and Magnetic Fields*

The wave equations for **E** and **B** can be solved to yield the plane wave solutions

$$\mathbf{E} = \mathbf{E}_0 \cos(\mathbf{k} \cdot \mathbf{r} - \omega t) \tag{5.36}$$

and

$$\mathbf{B} = \mathbf{B}_0 \cos(\mathbf{k} \cdot \mathbf{r} - \omega t) \tag{5.37}$$

where $\omega = ck$.

Consider a short antenna directed along the z axis with a current that oscillates at angular frequency ω. Let us take the midpoint of the antenna to be the origin. Consider the magnetic field at some distance in the x direction from the origin in the xy plane. From the equation $c^2\nabla \times \mathbf{B} = \frac{\mathbf{J}}{\epsilon_0}$ it is seen that the magnetic field curls in a circle with the antenna as an axis through the center. As the current changes, so does the magnetic field. A changing magnetic field induces an electric field perpendicular to the magnetic field. Thus, the electric field is perpendicular to the xy plane and is along the z axis. This oscillating electric field propagates radially from the oscillating current at the origin, and is everywhere directed perpendicular to the xy plane. The magnetic field is everywhere in the xy plane. Thus, the electric and magnetic fields are always perpendicular to each other. So $\mathbf{E} \cdot \mathbf{B} = 0$ for an electromagnetic wave generated by a source. The electric field is always in the z direction, and the magnetic field in the y direction. Both fields propagate in the x direction. What we have described is an example of a plane polarized wave. Real waves are superpositions of such waves.

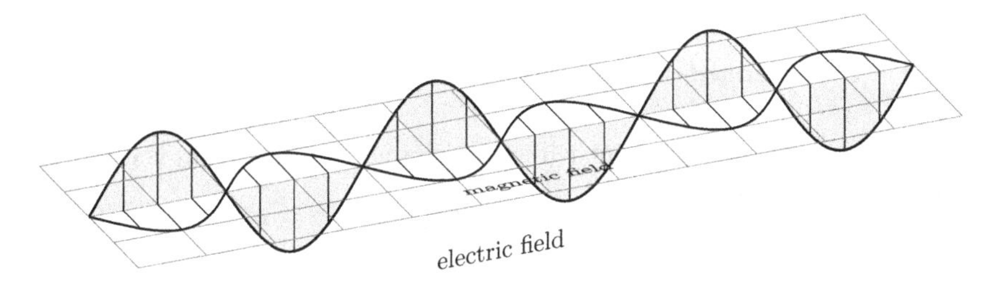

The graph illustrates a propagating electromagnetic field. The shaded portion indicates the plane of the electric field. The magnetic field is perpendicular to the electric field.

If we pursue the transmission of the wave along the x axis, this wave will have its electric field only in the z direction (positive or negative) and its magnetic field only in the y direction. Such a wave is said to be *polarized* in the xz plane, which is the plane of the electric field.

The electric field of a wave propagating along the x axis and polarized in the xz plane can be written as

$$\mathbf{E} = E_0 \hat{k} \cos(kx - \omega t) \tag{5.38}$$

and the magnetic field as

$$\mathbf{B} = B_0 \hat{\jmath} \cos(kx - \omega t) \tag{5.39}$$

For a wave propagating at speed c the following relation applies:

$$\frac{\omega}{k} = c \tag{5.40}$$

Substituting the solutions from Eqs. (5.38) and (5.39) into the Maxwell equations $\nabla \times \mathbf{E} = -\frac{\partial \mathbf{B}}{\partial t}$ or $c^2 \nabla \times \mathbf{B} = \frac{\partial \mathbf{E}}{\partial t}$ (since $\mathbf{J} = 0$ in empty space), we obtain the relationship between the amplitudes of the electric and magnetic fields:

$$E_0 = cB_0 \tag{5.41}$$

In general, where there are multiple sources, the time dependent electric field at any point is given by

$$\mathbf{E} = \sum_i \mathbf{E}_i \tag{5.42}$$

and likewise for the magnetic field.

5.3.6 *The Liénard-Wiechert Potentials*

The scalar potential at a point $\mathbf{r}$ due to a charge q at a point $\mathbf{r}_q$ is given by

$$\phi(\mathbf{r}) = \frac{q}{4\pi\epsilon_0 |\mathbf{r} - \mathbf{r}_q|} \tag{5.43}$$

The vector potential at $\mathbf{r}$ due to a charge q traveling with velocity $\mathbf{v}$ at $\mathbf{r}_q$ is

$$\mathbf{A}(\mathbf{r}) = \frac{1}{4\pi\epsilon_0 c^2} \frac{q\mathbf{v}_q}{|\mathbf{r} - \mathbf{r}_q|} \tag{5.44}$$

However, because the electric and magnetic fields propagate at the speed c, we need to calculate the potential at a point $\mathbf{r}$ at time t generated by a charge at the point $\mathbf{r}'$ at an earlier time t', and so $t - t'$ is the time taken for light to travel from the charge to the point $\mathbf{r}$.

The scalar potential at a point $\mathbf{r}$ at time t due to a charge q at a point $\mathbf{r}_{q(t')}$ at time t' is given by

$$\phi(\mathbf{r}, t) = \frac{q}{4\pi\epsilon_0} \int_{-\infty}^{\infty} \frac{1}{|\mathbf{r} - \mathbf{r}_q(t')|} \delta\left(\frac{|\mathbf{r} - \mathbf{r}_q(t')|}{c} - t + t'\right) dt' \qquad (5.45)$$

And the equation for the vector potential becomes

$$\mathbf{A}(\mathbf{r}, t) = \frac{q}{4\pi\epsilon_0 c^2} \int_{-\infty}^{\infty} \frac{\mathbf{v}_q(t')}{|\mathbf{r} - \mathbf{r}_q(t')|} \delta\left(\frac{|\mathbf{r} - \mathbf{r}_q(t')|}{c} - t + t'\right) dt' \qquad (5.46)$$

The delta function chooses what is called the retarded time t_r, which is the solution to the equation

$$t_r = t - \frac{1}{c}|\mathbf{r}(t) - \mathbf{r}_q(t_r)| \qquad (5.47)$$

We now make use of an identity:

$$\int_{-\infty}^{\infty} F(t')\delta[f(t')]dt' = \frac{F(t_r)}{|f'(t_r)|} \qquad (5.48)$$

where $f(t_r) = 0$ but $f'(t_r) \neq 0$.

We now set

$$f(t') = t' - t_r = t' - t + \frac{1}{c}|\mathbf{r}(t) - \mathbf{r}_q(t_r)| \qquad (5.49)$$

so that $f(t')$ vanishes at $t' = t_r$, but not $f'(t')$. We first set $t_r = t'$ and obtain the derivative

$$f'(t') = 1 - \frac{1}{c}|\mathbf{r}(t) - \mathbf{r}_q(t')|^{-1}(\mathbf{r} - \mathbf{r}_q(t')) \cdot \frac{d\mathbf{r}_q(t')}{dt'} \qquad (5.50)$$

and set $t' = t_r$ to obtain

$$f'(t_r) = 1 - \frac{1}{c}|\mathbf{r}(t) - \mathbf{r}_q(t_r)|^{-1}(\mathbf{r} - \mathbf{r}_q(t_r)) \cdot \frac{d\mathbf{r}_q(t_r)}{dt'} \qquad (5.51)$$

Exercise:

1. Prove the identity (5.48).
2. Derive Eq. (5.50).

Next, we define $\boldsymbol{\beta}_q = \mathbf{v}_q/c$, and $\mathbf{n}_q = \frac{\mathbf{r}-\mathbf{r}_q}{|\mathbf{r}-\mathbf{r}_q|}$.

$$f'(t') = 1 - \mathbf{n}_q \cdot \boldsymbol{\beta}_q$$

We next set $F(t')$ as $\frac{q}{4\pi\epsilon_0|\mathbf{r}-\mathbf{r}_q(t')|}$ for the scalar potential ϕ, and as $\frac{q\mathbf{v}_q}{4\pi\epsilon_0 c^2|\mathbf{r}-\mathbf{r}_q(t')|}$ for the vector potential. We set the time argument t' of $\mathbf{r}_q$ and $\boldsymbol{\beta}_q$ as the retarded time t_r, which is earlier than the time t at which the potentials are measured. Hence we write the Liénard-Wiechert potentials as

$$\phi(\mathbf{r}, t) = \frac{1}{4\pi\epsilon_0}\left(\frac{q}{(1-\mathbf{n}_q \cdot \boldsymbol{\beta}_q)|\mathbf{r}-\mathbf{r}_q|}\right) \tag{5.52}$$

$$\mathbf{A}(\mathbf{r}, t) = \frac{1}{4\pi\epsilon_0 c}\left(\frac{q\boldsymbol{\beta}_q}{(1-\mathbf{n}_q \cdot \boldsymbol{\beta}_q)|\mathbf{r}-\mathbf{r}_q|}\right) \tag{5.53}$$

5.4 Energy in a Magnetic Field

In the experiment shown in the following illustration the rod is mechanically moved to the right. In that process a current is set up in the closed circuit. The direction of the current is in the direction of the emf that is set up, which we can call the driving emf of the circuit.

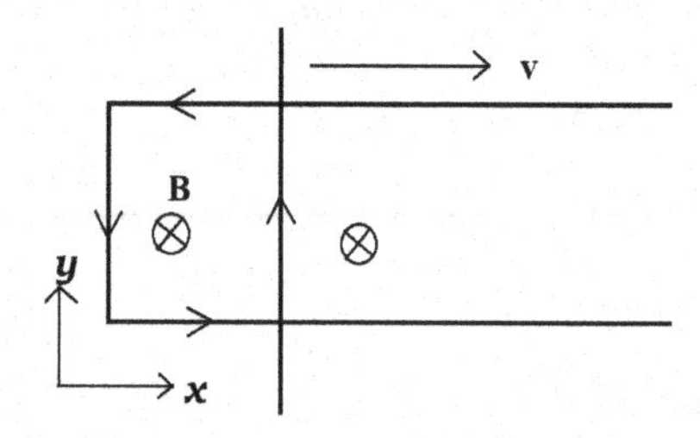

The direction of the current in the rod — moving along the positive y direction — is determined by the vector equation for the magnetic force $\mathbf{F} = q\mathbf{v} \times \mathbf{B}$ acting on a charge q on the rod. The potential difference set up across the ends of the rod will cause a current flow as shown in the counterclockwise direction.

This current in the rectangular loop will in turn generate a magnetic field through the loop. This induced magnetic field $\mathbf{B}_{ind}$ will be directed out of the page, and therefore in the opposite direction of the fixed magnetic field $\mathbf{B}$ that is directed into the page.

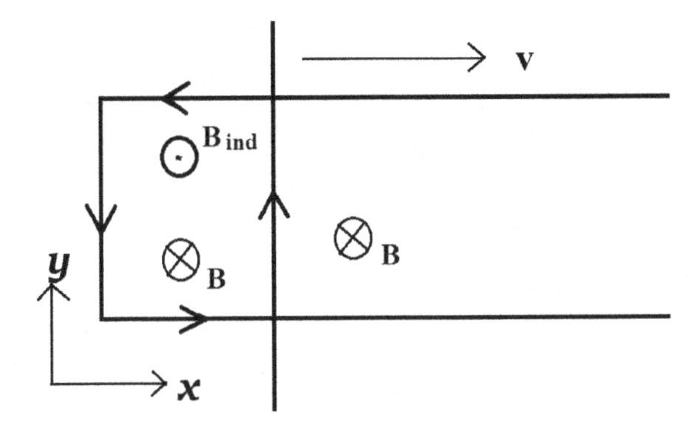

As we saw in Section 5.2.2, this is a property of induced currents, and is an example of Lenz's law. So, the induced magnetic field attempts to reduce the flux that is passing through the circuit. This is a sort of inertia of the magnetic field. And in order to overcome this inertia an external force must be applied and work must be done on the system.

The Biot-Savart law showed that a magnetic field is proportional to the current that generates the field. The flux generated within the circuit is therefore proportional to the current through the circuit. Self-inductance of a circuit is defined as the magnetic flux induced within the circuit divided by the current flowing through the circuit:

$$L = \frac{|\Phi_{ind}|}{|I|} \tag{5.54}$$

Since L by definition is positive, we are interested only in the magnitudes and not the directions of the flux or the current.

Now, by Faraday's law, the emf induced in the circuit is equal to the rate of change of flux through the circuit:

$$\mathcal{E}_{ind} = -\frac{\partial \Phi_{ind}}{\partial t} = L\frac{dI}{dt} \tag{5.55}$$

The change in current dI is opposite to the existing current I, and so dI has the opposite sign of I.

Thus the work done in a circuit in raising the current from 0 to I is

$$\int \mathcal{E}_{ind} I dt = \int LIdI = \frac{1}{2}LI^2 \tag{5.56}$$

This is the magnetic energy within an arbitrary circuit. In our experiment we generated the current by electromagnetic means, but the result will hold good even if we generate the current by other means, such as by a battery, or by a temporary current through discharge from a capacitor. Equation (5.56) does not include any energy due to the external magnetic field $\mathbf{B}$, but only the energy generated by the current through the circuit. We will next evaluate the magnetic energy within a solenoid generated by the current.

The magnetic field inside a solenoid has magnitude $B = \mu_0 nI$ where $n =$ number of turns per meter. $n = N/d$ where N is the number of turns in a length d of the solenoid. If A is the cross-section of the solenoid, the flux through N turns is $\Phi = BAN$, and therefore the self-inductance of a solenoid of length d is

$$L = \frac{BAN}{I} = \frac{\mu_0 NIAN}{Id} = \frac{\mu_0 N^2 A}{d} \tag{5.57}$$

So the magnetic energy stored within a solenoid of cross-section A and length d is equal to

$$U_m = \frac{1}{2}\frac{\mu_0 N^2 A}{d}\left(\frac{Bd}{\mu_0 N}\right)^2 = \frac{1}{2\mu_0}B^2 Ad \tag{5.58}$$

Setting $\mu_0 = \frac{1}{\epsilon_0 c^2}$, we obtain the magnetic energy per unit volume by dividing the right side of the equation by Ad, the volume of the space within the solenoid:

$$u_m = \frac{1}{2}\epsilon_0 c^2 B^2 \tag{5.59}$$

This result can be obtained more generally as follows:

The work done by an electromagnetic field on a small charge dq is given by

$$dW = \int \mathbf{F}\cdot d\mathbf{r} = \int dq(\mathbf{E} + \mathbf{v}\times\mathbf{B})\cdot d\mathbf{r} = \int dq\mathbf{E}\cdot\mathbf{v}dt \tag{5.60}$$

since $(\mathbf{v}\times\mathbf{B})\cdot\mathbf{v} = 0$. Replacing $dq\mathbf{v}$ by $\rho\mathbf{v}d\tau = \mathbf{J}d\tau$, we get an expression for the work done by the field on all the charges in the region

$$W = \int \mathbf{E}\cdot\mathbf{J}d\tau dt \tag{5.61}$$

Setting $\mathbf{J} = -\epsilon_0\frac{\partial\mathbf{E}}{\partial t} + \epsilon_0 c^2\nabla\times\mathbf{B}$, we obtain

$$W = \int \left(-\epsilon_0\mathbf{E}\cdot\frac{\partial\mathbf{E}}{\partial t} + \epsilon_0 c^2\mathbf{E}\cdot(\nabla\times\mathbf{B})\right)d\tau dt \tag{5.62}$$

Next, we use the identity

$$\mathbf{E} \cdot (\nabla \times \mathbf{B}) = \mathbf{B} \cdot (\nabla \times \mathbf{E}) - \nabla \cdot (\mathbf{E} \times \mathbf{B})$$

Hence

$$W = \int \left(-\epsilon_0 \mathbf{E} \cdot \frac{\partial \mathbf{E}}{\partial t} + \epsilon_0 c^2 \mathbf{B} \cdot (\nabla \times \mathbf{E}) - \epsilon_0 c^2 \nabla \cdot (\mathbf{E} \times \mathbf{B}) \right) d\tau dt \quad (5.63)$$

Now, if we choose the region over which the volume integral is evaluated to be approximately spherical with very large radius r, the electric and magnetic fields will become very small at every point on the boundary. We know that for arbitrary charge and current distributions the dominant term for the electric field is the monopole term which drops off as $\frac{1}{r^2}$, and the dominant term for the magnetic field is the dipole term which drops off as $\frac{1}{r^3}$. Thus the magnitude of $\mathbf{E} \times \mathbf{B}$ drops off at least as fast as $\frac{1}{r^5}$, whereas the area of the boundary increases only as r^2. Thus the volume integral of the third term becomes zero by Gauss's law for sufficiently large regions. Replacing $\nabla \times \mathbf{E}$ by $-\frac{\partial \mathbf{B}}{\partial t}$ we obtain

$$W = \int \left(-\epsilon_0 \mathbf{E} \cdot \frac{\partial \mathbf{E}}{\partial t} - \epsilon_0 c^2 \mathbf{B} \cdot \frac{\partial \mathbf{B}}{\partial t} \right) d\tau dt \quad (5.64)$$

We first perform the integral over time, using the fact that $\mathbf{E} \cdot \frac{\partial \mathbf{E}}{\partial t} = \frac{1}{2} \frac{\partial E^2}{\partial t}$ and likewise for the magnetic field, we obtain

$$W = -\frac{1}{2} \int \left(\epsilon_0 E^2 + \epsilon_0 c^2 B^2 \right) d\tau \quad (5.65)$$

Now, this is the expression for the work done *by* the electromagnetic field in order to generate a particular configuration of charges. The positive energy stored in the field is the work done by an external agent *on* the electromagnetic field, and has the opposite sign. This work done is the total energy stored in the field

$$U_{em} = \frac{1}{2} \int (\epsilon_0 E^2 + \epsilon_0 c^2 B^2) d\tau \quad (5.66)$$

The first term is the energy density for the electric field, and the second that of the magnetic field. Thus, the total energy per unit volume or energy density of the electromagnetic field is

$$u = \frac{\epsilon_0}{2} (E^2 + c^2 B^2) \quad (5.67)$$

We saw earlier (Eq. (5.41)) that in a plane electromagnetic wave the amplitude of the electric field E_0 is related to that of the magnetic field by $E_0 = cB_0$. The energy of a propagating wave is proportional to the square of the amplitude of the wave. So we find that in an electromagnetic wave described by the solutions given above, the electric and magnetic fields contribute equally to the total energy density.

5.5 Simple Circuits

5.5.1 *RC Circuits*

A capacitor can be charged by connecting the terminals of a battery of emf $\mathcal{E}$ across the plates of the capacitor in series with a resistance R, which safeguards the circuit from excessively high currents. A capacitor carrying charge Q_0 with potential V_0 can be discharged through the resistor after disconnection from the battery. A switch helps to toggle between the charging and the discharging processes as shown in Fig. 5.1.

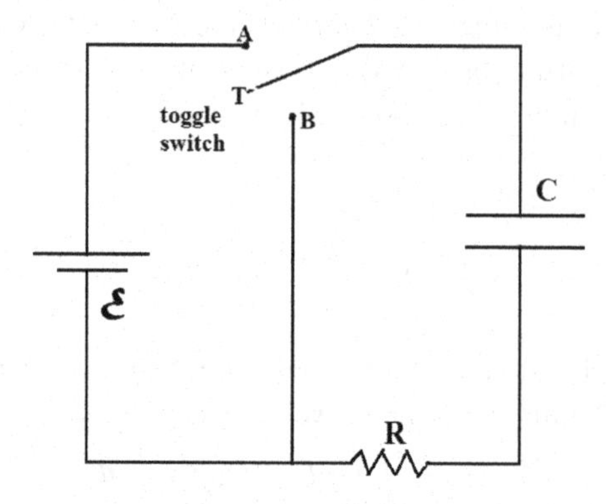

Fig. 5.1 A charging and discharging RC circuit.

It is an elementary exercise to derive the following expressions for the charge on the capacitor, the potential difference across the capacitor, and the current I through the resistor at any time during the charging:

$$
\begin{aligned}
Q &= Q_{max}(1 - e^{-\frac{t}{CR}}) \\
V &= \mathcal{E}(1 - e^{-\frac{t}{CR}}) \\
I &= \frac{\mathcal{E}}{R}e^{-\frac{t}{CR}}
\end{aligned}
\tag{5.68}
$$

It is equally straightforward to derive the equations for the charge and the potential of the capacitor, and the current I through the resistor during discharging:

$$
\begin{aligned}
Q &= Q_0 e^{-\frac{t}{CR}} \\
V &= V_0 e^{-\frac{t}{CR}} \\
I &= \frac{V_0}{R}e^{-\frac{t}{CR}}
\end{aligned}
\tag{5.69}
$$

> **Exercise:**
> Derive Eqs. (5.68) and (5.69).

5.5.2 *LC Circuits*

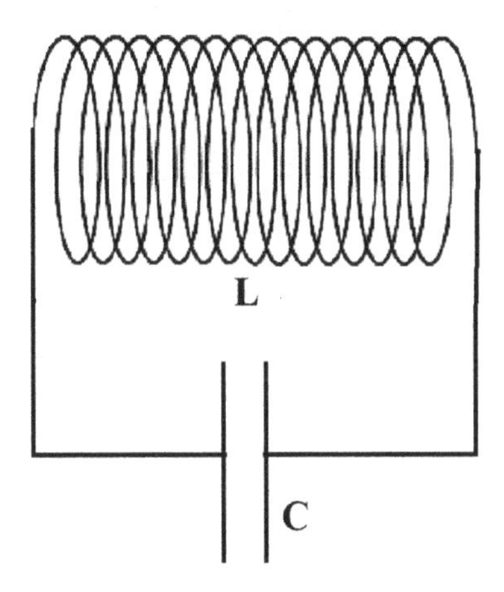

Consider a circuit consisting of a solenoid or inductor with self inductance L and a capacitor with capacitance C. Suppose initially the capacitor was given a charge and then it was discharged through the inductor. Let us neglect the resistance in the circuit.

At any time, the total energy in the circuit is given by

$$E = \frac{1}{2}LI^2 + \frac{1}{2}\frac{Q^2}{C} \tag{5.70}$$

As the capacitor discharges through the solenoid, there is a decrease of charge in the capacitor as this charge flows through the inductor as current. Since there is no resistance, there is no loss of energy in the circuit. Thus E is a constant. Taking the time derivative, we obtain

$$LI\frac{dI}{dt} + \frac{Q}{C}\frac{dQ}{dt} = 0 \tag{5.71}$$

Now, $\frac{dQ}{dt} = I$, and so

$$L\frac{dI}{dt} + \frac{Q}{C} = 0 \tag{5.72}$$

which can be written as

$$\frac{d^2Q}{dt^2} = -\frac{1}{LC}Q \tag{5.73}$$

This is a Simple Harmonic differential equation, and the solutions are

$$Q = A\cos\omega t + B\sin\omega t \tag{5.74}$$

where $\omega = \frac{1}{\sqrt{LC}}$.

The energy in the capacitor is due to the repulsion between the static charges, and may be called a potential energy. The energy in the inductor is due to flowing charges, and may be called kinetic energy. Thus the oscillating LC circuit is mathematically analogous to a pendulum or a vibrating spring. This analogy is important for subsequent discussions.

Chapter 6

Energy and Momentum of Fields

6.1 Tensor Analysis

6.1.1 *Transformation Laws*

A vector field is a set of three scalar fields that form the components of a vector whose direction and magnitude in general vary from point to point. A mathematical entity which can be considered as a sort of vector field for which every component is a set of three elements, and therefore a sort of nested vector with three components, is an example of a 3×3 tensor field. Such a tensor has 9 elements, and these elements can be written as a square array, or a square matrix. Not all matrices are tensors, because the elements of a tensor must follow definite rules for transformation from one coordinate system to another.

A vector field is a special tensor, called a tensor of rank 1. A tensor whose elements can be written as a square matrix is called a tensor of rank 2. A tensor of rank 3 can be pictured as a three-dimensional matrix, the elements arranged like atoms in a cubical crystal. We are concerned with tensors of rank 1 and 2 in this section.

A common example of a tensor of rank 1 is the displacement vector $d\mathbf{r} = \hat{i}dx + \hat{j}dy + \hat{k}dz$. From this we can construct the velocity vector $\mathbf{v} = \frac{d\mathbf{r}}{dt} = \hat{i}\frac{dx}{dt} + \hat{j}\frac{dy}{dt} + \hat{k}\frac{dz}{dt} = v_x\hat{i} + v_y\hat{j} + v_z\hat{k}$.

Suppose we change to a different coordinate system (u, v, w). This could be, say, the spherical (r, θ, φ) or the cylindrical (ρ, φ, z) system. The transformation equations can be expressed as

$$du = \frac{\partial u}{\partial x}dx + \frac{\partial u}{\partial y}dy + \frac{\partial u}{\partial z}dz \tag{6.1}$$

$$dv = \frac{\partial v}{\partial x}dx + \frac{\partial v}{\partial y}dy + \frac{\partial v}{\partial z}dz \tag{6.2}$$

$$dw = \frac{\partial w}{\partial x}dx + \frac{\partial w}{\partial y}dy + \frac{\partial w}{\partial z}dz \tag{6.3}$$

For the sake of economy of space, we use shorthand notation, and express the (x, y, z) coordinates by the symbol x^i where the superscript is not an exponent. Exponents are expressed as $(x^i)^2$, etc. We may express the (u, v, w) coordinates by the symbol u^i, and thus, the transformation equations can be expressed by a single line:

$$du^i = \sum_j \frac{\partial u^i}{\partial x^j}dx^j \tag{6.4}$$

Now, consider the gradient of a scalar field $\nabla\phi = (\frac{\partial\phi}{\partial x}, \frac{\partial\phi}{\partial y}, \frac{\partial\phi}{\partial z})$. The transformation equations are expressed simply as

$$\frac{\partial\phi}{\partial u^j} = \sum_i \frac{\partial\phi}{\partial x^i}\frac{\partial x^i}{\partial u^j} \tag{6.5}$$

The transformation coefficients are different in the two equations. The difference in the transformations reflects a basic difference in the two vectors. The displacement vector and vectors derived from the displacement such as the velocity vector transform one way. The gradient vector transforms in a different way. The former belong to a class of vectors called contravariant vectors, and the latter are called covariant vectors, though alternate terminology is available to distinguish these two classes of vectors. So, it is not uncommon to hear the former referred to as simply *vectors*, and the latter as *one-forms*.

The contravariant vectors are depicted with a superscript dx^i, and the covariant vectors are depicted with a subscript F_i. Coordinate systems are distinguished by appending a prime to one set: x' is different from x. So, the transformation equations can be written as:

For contravariant vectors

$$dx'^i = \sum_j \frac{\partial x'^i}{\partial x^j}dx^j \tag{6.6}$$

and for covariant vectors

$$F'_i = \sum_j \frac{\partial x^j}{\partial x'^i} F_j \tag{6.7}$$

A note on the transformation factors:
It is important to note that $\frac{\partial x^j}{\partial x'^i}$ is not the reciprocal of $\frac{\partial x'^i}{\partial x^j}$. As evidence for this statement, let $x = (x, y, z)$ and $x' = (r, \theta, \varphi)$. Let us select the factor for which $i = j = 1$, so that $x_1 = x$ and $x'_1 = r$. $\frac{\partial x^1}{\partial x'^1} = \frac{\partial x}{\partial r} = \sin\theta\cos\varphi$. $\frac{\partial x'^1}{\partial x^1} = \frac{\partial r}{\partial x} = \frac{x}{\sqrt{x^2+y^2+z^2}} = \sin\theta\cos\varphi$. Thus $\frac{\partial x^1}{\partial x'^1}$ is not the reciprocal of $\frac{\partial x'^1}{\partial x^1}$, but counter to algebraic intuition, $\frac{\partial x^1}{\partial x'^1} = \frac{\partial x'^1}{\partial x^1}$. But this equation also cannot be generalized into a rule, for it is evident that $\frac{\partial y}{\partial \theta}$ cannot equal $\frac{\partial \theta}{\partial y}$, since y has the dimension of length and θ has no dimension. Indeed, $\frac{\partial y}{\partial \theta} = r\cos\theta\sin\varphi$ and $\frac{\partial \theta}{\partial y} = \frac{\cos\theta\sin\varphi}{r}$.

6.1.2 *The Metric Tensor*

A vector — covariant or contravariant — is called a tensor of rank 1. A scalar is called a tensor of rank 0 or an invariant. So the distance ds between two points is an invariant or a scalar because it is an absolute quantity irrespective of the coordinate system.

In Cartesian coordinates

$$ds^2 = dx^2 + dy^2 + dz^2$$

and in spherical polar coordinates

$$ds^2 = dr^2 + r^2 d\theta^2 + r^2 \sin^2\theta d\varphi^2$$

The coefficients of these displacement squared terms are called the elements of the metric, which is a 3×3 matrix. So, for Cartesian coordinates the metric is simply the identity matrix:

$$g = \begin{pmatrix} 1 & 0 & 0 \\ 0 & 1 & 0 \\ 0 & 0 & 1 \end{pmatrix} \tag{6.8}$$

and for the spherical polar coordinates

$$g = \begin{pmatrix} 1 & 0 & 0 \\ 0 & r^2 & 0 \\ 0 & 0 & r^2\sin^2\theta \end{pmatrix} \tag{6.9}$$

Both the Cartesian and the spherical polar coordinates are examples of orthogonal coordinates, since the axes are mutually orthogonal at every point. In the more general case, where the coordinate axes may not be orthogonal, the square of the distance becomes

$$ds^2 = \sum_{i,j} g_{ij} dx^i dx^j \tag{6.10}$$

The elements are then written as

$$g = \begin{bmatrix} g_{11} & g_{12} & g_{13} \\ g_{21} & g_{22} & g_{23} \\ g_{31} & g_{32} & g_{33} \end{bmatrix} \tag{6.11}$$

It turns out that in most cases where a superscript index (contravariant index) is the same as a subscript index (contravariant index), there is also a summation over the repeated index. Accordingly, for the sake of economy of space, we use the convention introduced by Einstein whereby a repeated index implies a summation over that index. And so, we may drop the summation symbol in the equation

$$ds^2 = g_{ij} dx^i dx^j \tag{6.12}$$

The metric g_{ij} is a tensor with two subscripts. It is therefore a covariant tensor of rank 2. We notice that the sums over the indices i and j on the right side result in a scalar or invariant ds^2. The dx^i are contravariant tensors of rank 1, and the g_{ij} is a covariant tensor of rank 2. So the sum over the products of the contravariant and covariant tensors yields a scalar or a tensor of rank 0. This process whereby a combination of tensors yields a tensor of lower rank is called *contraction.* A covariant and a contravariant vector can be contracted to yield a scalar:

$$A_i B^i = k \tag{6.13}$$

So, the length or the magnitude A of a vector A_i can be expressed as

$$A_i A^i = A^2 \tag{6.14}$$

which is an invariant. The above equation is also called the inner product of the vectors A and B. In Cartesian coordinates $A_i = A^i$, that is, covariant and contravariant vectors have the same form. But, in general,

$$A_i = g_{ij} A^j \tag{6.15}$$

And so the inner product $A^2 = A_i A^i$ can be written as

$$A^2 = A_i A^i = g_{ij} A^i A^j \tag{6.16}$$

Summation indices are also called dummy indices, since the actual symbol used is irrelevant. So, in the equation above, the sum is invariant under interchange of i and j. Hence it is evident that $g_{ij} = g_{ji}$. This is an example of a *symmetric* tensor. If $A_{ij} = -A_{ji}$ then A is called a *skew-symmetric* or *antisymmetric* tensor.

A second rank tensor can be covariant, contravariant, or it can be a mixed tensor with one covariant and one contravariant index, e.g. A_i^j. Such a mixed tensor can be contracted by itself, so A_i^i is the sum of all the diagonal elements of the tensor, i.e. the sum $\sum_{i,j} \delta_{ij} A_i^j$, and is a scalar or invariant. The sum of the diagonal elements of a matrix is called the *trace* or *spur* of the matrix.

This brief introduction to the terminology of tensor analysis will be helpful for the discussions that follow.

6.2 Energy and Momentum Flow in Fields

6.2.1 *Poynting Vector*

Newton's Third Law states that when an object A applies a force **F** on an object B, the object B *simultaneously* applies a force −**F** on object A. This is evident for contact forces, but because electric and magnetic fields are transported at finite speeds through space, this Law needs to be clarified for forces between charges. Suppose we have a stationary charge A and a charge B in motion. Because the potential at A due to be B travels at a finite speed, by the time charge B is at a later position, the force experienced by A would be due to the charge B at an earlier position:

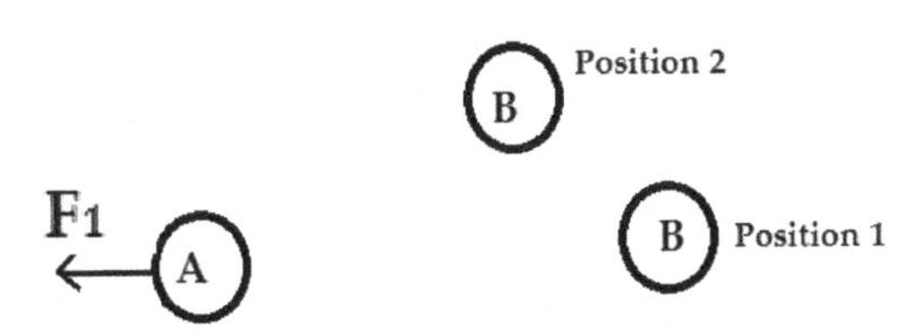

Charge A experiences a force F_1 due to charge B when it was in position 1

Thus, the charge A, which experiences a force $\mathbf{F}_1$, will apply an equal and opposite force not on the charge B, but on the field. As we saw in the previous chapter, an electromagnetic field carries energy. Now, we will see that a field also carries momentum. The force experienced by charge A is due to the momentum imparted per second to A by the field. We will now derive an expression for the momentum carried by a field.

The work done by an electromagnetic field on a small charge dq is given by

$$dW = \int \mathbf{F} \cdot d\mathbf{r} = \int dq(\mathbf{E} + \mathbf{v} \times \mathbf{B}) \cdot d\mathbf{r} = \int dq\mathbf{E} \cdot \mathbf{v}dt \tag{6.17}$$

since $(\mathbf{v} \times \mathbf{B}) \cdot \mathbf{v} = 0$. Replacing $dq\mathbf{v}$ by $\rho\mathbf{v}d\tau = \mathbf{J}d\tau$, we get an expression for the work done by the field on all the charges in the region

$$W = \iint \mathbf{E} \cdot \mathbf{J}d\tau dt \tag{6.18}$$

Setting $\mathbf{J} = -\epsilon_0 \frac{\partial \mathbf{E}}{\partial t} + \epsilon_0 c^2 \nabla \times \mathbf{B}$ from Maxwell's equation, we obtain

$$W = \iint \left(-\epsilon_0 \mathbf{E} \cdot \frac{\partial \mathbf{E}}{\partial t} + \epsilon_0 c^2 \mathbf{E} \cdot (\nabla \times \mathbf{B}) \right) d\tau dt \tag{6.19}$$

Next, we use the identity

$$\mathbf{E} \cdot (\nabla \times \mathbf{B}) = \mathbf{B} \cdot (\nabla \times \mathbf{E}) - \nabla \cdot (\mathbf{E} \times \mathbf{B})$$

Hence

$$W = \iint \left(-\epsilon_0 \mathbf{E} \cdot \frac{\partial \mathbf{E}}{\partial t} + \epsilon_0 c^2 \mathbf{B} \cdot (\nabla \times \mathbf{E}) - \epsilon_0 c^2 \nabla \cdot (\mathbf{E} \times \mathbf{B}) \right) d\tau dt \tag{6.20}$$

Applying Maxwell's equation $\nabla \times \mathbf{E} = -\frac{\partial \mathbf{B}}{\partial t}$ we obtain

$$W = \iint \left(-\epsilon_0 \mathbf{E} \cdot \frac{\partial \mathbf{E}}{\partial t} - \epsilon_0 c^2 \mathbf{B} \cdot \frac{\partial \mathbf{B}}{\partial t} - \epsilon_0 c^2 \nabla \cdot (\mathbf{E} \times \mathbf{B}) \right) d\tau dt \tag{6.21}$$

Let us consider the situation where the total work done by the field in the region is zero. So the integral is zero, which means that the integrand must be identically zero, since $d\tau$ and dt are both positive everywhere. If no work is done by the field, then no work is done by an external agent on the field either.

Hence, we obtain the important equation for a region where no work is done by an external agent on the field:

$$\epsilon_0 c^2 \nabla \cdot (\mathbf{E} \times \mathbf{B}) = -\epsilon_0 \mathbf{E} \cdot \frac{\partial \mathbf{E}}{\partial t} - \epsilon_0 c^2 \mathbf{B} \cdot \frac{\partial \mathbf{B}}{\partial t} = -\frac{1}{2} \frac{\partial}{\partial t} \left(\epsilon_0 E^2 + \epsilon_0 c^2 B^2 \right) \tag{6.22}$$

Now, if we consider an arbitrary region V enclosed by a surface S, we can write the equality of the integrals

$$\int_V \epsilon_0 c^2 \nabla \cdot (\mathbf{E} \times \mathbf{B}) d\tau = -\frac{\partial}{\partial t} \int_V (\epsilon_0 E^2 + \epsilon_0 c^2 B^2) d\tau \tag{6.23}$$

By Gauss's theorem this yields

$$\oint_S \epsilon_0 c^2 (\mathbf{E} \times \mathbf{B}) \cdot \hat{n} dA = -\frac{\partial}{\partial t} \int_V (\epsilon_0 E^2 + \epsilon_0 c^2 B^2) d\tau = -\frac{\partial U_{em}}{\partial t} \tag{6.24}$$

Thus the surface integral of $\epsilon_0 c^2 \mathbf{E} \times \mathbf{B}$ over a closed surface equals the rate of decrease of electromagnetic energy in the region enclosed by the surface.

Therefore the vector $\epsilon_0 c^2 \mathbf{E} \times \mathbf{B}$ must be rate of energy flow, expressing the conservation of electromagnetic energy in the field. This vector is called the Poynting Vector, and is commonly written as $\mathbf{S}$.

Poynting Vector: $\mathbf{S} = \epsilon_0 c^2 \mathbf{E} \times \mathbf{B}$

Since the electric and the magnetic fields propagate at speed c through space, that is also the speed of the Poynting Vector. Electromagnetic energy propagates through space at speed c.

The direction of the flow of energy is along the Poynting Vector $\epsilon_0 c^2 \mathbf{E} \times \mathbf{B}$ which is perpendicular to both the electric and the magnetic fields. Thus, the electric field propagates as a transverse wave insofar as the direction of the field at any point is perpendicular to the direction of the transport of the field energy. Likewise the magnetic field also propagates as a transverse wave.

6.2.2 *Maxwell Stress Tensor*

A vector field can be written in terms of its components as a row matrix or a column matrix: $\begin{bmatrix} E_x \; E_y \; E_z \end{bmatrix}$ or $\begin{bmatrix} B_x \\ B_y \\ B_z \end{bmatrix}$. Each of these vectors is a rank 1 tensor, and may be expressed in symbolic form as E_i or B_i, where $E_1 = E_x, E_2 = E_y$, etc.

We can obtain scalars — tensors of rank 0 — from these vectors by contraction, such as $E^2 = E_i E^i$, $B^2 = B_i B^i$, $\mathbf{E} \cdot \mathbf{B} = E_i B^i = E^i B_i$. In all these instances the scalars so obtained by contraction are invariant only under spatial transformations. (The definition of a scalar or invariant will be modified when we study transformations of space and time according to the rules of Special Relativity in a later chapter.)

One can also obtain tensors of rank 2 by suitable combinations of these vectors. So the symbol **EB** without a dot or a cross between them is a tensor of rank 2. It can be expressed in symbolic form and matrix form as

$$\mathbf{EB} = E_i B_j = \begin{bmatrix} E_x B_x & E_x B_y & E_x B_z \\ E_y B_x & E_y B_y & E_y B_z \\ E_z B_x & E_z B_y & E_z B_z \end{bmatrix} \tag{6.25}$$

Obviously, not all the possible tensors we could create from the components of the electric and magnetic fields have applications in physics. But one such tensor that has important physical meaning is the *Maxwell Stress Tensor.*

The word stress in an introductory study of elasticity is commonly defined as the ratio of applied force to cross sectional area. At the advanced level one makes a distinction between a traction vector $\mathbf{T}$, which is the force vector $\mathbf{F}$ divided by area A, and the second rank stress tensor $\boldsymbol{\sigma}$. The relationship between the two is given by the equation $\mathbf{T} = \mathbf{n} \cdot \boldsymbol{\sigma}$, where $\mathbf{n}$ is a unit vector perpendicular to some imaginary surface in the medium, and $\mathbf{T}$ is the traction force per unit area across that surface. $\mathbf{T}$ is in general not in the same direction as $\mathbf{n}$. The stress tensor so defined is also called the Cauchy Stress Tensor. An analogous stress tensor — called the Maxwell Stress Tensor — is important in electromagnetism.

Electromagnetic energy travels through space. Because it is in motion, it also has momentum. The rate of increase of the momentum of the field within a region of space is equal to the net force acting upon that region, by Newton's Second Law. And this net force divided by the area perpendicular to the force can be obtained from the electromagnetic Maxwell stress tensor.

6.2.3 *Momentum of a Field*

The force on a charge q moving with velocity $\mathbf{v}$ in an electromagnetic field is given by the Lorentz force equation

$$\mathbf{F} = q(\mathbf{E} + \mathbf{v} \times \mathbf{B}) \tag{6.26}$$

The total force acting on all the charges inside a volume V is given by Newton's Second Law

$$\frac{d\mathbf{P}_{mech}}{dt} = \int_V (\rho\mathbf{E} + \mathbf{J} \times \mathbf{B}) d\tau \tag{6.27}$$

where we have converted the sum over charges to an integral over charge and current densities. We can now apply Maxwell's equations to this integral.

We use the two equations

$$\nabla \cdot \mathbf{E} = \frac{\rho}{\epsilon_0}$$

and

$$\nabla \times \mathbf{B} = \frac{1}{\epsilon_0 c^2}\mathbf{J} + \frac{1}{c^2}\frac{\partial \mathbf{E}}{\partial t}$$

and write $\rho = \epsilon_0 \nabla \cdot \mathbf{E}$ and $\mathbf{J} = \epsilon_0 c^2 \nabla \times \mathbf{B} - \epsilon_0 \frac{\partial \mathbf{E}}{\partial t}$. Substituting these into the integrand above, and rearranging slightly, we get

$$\rho\mathbf{E} + \mathbf{J} \times \mathbf{B} = \epsilon_0 \left[\mathbf{E}(\nabla \cdot \mathbf{E}) + \mathbf{B} \times \frac{\partial \mathbf{E}}{\partial t} - c^2 \mathbf{B} \times (\nabla \times \mathbf{B})\right] \tag{6.28}$$

Now,

$$\frac{\partial}{\partial t}(\mathbf{E} \times \mathbf{B}) = \frac{\partial \mathbf{E}}{\partial t} \times \mathbf{B} + \mathbf{E} \times \frac{\partial \mathbf{B}}{\partial t}$$

Rearranging terms, and using Maxwell's equation $\nabla \times \mathbf{E} = -\frac{\partial \mathbf{B}}{\partial t}$, we obtain

$$\mathbf{B} \times \frac{\partial \mathbf{E}}{\partial t} = -\frac{\partial}{\partial t}(\mathbf{E} \times \mathbf{B}) - \mathbf{E} \times (\nabla \times \mathbf{E})$$

Substituting into Eq. (6.28),

$$\rho\mathbf{E} + \mathbf{J} \times \mathbf{B} = \epsilon_0 \left[\mathbf{E}(\nabla \cdot \mathbf{E}) - \frac{\partial}{\partial t}(\mathbf{E} \times \mathbf{B}) - \mathbf{E} \times (\nabla \times \mathbf{E}) - c^2 \mathbf{B} \times (\nabla \times \mathbf{B})\right]$$

$$= \epsilon_0 \left[\mathbf{E}(\nabla \cdot \mathbf{E}) - \mathbf{E} \times (\nabla \times \mathbf{E}) - c^2 \mathbf{B} \times (\nabla \times \mathbf{B})\right] - \epsilon_0 \frac{\partial}{\partial t}(\mathbf{E} \times \mathbf{B})$$

The expression inside the brackets can be given a symmetry by adding $c^2\mathbf{B}(\nabla \cdot \mathbf{B})$ which has the value 0. And so we get

$$\rho\mathbf{E} + \mathbf{J} \times \mathbf{B} + \epsilon_0 \frac{\partial}{\partial t}(\mathbf{E} \times \mathbf{B})$$

$$= \epsilon_0 \left[\mathbf{E}(\nabla \cdot \mathbf{E}) - \mathbf{E} \times (\nabla \times \mathbf{E}) + c^2 \mathbf{B}(\nabla \cdot \mathbf{B}) - c^2 \mathbf{B} \times (\nabla \times \mathbf{B})\right]$$

From Eq. (6.27) above $\frac{d\mathbf{P}_{mech}}{dt} = \int_V (\rho\mathbf{E} + \mathbf{J}\times\mathbf{B})d\tau$ and so

$$\frac{d\mathbf{P}_{mech}}{dt} + \frac{d}{dt}\int_V \epsilon_0(\mathbf{E}\times\mathbf{B})d\tau$$

$$= \epsilon_0 \int_V \left[\mathbf{E}(\nabla\cdot\mathbf{E}) - \mathbf{E}\times(\nabla\times\mathbf{E}) + c^2\mathbf{B}(\nabla\cdot\mathbf{B}) - c^2\mathbf{B}\times(\nabla\times\mathbf{B})\right] d\tau$$

In the second term on the left a partial derivative with respect to time has been replaced by a total derivative with respect to time. The partial time derivative meant that the derivative was taken only with respect to time t, and not the space variables (x, y, z), but when we integrate the expression over the volume V the space variables have disappeared and the only variable is time.

The terms in $\mathbf{E}$ and $\mathbf{B}$ in the integrand on the right side of the equation have a similar mathematical form. Let us examine the $\mathbf{E}$ terms:

We will use the abbreviation $i = 1, 2, 3$ to denote the subscripts for x, y and z respectively. The cross product $-\mathbf{E}\times(\nabla\times\mathbf{E})$ can be expressed in determinant form as

$$-\begin{vmatrix} \hat{i} & \hat{j} & \hat{k} \\ E_1 & E_2 & E_3 \\ \left(\frac{\partial E_3}{\partial x_2} - \frac{\partial E_2}{\partial x_3}\right) & \left(\frac{\partial E_1}{\partial x_3} - \frac{\partial E_3}{\partial x_1}\right) & \left(\frac{\partial E_2}{\partial x_1} - \frac{\partial E_1}{\partial x_2}\right) \end{vmatrix}$$

$= -\hat{i}[E_2(\frac{\partial E_2}{\partial x_1} - \frac{\partial E_1}{\partial x_2}) - E_3(\frac{\partial E_1}{\partial x_3} - \frac{\partial E_3}{\partial x_1})] + ...$ similar terms for $\hat{j}$ and $\hat{k}$. To these we will add $\mathbf{E}(\nabla\cdot\mathbf{E})$.

And so we obtain the x component of the terms containing $\mathbf{E}$ as

$$E_1\left(\frac{\partial E_1}{\partial x_1} + \frac{\partial E_2}{\partial x_2} + \frac{\partial E_3}{\partial x_3}\right) - E_2\left(\frac{\partial E_2}{\partial x_1} - \frac{\partial E_1}{\partial x_2}\right) + E_3\left(\frac{\partial E_1}{\partial x_3} - \frac{\partial E_3}{\partial x_1}\right)$$

The above expression can be rewritten as

$$\frac{\partial}{\partial x_1}(E_1^2) + \frac{\partial}{\partial x_2}(E_1E_2) + \frac{\partial}{\partial x_3}(E_1E_3) - \frac{1}{2}\frac{\partial}{\partial x_1}(E_1^2 + E_2^2 + E_3^2)$$

Thus the ith component of the $\mathbf{E}$ terms are expressible as

$$[\mathbf{E}(\nabla\cdot\mathbf{E}) - \mathbf{E}\times(\nabla\times\mathbf{E})]_i = \sum_j \frac{\partial}{\partial x_j}(E_iE_j - \frac{1}{2}\mathbf{E}\cdot\mathbf{E}\,\delta_{ij}) \tag{6.29}$$

The term inside the brackets on the right side is a tensor. When we add to it the corresponding magnetic field terms, we obtain the Maxwell Stress Tensor:

$$T_{ij} = \epsilon_0[E_iE_j + c^2B_iB_j - \frac{1}{2}(\mathbf{E}\cdot\mathbf{E} + c^2\mathbf{B}\cdot\mathbf{B})\delta_{ij}] \tag{6.30}$$

The Maxwell Stress Tensor can be written in matrix form as $\frac{1}{\epsilon_0}\mathbf{T} =$

$$\begin{bmatrix} E_x^2 - \frac{1}{2}E^2 + c^2(B_x^2 - \frac{1}{2}B^2) & E_xE_y + c^2B_xB_y & E_xE_z + c^2B_xB_z \\ E_yE_x + c^2B_yB_x & E_y^2 - \frac{1}{2}E^2 + c^2(B_y^2 - \frac{1}{2}B^2) & E_yE_z + c^2B_yB_z \\ E_zE_x + c^2B_zB_x & E_zE_y + c^2B_zB_y & E_z^2 - \frac{1}{2}E^2 + c^2(B_z^2 - \frac{1}{2}B^2) \end{bmatrix} \tag{6.31}$$

The inner product of a tensor and a vector is another vector: $\mathbf{Tx} = \mathbf{y}$. Suppose we rotate the coordinate system, so that the individual elements of $\mathbf{T}$, $\mathbf{x}$ and $\mathbf{y}$ all change. So the equation now becomes $\mathbf{T}'\mathbf{x}' = \mathbf{y}'$. The relationship between the new and the old elements is expressible in terms of the rotation matrix $\mathbf{R}$ as $\mathbf{x}' = \mathbf{Rx}$, $\mathbf{y}' = \mathbf{Ry}$, and $\mathbf{R}(\mathbf{Tx}) = \mathbf{Ry}$.

Hence the equation for the transformation of the tensor: $\mathbf{T}' = \mathbf{RTR}^{-1}$.

The trace of a square matrix A is the sum of its diagonal elements: $\text{Tr}A = \sum_{ii} A_{ii}$.

It is easy to show that Tr $(AB) = \text{Tr}(BA)$:

An element of AB can be written as $\sum_j A_{ij}B_{jk}$. Hence Tr $(AB) = \sum_i \sum_j A_{ij}B_{ji} = \sum_j \sum_i B_{ji}A_{ij} =$Tr (BA). Thus, Tr $(\mathbf{T}')$ = Tr $(\mathbf{RTR}^{-1})$ = Tr $(\mathbf{R}^{-1}\mathbf{RT})$ = Tr $\mathbf{T}$.

Thus the trace of a matrix remains invariant under rotations of the coordinate system. Let us examine the physical significance of this result for the Maxwell Stress Tensor.

The trace of the stress tensor $= -\frac{\epsilon_0}{2}(E^2+c^2B^2)$ which is simply the negative of the energy density of the electromagnetic field. We see that it is invariant under coordinate rotations. This is an indicator that space is isotropic. There is no preferred direction in space.

An inner product between a second rank tensor and a vector is another vector. Such an inner product is obtained by contraction. We can write such a dot product in component form as $\Sigma_j T_{ij}E^j$. (Of course, for three-dimensional Cartesian spatial coordinates there is no difference between E_i and E^i, but we will use conventional tensor notation for the sake of consistency.) For $i = 1$, this sum written out in full form becomes $T_{11}E^1 + T_{12}E^2 + T_{13}E^3$. We can write similar sums for $i = 2$ and $i = 3$. When we carry out the sum $\sum_j T_{ij}E^j$ over j the only subscript or superscript that remains is i. Thus the result of this inner product or contraction is a vector.

One can also perform an inner product between the vector operator ∇ and the stress tensor, and the result will be a vector.

The term on the right side of Eq. (6.29) is the inner product of the operator ∇ and the stress tensor. When we take the divergence of a vector we obtain a scalar. But when we take the divergence of the stress tensor, we obtain another vector. So, in Eq. (6.29) the left side is the ith component of a vector, and we can see that on the right side, after the sum is made over the index j, we have only the ith component of some vector.

The stress tensor is sometimes written as $\overleftrightarrow{\mathbf{T}}$ with a double arrow over the head, showing that one can form a dot product either from the left or from the right. In our case we are doing a dot product from the left by taking the divergence.

$$\sum_j \frac{\partial}{\partial x_j} T_{ij} \equiv (\nabla \cdot \overleftrightarrow{\mathbf{T}})_i \tag{6.32}$$

The symbol $\frac{\partial}{\partial x_j}$ is the contravariant form of the gradient or del operator $\frac{\partial}{\partial x^j}$. The two are related via the metric tensor

$$g_{ij} \frac{\partial}{\partial x_j} = \frac{\partial}{\partial x^i}$$

where the summation is implied over the repeated indices. This equation can be expressed equivalently as

$$g_{ij} \nabla^j = \nabla_i$$

or as

$$g_{ij} \partial^j = \partial_i$$

Hence we can write, using the Maxwell Stress Tensor,

$$\frac{d\mathbf{P}_{mech}}{dt} + \frac{d}{dt} \int_V \epsilon_0 (\mathbf{E} \times \mathbf{B})\, d\tau = \int_V \nabla \cdot \overleftrightarrow{\mathbf{T}}\, d\tau \tag{6.33}$$

One can apply the divergence theorem to the integral on the right side:

$$\int_V \nabla \cdot \overleftrightarrow{\mathbf{T}}\, d\tau = \oiint_S \overleftrightarrow{\mathbf{T}} \cdot \hat{n}\, dA \tag{6.34}$$

This would yield the equation

$$\frac{d\mathbf{P}_{mech}}{dt} + \frac{d}{dt} \int_V \epsilon_0 (\mathbf{E} \times \mathbf{B}) d^3x = \oiint_S \overleftrightarrow{\mathbf{T}} \cdot \hat{n} dA \tag{6.35}$$

The first term on the left is the rate of change of the momentum of the charges inside the volume V. We need to understand the second term.

The rate of flow of electromagnetic energy through space is given by the Poynting Vector $\mathbf{S} = \epsilon_0 c^2 \mathbf{E} \times \mathbf{B}$. The magnitude of this vector equals the amount of electromagnetic energy crossing unit area in unit time.

Consider a region of space in the shape of a cylinder of length c and cross-section 1 m^2. The volume of this region is $V = c$ m^3. We now consider an electromagnetic field propagating along the length of this cylinder.

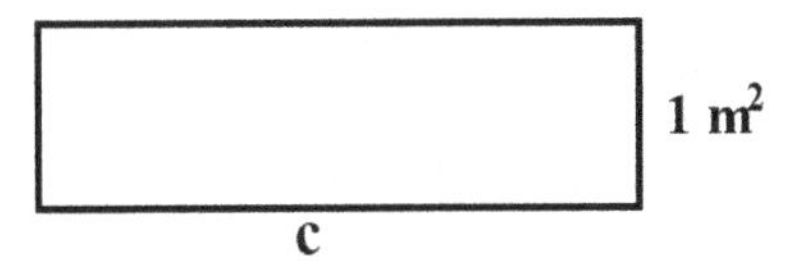

Electromagnetic fields travel at the speed c through space. So, the average value of the magnitude of the Poynting vector within this region equals the magnitude of the electromagnetic energy contained within this cylinder of length c and cross-section 1 m^2. Since the energy travels at speed c, all the energy contained in this cylinder will cross the area of 1 m^2 in one second. So, the amount of energy present inside this cylinder at any given instant of time is simply

$$W = |\epsilon_0 c^2 \mathbf{E} \times \mathbf{B}| \tag{6.36}$$

Let us picture the cylinder in the figure above as empty space into which the electromagnetic field is being introduced. By the Work Energy Theorem, the energy received by this cylinder equals the work done on the space within the cylinder. This work is given by $W = F\Delta x$ where F is the average force acting on the space and Δx the displacement of the force. By the Impulse Momentum Theorem, which is essentially Newton's Second Law, $F = \frac{P}{t}$ where P is the momentum communicated to the space within the cylinder, and t is the time over which this momentum is imparted. Since this momentum is carried entirely by the electromagnetic field, and not by material particles, we will denote it as P_{field}.

So $W = F\Delta x = \frac{P_{field}}{t}\Delta x = P_{field}\frac{\Delta x}{t}$.

Now, $\frac{\Delta x}{t} = c$, the rate of transport of electromagnetic energy. So $W = cP_{field}$.

Hence the momentum communicated by the field to the space within the cylinder is $P_{field} = \frac{1}{c}W$. This is therefore the magnitude of the momentum within the volume of the cylinder at any given instant of time.

Writing in vector form, the total momentum of the field within the cylinder is

$$\mathbf{P}_{field} = \epsilon_0 c\mathbf{E} \times \mathbf{B} \tag{6.37}$$

We will denote the momentum per unit volume as $\mathbf{p}_f$. Since the volume of the cylinder is c, the momentum of the field per unit volume is $\mathbf{p}_f = \epsilon_0 \mathbf{E}\times\mathbf{B}$.

Thus the integrand of the second integral on the left side of Eq. (6.35) is the momentum per unit volume, or the momentum density, which is being integrated over the entire volume, and hence becomes the total field momentum within the volume. The time derivative of the field momentum is the force acting on the field, by Newton's Second Law.

Thus, Eq. (6.35) can be written as

$$\frac{d\mathbf{P}_{mech}}{dt} + \frac{d\mathbf{P}_{field}}{dt} = \oint\!\!\!\oint_S \overleftrightarrow{\mathbf{T}} \cdot \hat{n}\, dA \tag{6.38}$$

The left side is the sum of the rate of change of momentum of the charges and the rate of change of momentum of the field. This should equal the net force acting on the volume. Thus the quantity $\overleftrightarrow{\mathbf{T}} \cdot \hat{n}dA$ can be interpreted as force acting on an area dA on the surface of the region and directed into the region. The integral of this quantity over the entire closed surface is equal to the net force acting on the region. Thus $\overleftrightarrow{\mathbf{T}}$ has the dimension of pressure or stress, and is consequently known as the Maxwell Stress Tensor. Notice that the unit vector $\hat{n}$ is directed outwards, but the stress is directed inwards.

Now, let us move the boundaries so that the integral on the right is over a sphere of very large radius r. Since the stress tensor contains fields in the second power, and the electric field drops off as the inverse square of r, the integral vanishes for large r. Thus, for a sufficiently large region,

$$\frac{d\mathbf{P}_{mech}}{dt} = -\frac{d\mathbf{P}_{field}}{dt} \tag{6.39}$$

The net force on the field is equal and opposite to the net force acting on the particles. This is the electromagnetic field equivalent of Newton's Third Law.

6.2.4 *Momentum and Energy of a Field*

Comparing Eqs. (6.36) and (6.37) we find that the momentum of the radiation equals its energy divided by the speed c:

$$P = E/c \tag{6.40}$$

This equation enables us to calculate the momentum imparted to a mechanical surface by the impact of electromagnetic radiation on the surface.

In classical mechanics one makes a distinction between mass and energy. Classical dynamics deals with individual objects endowed with individual and unchangeable mass which can exchange energy with other such individual objects. Thus mass is fixed, and energy is communicable. The momentum of an object is closely related to its velocity, and in Newtonian kinematics the momentum has magnitude $p = mv$, where m is the mass and v the speed of the object. Thus, if both the momentum and the speed could be measured, the mass of an object could be calculated as $m = p/v$. When measured thus, the mass of the object is called its *inertial mass.* The inertial mass is also expressible in terms of Newton's Second Law as

$$m = \frac{F}{a} \tag{6.41}$$

If a charged solid object has a charge q and it is placed in an electric field E we can calculate its mass by calculating the force $F = qE$ and by measuring its acceleration a. The mass of an object can also be calculated from the force $F = mg$ it experiences when measured by a spring scale. The mass so measured is called the *gravitational mass* of the body. Newton did not make a distinction between the inertial and gravitational mass. Einstein made the equivalence of the two masses a principle of his General theory of Relativity.

We can define an inertial mass of the electromagnetic field in terms of its momentum and velocity. So, by extrapolating the equations $P = E/c$ and $m = p/v$ to the electromagnetic field, we obtain the mass m of the field within a region of space that is related to the momentum of the field within that space by $m = p/c$. Thus, we would obtain the equation for the mass of a region of space containing nothing but electromagnetic fields as

$$m = \frac{E}{c^2} \tag{6.42}$$

For the present we will treat this mass as purely inertial, with measurable consequences.

One such consequence is that an electromagnetic field carrying momentum also has a measurable angular momentum.

The angular momentum $\mathbf{L}$ of a body of mass m with momentum $\mathbf{p}$ about some point is defined as

$$\mathbf{L} = \mathbf{r} \times \mathbf{p} \tag{6.43}$$

where $\mathbf{r}$ is the displacement vector drawn from the point to the center of mass of the body. Using the equation $\mathbf{p} = \epsilon_0 \mathbf{E} \times \mathbf{B}$ for the momentum density of a field, we can obtain an expression for the angular momentum of an electromagnetic field as follows:

$$\mathbf{L} = \epsilon_0 \iiint \mathbf{r} \times (\mathbf{E} \times \mathbf{B}) d\tau \tag{6.44}$$

Now, the field is constantly interacting with the charged particles that are the source of the field. During this process the particles exchange angular momentum with each other and with the field. Conservation of global angular momentum implies the conservation of the sum of the angular momenta of the field and the particles.

We saw earlier that the net force applied by the charges to the field is equal and opposite to the net force applied by the field to the charges:

$$\frac{d\mathbf{P}_{mech}}{dt} = -\frac{d\mathbf{P}_{field}}{dt}$$

Now, if $\mathbf{r}$ is the position vector of any point in space relative to some point of origin, then we obtain

$$\mathbf{r} \times \frac{d\mathbf{P}_{mech}}{dt} = -\mathbf{r} \times \frac{d\mathbf{P}_{field}}{dt} \tag{6.45}$$

Since $\mathbf{r}$ is constant,

$$\frac{d}{dt}(\mathbf{r} \times \mathbf{P}_{mech} + \mathbf{r} \times \mathbf{P}_{field}) = 0 \tag{6.46}$$

The total angular momentum of the particles and the field about any point remains constant. In a space far away from any charges the angular momentum of the field is constant. Alternatively, we could say the net torque of the electromagnetic field in empty space is zero:

$$\mathbf{r} \times \frac{d\mathbf{P}_{field}}{dt} = \mathbf{r} \times \mathbf{F}_{field} = \tau_{field} = 0 \tag{6.47}$$

6.3 Energy Flow in Simple Circuits

Consider a simple RC circuit as shown below:

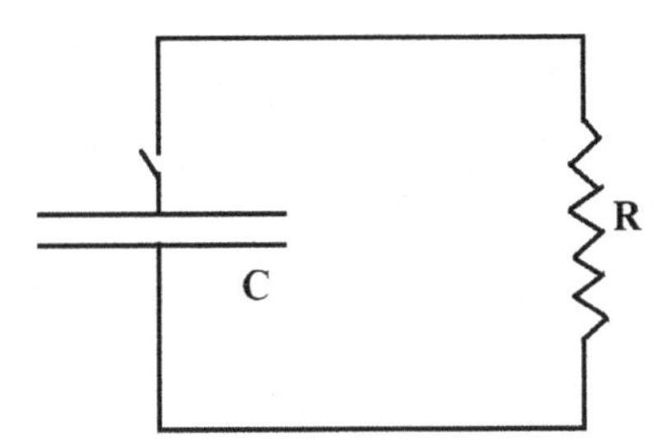

A parallel plate capacitor C consisting of circular plates of radius a each and separation d between the plates is connected to a resistor R through an open/shut key. The capacitor is initially charged and then the key is closed so that the charge flows through the resistor as the capacitor is discharged.

As the capacitor is being discharged, there is a change in the electric field between the plates. So, we apply Maxwell's equation for the space between the plates. The current density in this space is 0, and so the equation becomes

$$c^2\nabla \times \mathbf{B} = \frac{\partial \mathbf{E}}{\partial t} \tag{6.48}$$

We take the surface integral of both sides across the cross-section of the space between the capacitor plates:

$$\iint c^2\nabla \times \mathbf{B} \cdot \hat{n} dA = \frac{\partial}{\partial t} \iint \mathbf{E} \cdot \hat{n} dA \tag{6.49}$$

By applying Stokes's theorem, the left-hand side becomes a line integral over the circular perimeter of the cross-section:

$$c^2 \oint \mathbf{B} \cdot d\mathbf{r} = 2\pi a c^2 B = \frac{\partial E}{\partial t} \pi a^2 \tag{6.50}$$

Now, $E = \frac{\sigma}{\epsilon_0} = \frac{Q}{\pi a^2 \epsilon_0}$, and so

$$2\pi a c^2 B = \frac{1}{\epsilon_0}\frac{dQ}{dt} = \frac{I}{\epsilon_0} \tag{6.51}$$

Hence

$$B = \frac{I}{2\pi a \epsilon_0 c^2}$$

Thus at every point in the space between the capacitor plates there is an electric field perpendicular to the plates and a circling magnetic field parallel to the plates. An examination of the directions of these fields shows that the Poynting vector $\epsilon_0 c^2 \mathbf{E} \times \mathbf{B}$ is everywhere directed radially outwards. The flow of electromagnetic energy outwards through the sides of the capacitor plates is therefore $\epsilon_0 c^2 EB2\pi ad$. Thus the rate of flow of electromagnetic energy from between the plates is EdI which equals VI where V is the potential difference across the plates.

Now, if the conducting wires have zero resistance, the potential difference across the plates equals the potential difference across the resistor. So, if the resistor has length L, the electric field across the resistor is V/L. The magnetic field that curls around the resistor has magnitude $B = \frac{I}{2\pi r \epsilon_0 c^2}$ where the field is measured at a distance r from the resistor.

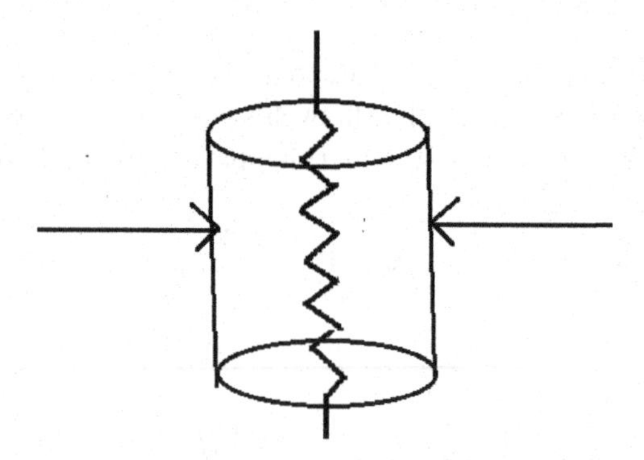

The influx of electromagnetic energy into a cylinder of length L and radius r around the resistor is therefore $\epsilon_0 c^2 EB2\pi rL = ELI = VI$. This energy flows into the resistor from all directions perpendicular to the length of the resistor wire. The power generated in the resistor — i.e. the amount of electrical energy converted to heat every second, is simply VI. And so, we get this somewhat counterintuitive result that the amount of energy radiated away equally in all directions from the space between the capacitor plates equals the amount of electromagnetic energy flowing into the resistor per second.

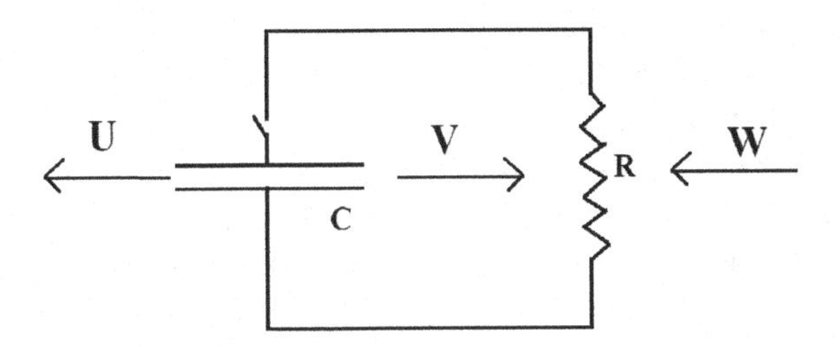

In the figure above, U represents the energy flowing away from the circuit to the left, V the energy flowing to the right, within the space of the circuit, and W the energy flowing from the right into the resistor. Since $U + V = V + W$, it follows that $U = W$. Electromagnetic energy flows from the capacitor into space in a direction away from the resistor, and an equal amount of energy flows from space into the resistor from a direction opposite to that of the capacitor. This diagram illustrates the importance of the role played by the field in the flow of electromagnetic energy. The field is like a bank. The capacitor makes a deposit into this bank, and the resistor makes a withdrawal of an equal amount from the bank. The energy flows out from between the plates of the capacitor, and enters the resistor where it is converted to heat. But the energy does not flow exclusively through the wires. The energy flows primarily through empty space. This is important for understanding a puzzling phenomenon regarding the rates of charge and energy flow in a circuit.[1]

Consider a conductor along which charges of magnitude q are transported at an average speed v_d. If n is the number of charges per unit volume, and A the cross-section of the wire, then the current I through the wire is given by

$$I = \frac{dQ}{dt} = nqv_d A \tag{6.52}$$

Suppose we have a copper wire of cross-section $A = 2.45 \times 10^{-6}\ \text{m}^2$ carrying a current of $I = 1.00$ A. Given the density of copper at room temperature to be $\rho = 8920\ \text{kg/m}^3$, we will calculate the drift speed v_d of electrons through the wire.

[1]cf. B.S. Davis and L. Kaplan, "Poynting Vector Flow in a Circular Circuit," *American Journal of Physics*, **79**, 1155 (2011).

The atomic weight, or molar mass of copper is $M = 63.5$. So the mass of 1 mole of copper is 0.0635 kg. The volume of 1 mole of copper is therefore $0.0635/8920 = 7.12 \times 10^{-6}$ m^3. The number of copper atoms in 1 mole is Avogadro's number 6.02×10^{23}. Assuming that each copper atom contributes 1 free electron to the current, the number of free electrons per cubic meter $n = 6.03 \times 10^{23}/7.12 \times 10^{-6}$ m^{-3}. The magnitude of the charge of an electron is $q = 1.60 \times 10^{-19}$ C. Hence we find the value of the drift speed of electrons in the wire to be

$$v_d = \frac{I}{nqA} = \frac{7.12 \times 10^{-6}}{6.02 \times 10^{23} \times 1.60 \times 10^{-19} \times 2.45 \times 10^{-6}} = 3.02 \times 10^{-5} \text{ m/s} \tag{6.53}$$

The charges travel rather slowly through this wire carrying a current of 1 A. If the distance from the capacitor to the resistor is 1 meter, it would take over 9 hours for the current to reach the resistor from the capacitor. But our experience with household circuits suggests that the current should reach the resistor within a fraction of a second after the key is closed near the capacitor. The explanation is that the energy does not flow through the wires along with the charges, but travels through the intervening space at the speed of light, taking less than 10^{-8} s to travel from the capacitor to the resistor.

Chapter 7

Special Relativity and Electromagnetism

7.1 Detection of the Ether

7.1.1 *Medium of Electromagnetic Waves*

An electromagnetic wave propagates in a direction perpendicular to the electric and magnetic fields. The direction of propagation is given by the Poynting Vector

$$\mathbf{S} = \epsilon_0 c^2 \mathbf{E} \times \mathbf{B}$$

A wave is understood as a disturbance that is propagated along some medium. In the course of the propagation the medium is not transported. But the shape of the wave and the energy of the wave are displaced from point to point along the path of propagation.

Since the medium undergoes local oscillation or vibration during wave propagation, it was believed that electromagnetic waves could not travel in a space that was absolutely empty. It was believed that there had to be an invisible physical medium whose local vibration provided the energy that was transmitted by an electromagnetic wave.

The medium through which electromagnetic waves propagate was called ether. This was the term used by the ancient Greek philosophers for the medium in which the heavenly bodies moved. Naturally, scientists of the nineteenth century were not particularly interested in the ancient philosophical attributes of ether. They found the term convenient for denoting the invisible stuff that permeated all of otherwise empty space.

So, material objects move through the ether, like fishes swimming under water. However, unlike the case of water and air, nobody had ever detected any ether currents or ether eddies due to the motion of planets, asteroids or meteorites through space. Water in turbulent motion does distort the view of objects under the water, and the hot gases rising from a flame do distort the view of objects in the air. However, no such distortion of light was detected as a result of the relative motion of objects through ether. The motion of the moon through space did not produce any perceptible distortion of the light coming from distant stars. However, this negative observation did not seem significant. It was entirely possible that ether is an ideal fluid with zero viscosity, and hence no turbulence or frictional forces are generated through a motion of a solid object through the ether.

7.1.2 *Motion of Detector Relative to the Medium*

However, there is one effect which should be perceptible even in a non-viscous ether. Suppose we have a stationary source of sound such as a siren and a speeding car that approaches the siren. And let us say we have an apparatus fixed in the car that is capable of measuring the speed of the sound waves emitted by the siren. A simple calculation shows that the speed of the sound as measured by the car approaching at speed u relative to the road is

$$w = v + u$$

where v is the speed of sound waves relative to the air. Similarly, if the speed of the sound is measured by a car receding from the source at speed u, we would get

$$w = v - u$$

Suppose we have a long ship that is capable of moving at very high speed over the water, say at about 5 to 10 percent of the speed of sound. Let us say two observers are situated on the deck at a distance of L (about 300 m) from each other along the length of the ship. The time taken for a sound wave to travel from one observer to the other and back again would be

$$T_1 = \frac{L}{v-u} + \frac{L}{v+u} = \frac{2L}{v} \frac{1}{\left(1 - \frac{u^2}{v^2}\right)} \tag{7.1}$$

If we know the speed of sound in air v, and measure the time taken for the sound to make the round trip, then we can calculate the speed of the ship relative to the air.

Next, consider a ship with a wide deck moving at speed u relative to the ground, on a still day. Consider two observers O_1 and O_2 facing each other at a distance L perpendicular to the direction of motion of the ship. Let T_2 be the time taken for a sound wave emitted by O_1 to return to O_1 after reflection from O_2. See Fig. 7.1.

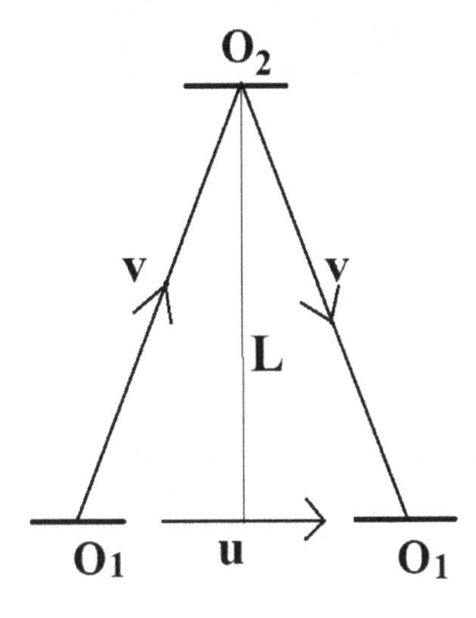

Fig. 7.1

The total distance traveled by the sound during the time interval T_2 taken to go from O_1 to O_2 and back to O_1 is given by the equation

$$\frac{vT_2}{2} = \sqrt{(uT_2/2)^2 + L^2} \tag{7.2}$$

Solving, we obtain

$$T_2 = \frac{2L}{v}\frac{1}{\sqrt{1 - \frac{u^2}{v^2}}} \tag{7.3}$$

There is a difference in the round trip times when the motion of the ship is in the same direction as the sound, and when it is perpendicular to the sound.

7.1.3 *Michelson-Morley Experiment*

Michelson and Morley set out to measure the speed of the earth relative to the ether as it moves in its orbit round the sun. One can calculate this speed knowing the radius of the earth's orbit and the time taken for one orbit (= one year). This calculation yields a speed of 2.99×10^4 m/s, which is about 0.0001 times the speed of light. This ratio is small, but not negligible.

Michelson and Morley split a beam into two parts, one traveling in the same direction as the earth's motion, and the other perpendicular to the earth's motion. The distances traveled by the beams were made equal. The two beams would — by analogy to the sound wave experiment — take slightly different times, and the time difference would come to

$$T_2 - T_1 = \frac{2L}{c}\left[\frac{1}{\sqrt{1-\frac{u^2}{c^2}}} - \frac{1}{1-\frac{u^2}{c^2}}\right] \tag{7.4}$$

The time difference between the paths of the two beams would be measurable as a phase difference. The experiment was conducted using what is known as the Michelson Interferometer. The experiment was repeated rotating the apparatus through ninety degrees, so that the paths parallel to and perpendicular to the direction of the earth's motion were interchanged. After taking into account every possible experimental error, the calculations showed that there should be a measurable phase difference between the two beams. However, the result of the experiment was found to be negative every time. There was apparently no time difference between the paths of the light beams traveling parallel and perpendicular to the direction of the earth's motion through space. Thus, it was impossible to measure the speed of the earth relative to the ether using this method.

Fitzgerald and Lorentz argued that the negative result obtained by Michelson and Morley could be explained away by making the hypothesis that material objects undergo a contraction as they move through ether, so that if the length of a stationary object is L, its length L' when it is moving at speed v relative to ether becomes

$$L' = L\sqrt{1-\frac{v^2}{c^2}} \tag{7.5}$$

Hence we obtain the value of T_1 as

$$T_1 = \frac{2L}{c}\frac{\sqrt{1-\frac{u^2}{c^2}}}{1-\frac{u^2}{c^2}} = \frac{2L}{c}\frac{1}{\sqrt{1-\frac{u^2}{c^2}}} = T_2 \tag{7.6}$$

Hence the times taken by light to travel parallel and perpendicular to the earth's motion become equal, and so there is no phase difference between the two beams of light.

The contraction of material objects as a result of their motion came to be called the *Fitzgerald contraction* or more commonly the *Lorentz contraction*, and less commonly as the *Lorentz-Fitzgerald contraction.*

Whatever be the source or cause of the Lorentz contraction, the consequence of this contraction is that it is impossible to measure the speed of an object relative to the ether. And since ether did not exhibit any viscosity or turbulence, the inability to detect a speed relative to the ether simply confirmed the suspicion that the ether was too elusive to be detected. Thus, there is a fundamental difference between the medium of a wave such as sound and the medium of electromagnetic waves.

7.2 Einstein's Theory of Special Relativity

7.2.1 *Postulates of Special Relativity*

Newton's first law of motion states that every body remains at rest or moves in a straight line with constant speed unless compelled to do otherwise by an external agent. Even though inertia could be manifested either by rest or by uniform motion, the states of rest and motion were still treated as fundamentally different. The distinction was most obvious in astronomy. Prior to Copernicus it was believed that the earth was at rest, and the stars, the sun, the moon and the planets moved round the earth. After Copernicus it became convenient to think of the sun as an object whose center of mass was at rest and all the planets were in motion relative to the sun. However, when it became known that the sun itself revolves around the center of our Milky Way galaxy, and that galaxies themselves are in relative motion, it became difficult to pinpoint any one object in the sky as something that was at rest. Nevertheless, the distinction between rest and motion, being so deeply ingrained in human consciousness, remained a fundamental notion in physics. Thus, physicists — both subconsciously and consciously — thought that the concept of absolute rest was as real as any other concept in physics. Absolute rest meant at rest relative to empty space, or more precisely, at rest relative to the ether. The ether was the one entity that was at absolute rest, and all motion was defined relative to this invisible ether.

But the failure to detect this ether raised problems. Fitzgerald and Lorentz had tried to rescue the concept of ether by positing a contraction undergone by bodies in motion relative to the ether. But Einstein was not satisfied with this ad hoc solution. Einstein interpreted the failure to detect the ether — a substance that was supposed to define absolute rest — as a signal to abandon the time honored concept of absolute rest that was distinct from motion of any kind.

Einstein's hypothesis was that there was no difference between rest and motion with constant velocity. A laboratory that was moving with constant velocity (i.e. not accelerating) and where no external forces act on any of the bodies — called an inertial frame of reference — was fully equivalent to any other inertial frame of reference. A law of physics that is valid in one inertial frame would have to be valid in any other inertial frame. This also means that any measurement restricted exclusively to objects within a laboratory in one inertial frame would yield the same results as an identical measurement in an identical laboratory in a different inertial frame. Thus, it is impossible to measure the velocity of an inertial laboratory if the measurements are limited to objects within that laboratory. And as a corollary, there will be no difference in the speed of light as measured by observers in two different inertial frames.

The hypothesis and the corollary of the last paragraph express the two postulates of Special Relativity enunciated by Einstein:

Postulate 1:

The laws of physics take the same form in every inertial frame of reference.

Postulate 2:

The speed of light in vacuum is the same when measured in any inertial frame of reference.

The first postulate eliminates the notion of a preferred frame of reference. All inertial frames are fully equivalent. Thus, there is no such thing as absolute rest, a concept which implies a preferred frame of reference against which all other frames are measured. So we can no longer talk about the absolute velocity of an object, only a velocity in some frame of reference.

The equivalence of all inertial frames and the invariance of the speed of light in all frames provide the axiomatic bases for the mathematical foundation of Special Relativity.

7.3 The Geometry of Special Relativity

7.3.1 *World Lines*

In the study of motion and velocity we use displacement-time graphs. Consider a straight road that runs west to east. Suppose a car is moving at a constant speed on this road towards east. A stationary observer called Alice draws a graph of the car's position at different times and obtains something like this:

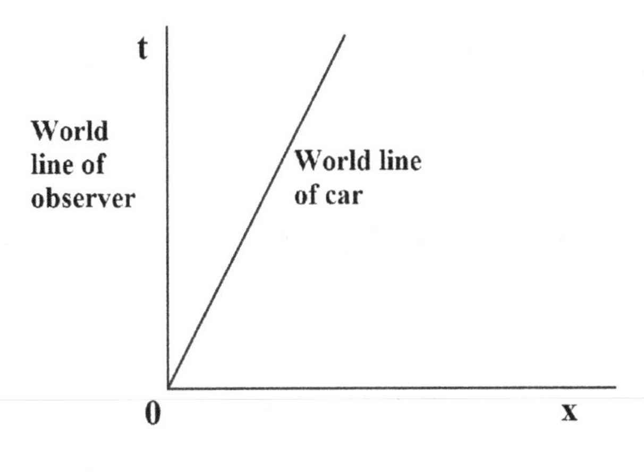

Fig. 7.2

The horizontal space axis — labeled x — represents the road along which the car travels from west to east. The vertical time axis labeled t represents the passage of time. The observer Alice sees the car at different positions on the road at different times. If the car is at a point labeled $x = 0$ at the starting moment when time $t = 0$ then the oblique line is the graph of the motion of the car. We call the oblique line the *world line* of the car. Since Alice herself is traveling forwards not in space but in time in this reference frame, the time axis in this graph is Alice's own world line.

We now apply the postulate of the constancy of the speed of light in all reference frames. In nonrelativistic kinematics we can identify a point in space and imagine that this point will not change its position through the passage of time. However, such a notion of an absolutely immovable fixed point in space has to be thrown out along with the concept of a universal ether. So, instead of points in space, we will talk about points in space-time. So, we will use the concept of a point to identify *events*. An event occurs at a specific point at a specific moment of time. Thus an event has both spatial and time coordinates. Suppose we are interested in two events: the emission of light by a source and the absorption of the same light by a detector. For the sake of simplicity we shall limit our discussion to one spatial dimension, so that the light travels along the x axis. So we shall call the coordinates of the emission (x_1, t_1) and those of the detection (x_2, t_2). Clearly,

$$\frac{x_2 - x_1}{t_2 - t_1} = c \tag{7.7}$$

Writing $\Delta x = x_2 - x_1$ and $\Delta t = t_2 - t_1$, we write $\Delta x = c\Delta t$. Allowing for positive and negative directions of motion — with positive and negative velocities — we can write

$$\frac{\Delta x^2}{\Delta t^2} = c^2 \tag{7.8}$$

The invariance of the speed of light in any reference frame means that

$$\Delta x^2 - c^2\Delta t^2 = \Delta x'^2 - c^2\Delta t'^2 \tag{7.9}$$

This equation can be written as

$$\Delta x^2 + (ic\Delta t)^2 = \Delta x'^2 + (ic\Delta t')^2 \tag{7.10}$$

We shall now generalize the direction of the motion of the light beam, and write

$$\Delta x^2 + \Delta y^2 + \Delta z^2 + (ic\Delta t)^2 = \Delta x'^2 + \Delta y'^2 + \Delta z'^2 + (ic\Delta t')^2 \tag{7.11}$$

This equation suggests that we can define a four-dimensional space including the three spatial dimensions and a time dimension, and render the time dimension spatial by the factor ic. Such a four-dimensional space is called a *Minkowski space*. A four-dimensional interval Δs in Minkowski space can be defined as

$$\Delta s^2 = \Delta x^2 + \Delta y^2 + \Delta z^2 + (ic\Delta t)^2 \tag{7.12}$$

In three-dimensional space, a spatial interval is defined as $\Delta r^2 = \Delta x^2 + \Delta y^2 + \Delta z^2$. Such an interval is invariant under translation, rotation and reflection of coordinates. Equation (7.12) shows that an interval in four-dimensional space has the components $\Delta x, \Delta y, \Delta z, ic\Delta t$. The factor ic has the effect of converting time into a spatial dimension. So this is a result we obtained from the postulate that the speed of light is invariant in any inertial reference frame. We will next apply the second postulate — the invariance of the laws of physics — to obtain another geometrical principle of special relativity. In the observer's reference frame the time axis is the observer's world line. In this frame the world line of a car moving away from the observer is indicated by an oblique straight line. Now, in the reference frame of the car, this line becomes the time axis. The invariance of physical laws under a change of reference frame implies that the space axis of the car's reference frame should be perpendicular to the car's time axis or world line. Thus, the reference frame of the car is *rotated* relative to the reference frame of the observer. A rotation in Minkowski space is also called a *Lorentz transformation* or a *Lorentz boost.*

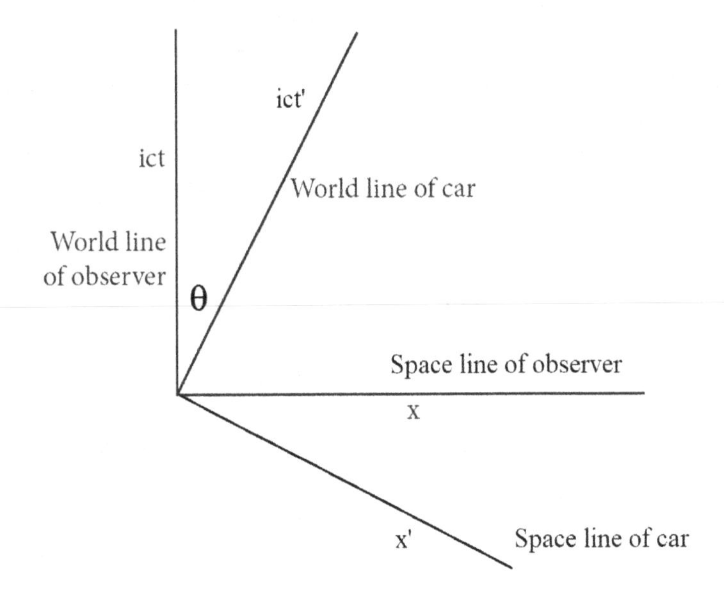

Fig. 7.3 Rotation in Minkowski Space.

Thus, using the postulates of special relativity we have obtained two important geometrical results. The constancy of the speed of light enabled us to convert time t into a spatial coordinate by multiplying by ic. The equivalence of all inertial frames led to the geometrical description of a transformation of reference frame as a rotation in Minkowski space.

In Fig. 7.3 the primed coordinates represent a frame that is in relative uniform motion with respect to the unprimed coordinates. Evidently the primed frame is moving in the positive x direction as measured in the unprimed frame. Let the velocity of the primed frame relative to the unprimed frame be v. Clearly $v = \frac{\Delta x}{\Delta t}$.

Now, from the graph, $\tan\theta = \frac{\Delta x}{ic\Delta t}$. And so we obtain the important result that $\tan\theta = \frac{v}{ic}$. Equivalently, $v = ic\tan\theta$. Since v and c are real, it follows that $\tan\theta$ must be imaginary, and since $\tan\theta$ is an odd function of θ, it is reasonable to infer that θ is imaginary. Dealing with imaginary angles is problematic, and we will find a way of eliminating imaginary numbers from special relativity, but for now we will continue with the procedure followed by Einstein, since the algebra is simple, and helps us to obtain some very important physical results.

An interval Δs in Minkowski space is invariant under a transformation from one inertial frame to another. Such a transformation is a rotation in Minkowski space. And since an interval in three-space is invariant under spatial rotation, an interval in Minkowski space is invariant under a four-dimensional rotation. So, the quantity $\Delta x^2 + \Delta y^2 + \Delta z^2 + (ic\Delta t)^2$ is an invariant that has the same value in all inertial frames. It is clearly a real number, which could be positive, negative or zero.

7.3.2 *Space-like, Time-like and Light-like Intervals*

Suppose we have an interval of space Δr of the order of the distance between the earth and the sun (about 150 million km) and an interval of time Δt equal to 3 minutes. Clearly, it takes longer than 3 minutes (about 8 minutes) for light to travel 150 million km. So if an event occurred at 8:00 am US Central Time on the sun and another event occurred at 8:03 am US Central Time on the earth (on the same day) these two events are separated by a spatial distance that is greater than the distance traveled by light in 3 minutes. We say these two events are separated by a *space-like interval*. So for this interval $\Delta r > c\Delta t$ implying that $\Delta r^2 > c^2\Delta t^2$ and therefore $\Delta x^2 + \Delta y^2 + \Delta z^2 - c^2\Delta t^2 = k^2$ where k is some real number. We could write this equation as

$$(\Delta x)^2 + (\Delta y)^2 + (\Delta z)^2 + (ic\Delta t)^2 = k^2 \tag{7.13}$$

The number k^2 is invariant under inertial frame transformations. This means that if two events are separated by a space-like interval in one inertial frame of reference, then they are separated by a space-like interval in any other inertial frame.

Now, if we consider an event on the sun at 10:00 am US Central Time, and another on the same day on earth at 11:00 am US Central Time, the spatial and temporal intervals are now related by the inequality $\Delta r < c\Delta t$, because the time difference between the events is greater than the time taken for light to move from the location of one event to the location of the other. Such an interval is called a *time-like interval.* This would lead to the equation

$$(\Delta x)^2 + (\Delta y)^2 + (\Delta z)^2 + (ic\Delta t)^2 = -k^2 \tag{7.14}$$

where k is a real number. It is evident that if two events are separated by a time-like interval in one inertial frame, they have a time-like interval in any other inertial frame.

One could also consider two events for which $k = 0$. Such an interval is called a *light-like* interval, because the distance between the events is equal to the distance traveled by light during that time interval. Two events separated by a light-like interval in one frame will have a light-like interval in any other inertial frame. This expresses the principle that the speed of light is the same in any inertial frame.

Hence we can enunciate an important principle concerning intervals in Special Relativity:

Interval Principle:

The type of an interval (space-like, light-like or time-like) between any two events will not change under a transformation between inertial frames of reference in which the events are measured.

7.3.3 *Limiting Cases of Time-like and Space-like Intervals*

Figure 7.4 represents the inertial reference frame of an observer. The points A and B represent two points on the observer's world line. For the interval AB the spatial separation $\Delta x = 0$ and so the interval is time-like. The points C and D represent two events that occur at time $t = 0$ at different spatial locations in the observer's frame. The temporal separation $\Delta t = 0$ for the interval CD, and so CD is a space-like interval. AB and CD represent limiting cases of time-like and space-like intervals. It is not hard to show — e.g. using Eqs. (7.13) and (7.14) that the value of k^2 is equal to the limiting Δx^2 for a space-like interval, and equal to the limiting $c^2\Delta t^2$ for a time-like interval.

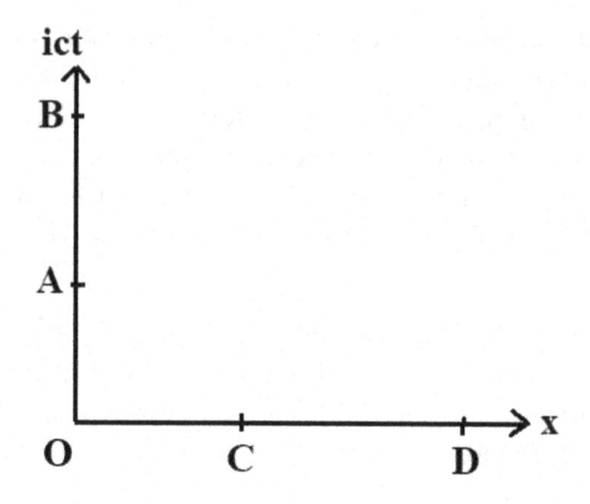

Fig. 7.4 AB is a time-like interval; CD is a space-like interval.

Suppose two events have a space-like separation in some reference frame. It is possible to find some inertial frame of reference in which these two events will occur simultaneously at two different points in space. This leads to an important inference. It is impossible for two events separated by a space-like separation to influence one another. So, events with a space-like separation are mutually acausal — one cannot be the cause of the other, and one cannot be the result of the other.

Suppose two events have a time-like separation in some reference frame. It is possible to find some inertial frame of reference in which these two events will occur at the same place but at two different times. Also, the time order of the events will not change, and so the earlier event in one frame will remain the earlier event in any other frame. Hence, causality is preserved. Events separated by a time-like separation can have a causal relationship.

In Fig. 7.5 an observer Simon is represented at an instant of time at the origin O. Simon moves along his own time axis ict, which is his world line. OC is the world line of a light ray emitted by him at time t = 0, i.e. when he was at O. Now, consider a traveler Alice moving with a speed v less than c, whose world line is represented by OA. Clearly, the interval O-A is time-like as measured by Simon. For Alice the interval O-A, which runs along her world line, would be a limiting time-like interval. The space-like interval OB is shown for purposes of comparison.

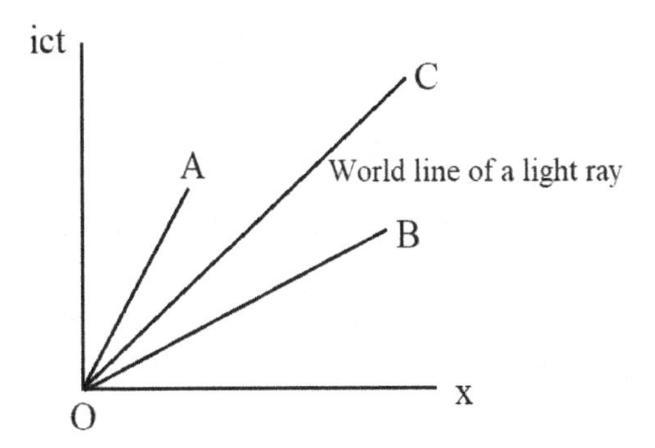

Fig. 7.5 Intervals: OA = time-like; OC = light-like; OB = space-like.

Suppose Simon were to move in a direction opposite to that of Alice. Alice's world line would change in Simon's frame, but it would still remain time-like. No matter how fast Simon could move, he would never observe Alice moving at light speed, because Alice's world line cannot change from time-like to light-like. And Alice would never observe Simon moving at light speed. Thus, no matter how fast Simon moved, he could never reach the speed of light as measured in any frame of reference.

Corollary 1:

No object can be accelerated to reach the speed of light.

Of course, light itself does travel at light speed, but then it was never accelerated. Light can travel only at the speed of light. It cannot travel at any other speed. (The average speed of light within a medium is less than c, but that is because light is absorbed and re-emitted by the atoms of the medium. Light travels at speed c from one atom to another.) It is impossible to find a frame of reference in which light is observed at any speed other than c. This also holds for any entity that travels at speed c in some reference frame.

Corollary 2:

Any physical entity, initially traveling at speed c, cannot be slowed down or sped up.

Electromagnetic waves travel at speed c in vacuum. At one time it was thought that neutrinos also travel at speed c, but now it is known that neutrinos travel slower than light.[1]

In Fig. 7.5 the interval O-B is space-like, since the spatial separation between B and O is greater than the distance that light would travel during the temporal separation between B and O. Thus an object traveling at light speed cannot be found traveling faster than light in any reference frame. Thus, it is impossible to send a message by an electromagnetic wave at any speed greater than c. But that still does not preclude the possibility that there may be some entities that always travel faster than light, and which cannot travel slower than light. But we can dismiss this possibility without too much difficulty.

Suppose it were possible for some object to travel faster than light. If so, OB would be a physically possible world line. The interval O-B is space-like, and hence it is possible, through a suitable transformation of reference frame, to render this interval into a purely spatial interval for some observer, along the observer's x axis. Then the events O and B would now lie at different points on the x axis, or at different spatial points at the same time. But O and B represent the same object at different times in the object's own reference frame. So, our transformation of frames has yielded the bizarre result that the same object is now detected at two different points at some instant of time by an observer in a different frame. Moreover, proceeding with this line of reasoning we can show that the object is detected not merely at two different points, but at every point on the observer's x axis!

Another problem with the possibility of faster than light motion is illustrated by the graph in Fig. 7.6. Suppose an object W were to travel at speed w relative to a frame of reference indicated by the x, ict axes as shown. Next, we consider an object V moving at speed v in the negative x direction. In the reference frame of object V it will be seen that object W is moving backwards in time! So, if V and W are twins who take off in opposite directions, with V traveling slower than light and W traveling faster than light, then W would be moving backwards in time as observed by V!

[1]cf., J. Beringer (Particle Data Group) *et al.* (2012). "Neutrino Properties — Review of Particle Physics," *Physical Review D*. 86 (1): 010001.

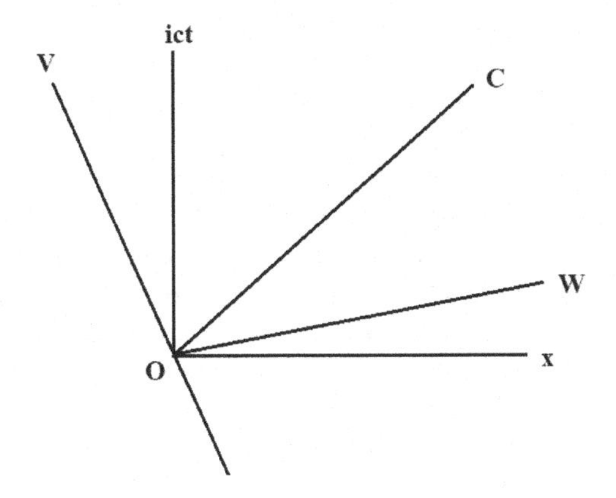

Fig. 7.6 Object moving faster than light is seen moving backwards in time.

These arguments dismiss the possibility of existence of any faster than light particles or *tachyons* which were postulated in the 20th century. It was believed that tachyons always traveled faster than light, and the faster the tachyon the lower its kinetic energy.[2]

Corollary 3:

No physical entity can travel with a speed greater than c.

Equation (7.13) is the equation for a space-like interval. Two events separated by a space-like interval are truly independent of each other. One event cannot affect the other. It would take longer for a message at light speed to travel from one place to the other than the time interval between these two events in the two places. So people who are present at one event cannot communicate the news of that event to the other place before the second event takes place. Events that have a time-like interval may not always be independent, because it could be possible for a message to be sent from one event which arrives at the other place before the other event occurs.

Are events separated by a light-like interval independent of each other? A message can in principle be sent from one event to the other at the speed of light, but since it takes some time — however small — for the message

[2]Feinberg, G. (1967). "Possibility of Faster-than-light Particles," *Physical Review.* 159 (5): 1089–1105. Bilaniuk, O-M.P., Sudarshan, E.C.G. (1969). "Particles beyond the Light Barrier," *Physics Today.* 22 (5): 43–51.

to be read and interpreted, events separated by an exact light-like interval are independent.

Exercise:
A distant star has two planets. A volcano erupts at 02:12:34 AM (hours:minutes:seconds) according to some interstellar standard time on one planet. A meteor hits the surface of the other planet at 11:56:43 AM the same day. If these planets are 1200 million kilometers apart, determine if the interval between the two events is space-like, light-like or time-like.

7.4 The Algebra of Special Relativity

Consider three observers A, B and C. A is stationary relative to the earth. B moves in the positive x direction with speed u relative to A. C moves in the positive x direction with speed v relative to B.

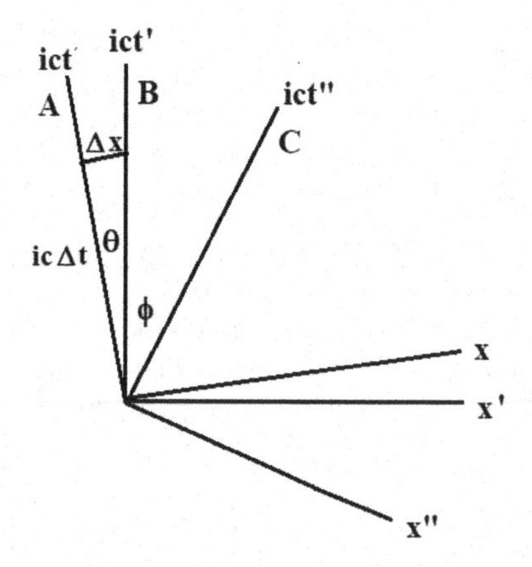

Fig. 7.7 A, B and C in relative motion in the x direction.

The coordinates x, t are related to the coordinates x', t' by the transformation equations

$$x' = x\cos\theta - ict\sin\theta$$

$$ict' = x\sin\theta + ict\cos\theta \tag{7.15}$$

From A's reference frame, B makes a displacement Δx in time Δt. So, B's velocity as measured by A is $\frac{\Delta x}{\Delta t} = u$. From the graph $\tan\theta = \frac{\Delta x}{ic\Delta t} = \frac{u}{ic}$. So $\sec\theta = \sqrt{1+\tan^2\theta}$, $\cos\theta = \frac{1}{\sec\theta}$, and $\sin\theta = \tan\theta\sec\theta$.

Therefore $\cos\theta = \frac{1}{\sqrt{1-\frac{u^2}{c^2}}}$ and $\sin\theta = \frac{u}{ic\sqrt{1-\frac{u^2}{c^2}}}$.

And thus we get the transformation equations:

$$x' = \frac{1}{\sqrt{1-\frac{u^2}{c^2}}}(x - ut)$$

$$t' = \frac{1}{\sqrt{1-\frac{u^2}{c^2}}}(-ux/c^2 + t) \tag{7.16}$$

For a relative motion along the x direction, the other two spatial coordinates remain the same: $y' = y$ and $z' = z$. Equations (7.16) are the algebraic forms of the *Lorentz transformation* equations, or *Lorentz boosts*, that we had previously defined geometrically. We can write these equations in compact form by using the symbol $\gamma = \frac{1}{\sqrt{1-\frac{v^2}{c^2}}}$. The matrix form of the Lorentz transformation for relative motion along the x direction is

$$\Lambda = \begin{bmatrix} \gamma & -\frac{\gamma u}{c} & 0 & 0 \\ -\frac{\gamma u}{c} & \gamma & 0 & 0 \\ 0 & 0 & 1 & 0 \\ 0 & 0 & 0 & 1 \end{bmatrix} \tag{7.17}$$

so that the transformation itself is written as

$$\begin{bmatrix} ct' \\ x' \\ y' \\ z' \end{bmatrix} = \begin{bmatrix} \gamma & -\frac{\gamma u}{c} & 0 & 0 \\ -\frac{\gamma u}{c} & \gamma & 0 & 0 \\ 0 & 0 & 1 & 0 \\ 0 & 0 & 0 & 1 \end{bmatrix} = \begin{bmatrix} ct \\ x \\ y \\ z \end{bmatrix} \tag{7.18}$$

It is easy to show that

$$\Lambda^{-1} = \begin{bmatrix} \gamma & \frac{\gamma u}{c} & 0 & 0 \\ \frac{\gamma u}{c} & \gamma & 0 & 0 \\ 0 & 0 & 1 & 0 \\ 0 & 0 & 0 & 1 \end{bmatrix} \tag{7.19}$$

and hence

$$\Lambda^{-1}(u) = \Lambda(-u) \tag{7.20}$$

This states that a transformation to a frame moving at velocity u can be reversed by a transformation to a frame moving at velocity $-u$. Two such transformations applied in succession bring us to the original frame: $\Lambda(u)\Lambda(-u) = I$ (the unit matrix of order 4).

However, it is not obvious that $\Lambda(u)\Lambda(v) = \Lambda(u+v)$. We will next see why this is equation is not true.

In Fig. 7.7 the traveler B is in motion with velocity u relative to A, and C is moving with velocity v relative to B, both along the common x direction of A. According to classical kinematics the relative velocity of C with respect to A is $u + v$. This cannot be true in relativity, because if $u = 0.9c$ and $v = 0.9c$ then we would get the value of $1.8c$ for the speed of C measured by A. This clearly violates Corollary 3. The calculation of relative velocity is different in relativistic kinematics.

Let the relative velocity of C with respect to A be w. So, according to the formula we obtained earlier, $w = ic\tan(\theta + \phi)$, and $u = ic\tan\theta$ and $v = ic\tan\phi$.

Applying the rules of trigonometry,

$$w = ic\left(\frac{\tan\theta + \tan\phi}{1 - \tan\theta\tan\phi}\right) = \frac{u+v}{1+\frac{uv}{c^2}} \tag{7.21}$$

So $\Lambda(w) = \Lambda(v)\Lambda(u) = \Lambda(u)\Lambda(v)$.

But since $w \neq u + v$, it follows that $\Lambda(v)\Lambda(u) \neq \Lambda(u+v)$.

It can be easily shown that for all $u, v < c$, $w < c$. If either u or v equals c then $w = c$. This expresses the postulate that the speed of light is the same when measured by any observer.

7.4.1 *Transformation of Space and Time Intervals*

7.4.1.1 *Lorentz contraction*

Suppose spaceship B is moving with a velocity v relative to an identical spaceship A, each having length L when they are stationary relative to each other. What is the length of B as measured by A when they are in relative motion?

This question needs to be stated with clarity. What we mean is, if an observer in A notes the positions of the front end and the rear end of B *at the same time* in A's reference frame, what will be the length of B when measured thus?

The measurement of the position of the front end of A's space ship is one event, and the measurement of the position of the rear of A's ship is a different event. If this measurement is done by A on A's own ship, we would get $\Delta x = L$. If this measurement is done by A on B's ship we would get $\Delta x'$, and since both events occur at the same time for A, we must set $\Delta t' = 0$. The Lorentz transformation equations yield

$$\Delta x' = \Delta x\sqrt{1 - \frac{v^2}{c^2}} \tag{7.22}$$

The algebraic steps are detailed in the box below:

We write the Lorentz transformation equations in terms of differences thus:

$$\Delta x' = \gamma(\Delta x - u\Delta t)$$

$$\Delta t' = \gamma(\Delta t - u\Delta x/c^2)$$

Setting $\Delta t' = 0$, we obtain $\Delta t = u\Delta x/c^2$. And so,

$$\Delta x' = \gamma(\Delta x - u^2\Delta x/c^2) = \Delta x\sqrt{1 - \frac{u^2}{c^2}}$$

Thus B's ship will be contracted when viewed in A's frame. This, then, is the true Lorentz contraction. The negative result of Michelson and Morley is explained in terms of an equation derived on the basis of the postulates of Einstein's special relativity. The Lorentz contraction is not caused by any compression brought about through the motion of a solid through ether. It is a consequence of the relativity of space and time. Indeed, a calculation shows that while A observes B's ship contracted, B will observe the same thing about A. A's ship will be contracted by the exact amount when observed by B.

7.4.1.2 *Time dilatation*

Consider the following scenario. Twins Molly and Jane decide to be apart for some time. Molly stays on earth, while Jane takes a trip on a spaceship traveling at a speed close to that of light. Molly measures the time Δt that elapses while Jane is away. Jane measures the time $\Delta t'$ for her journey. Now, since Jane's spatial position does not change relative to her own frame of reference, $\Delta x' = 0$. Using the Lorentz transformation equations, we obtain

$$\Delta t' = \Delta t\sqrt{1 - \frac{v^2}{c^2}} \tag{7.23}$$

A shorter time has elapsed on Jane's spaceship than on earth where Molly is waiting. This phenomenon is called *time dilatation.* Actually the situation described in this process is not a pure special relativity effect. In order to return to earth Jane would have to turn around or slow down, stop and retrace her path. Either way, she would need to accelerate. And acceleration is not included in special relativity. But if the distance traveled by Jane is sufficiently great, then the amount of time spent on acceleration can be made proportionately negligible.

7.4.1.3 *Proper time*

One can divide up Jane's travel into a large number of infinitesimal steps, so that the velocity remains more or less constant during each interval. This way one could apply special relativity to each such infinitesimal step. The time measured by Jane in one such infinitesimal step would be dt', and the duration of this step would be measured by Molly as dt so that $dt' = dt\sqrt{1 - \frac{v^2}{c^2}}$. The total time experienced by Jane on her voyage can be calculated as

$$T = \int dt' = \int \sqrt{1 - \frac{v^2}{c^2}}dt \tag{7.24}$$

where v is a function of t. This is clearly less than the time that has passed on the earth, which is simply $\int dt$. The time interval $dt\sqrt{1 - \frac{v^2}{c^2}}$ is called Jane's *proper time interval,* signifying that this duration of time is proper to Jane's own reference frame. In general, we can define a *proper time* τ according to

$$\tau = t\sqrt{1 - \frac{v^2}{c^2}} \tag{7.25}$$

($d\tau$ as a proper time interval should not be confused with our earlier use of $d\tau$ as an element of volume.) The definition of proper time enables us to define a proper velocity as

$$\mathbf{U} = \frac{\mathbf{v}}{\sqrt{1 - \frac{v^2}{c^2}}} \tag{7.26}$$

An important application of proper velocity is the definition of relativistic momentum. For small velocities (much less than that of light) we define the momentum of an object as $\mathbf{p} = m\mathbf{v}$. But for arbitrary velocities this definition needs to be replaced by the relativistic equation

$$\mathbf{p} = m\mathbf{U} = \frac{m\mathbf{v}}{\sqrt{1 - \frac{v^2}{c^2}}} \tag{7.27}$$

7.4.2 *Energy and Mass*

The Work Kinetic Energy theorem of classical mechanics states that the work done by some force on an object equals the change in kinetic energy of that object. So, the work done by a force $\mathbf{F}$ on an object which undergoes a displacement is given by

$$W = \int \mathbf{F} \cdot d\mathbf{r} = T_f - T_i \tag{7.28}$$

By Newton's Second Law, $\mathbf{F} = \frac{d\mathbf{p}}{dt}$. And so

$$W = \int \frac{d\mathbf{p}}{dt} \cdot d\mathbf{r} = T_f - T_i \tag{7.29}$$

If the motion is confined to one direction, which we call x, we can evaluate the work needed to accelerate the object from rest (so $T_i = 0$)to speed v by

$$W = \int \frac{dp}{dt} dx = \int dp \frac{dx}{dt} = \int_0^v u dp = [up]_0^v - \int_0^v p du \tag{7.30}$$

Substituting $p = \frac{mu}{\sqrt{1-\frac{u^2}{c^2}}}$ and carrying out the integration, we obtain

$$T_f = \frac{mc^2}{\sqrt{1 - \frac{v^2}{c^2}}} - mc^2 \tag{7.31}$$

The first term on the right is called the *relativistic energy* E of the object and the second term is called its *rest energy.*

The quantity $\frac{m}{\sqrt{1-\frac{v^2}{c^2}}}$ is called the relativistic mass of the object. So the relativistic energy of a moving object can be expressed in terms of its relativistic mass m_r as

$$E = m_r c^2 \tag{7.32}$$

It is also customary to denote the relativistic mass as m and the rest mass as m_0. In this book we will refer to the rest mass simply as m in keeping with the preferred usage in current literature. The energy and the momentum are calculated from the relativistic mass of an object, but the rest mass is important because it is an invariant quantity. Using some elementary algebra it is easy to show that the relativistic energy and the momentum are related by the equation

$$E^2 - p^2c^2 = m^2c^4 \tag{7.33}$$

This equation shows that whereas E and $\mathbf{p}$ vary with motion, and hence these quantities take on different values in different reference frames, $E^2 - p^2c^2$ is an invariant. This suggests that the four quantities E, p_x, p_y, p_z could be thought of as components of a space-time vector, similar to dt, dx, dy, dz, which are the components of a four-dimensional displacement with invariant magnitude

$$ds^2 = dx^2 + dy^2 + dz^2 - c^2dt^2 \tag{7.34}$$

We may now dispense with imaginary numbers by defining a metric tensor according to the equation for four-dimensional displacement as

$$ds^2 = g_{\mu\nu}dx^\mu dx^\nu \tag{7.35}$$

where the repeated indices signify addition over the index.

Various conventions are used for the elements of the metric tensor, all leading to the same physical results. We will use the following convention, which is also widely employed in the literature:

$$g_{\mu\nu} = \begin{bmatrix} -1 & 0 & 0 & 0 \\ 0 & 1 & 0 & 0 \\ 0 & 0 & 1 & 0 \\ 0 & 0 & 0 & 1 \end{bmatrix} \tag{7.36}$$

There are also various different conventions for numbering the rows and columns. We will adopt a convention that has a respectable following: The first row and the first column are given the cardinal number 0, and the other rows and columns are numbered 1, 2, and 3. So $g_{00} = -1$ and $g_{11} = g_{22} = g_{33} = 1$. All other elements are zero, so $g_{\mu\nu} = 0$ if $\mu \neq \nu$. One important property of tensors in Minkowski space is that covariant and contravariant elements are identical.

Following the rules of the conventions outlined above, $dx^0 = cdt, dx^1 = dx, dx^2 = dy$, and $dx^3 = dz$. To avoid confusion between superscripts and exponents, we will write dy^2 as $(dx^2)^2$, etc.

The elements of the energy-momentum four-vector — which we express by the symbol p^μ — can be chosen in a variety of ways. One valid choice would be $(E, p_x c, p_y c, p_z c)$. Another equally valid choice is $(E/c, p_x, p_y, p_z)$. We will select the latter, since it leaves three of the components unchanged.

So according to our choice $p^0 = E/c, p^1 = p_x, p^2 = p_y, p^3 = p_z$. The invariant equation becomes

$$g_{\mu\nu}p^\mu p^\nu = p^\mu p_\mu = -m^2c^2 \tag{7.37}$$

Exercises:

1. Two Deuterium (H_1^2) nuclei (each containing one proton and one neutron — also called a deuteron) combine to form a Helium 4 (He_2^4) nucleus (containing two protons and two neutrons — also called an alpha particle). The mass of a Deuterium nucleus is $3.34358348 \times 10^{-27}$ kg and the mass of a Helium 4 nucleus is $6.64465675 \times 10^{-27}$ kg. Take $c = 2.99792458 \times 10^8$ m/s. How much energy is released in this nuclear fusion reaction?
2. In laboratory X a free neutron is observed to decay into a proton, electron and neutrino 9.30 minutes after it was produced. How long was this neutron observed to be in existence before decaying when observed in laboratory Y that is moving with speed 0.900c relative to X?

7.5 Electromagnetic Fields

7.5.1 *Covariant Forms*

Invariants can also be obtained by taking the inner product of two four-vectors. Inner product or contraction renders two vectors into a scalar. The equation of continuity in electrodynamics runs as follows:

$$\nabla \cdot \mathbf{J} + \frac{\partial \rho}{\partial t} = \frac{\partial J_x}{\partial x} + \frac{\partial J_y}{\partial y} + \frac{\partial J_z}{\partial z} + \frac{\partial \rho}{\partial t} = 0 \tag{7.38}$$

Writing the equation in shorthand form, we get $\partial^\mu J_\mu = 0$, where $\partial^\mu = \frac{\partial}{\partial x_\mu}$ and $J_0 = c\rho$. Thus, the inner product of ∂_μ and J_μ is an invariant under

Lorentz transformations. So, it follows that $(c\rho, \mathbf{J})$ is a relativistic four-vector.

Thus, the sources of the electromagnetic fields transform as a four-vector. We will now examine the potentials ϕ and $\mathbf{A}$.

Using the equations for the fields $\mathbf{E} = -\nabla\phi - \frac{\partial \mathbf{A}}{\partial t}$ and $\mathbf{B} = \nabla \times \mathbf{A}$ and the Maxwell equations $c^2\nabla \times \mathbf{B} = \frac{1}{\epsilon_0}\mathbf{J} + \frac{\partial \mathbf{E}}{\partial t}$ and $\nabla \cdot \mathbf{E} = \frac{\rho}{\epsilon_0}$ we obtain the following:

$$\nabla\left(c^2\nabla \cdot \mathbf{A} + \frac{\partial \phi}{\partial t}\right) - c^2\nabla^2\mathbf{A} + \frac{\partial^2 \mathbf{A}}{\partial t^2} = \frac{1}{\epsilon_0}\mathbf{J} \tag{7.39}$$

and

$$-\frac{\partial}{\partial t}(\nabla \cdot \mathbf{A}) - \nabla^2\phi = \frac{\rho}{\epsilon_0} \tag{7.40}$$

The potentials ϕ and $\mathbf{A}$ have a certain arbitrariness to them. Since $\mathbf{B} = \nabla \times \mathbf{A}$, we can add a gradient of a function $\nabla\chi$ to $\mathbf{A}$ without altering the value of $\mathbf{B}$ provided we add $-\frac{\partial \chi}{\partial t}$ to ϕ. This freedom in the choice of the potentials is called gauge invariance. Suppose we select a gauge in which $c^2\nabla \cdot \mathbf{A} + \frac{\partial \phi}{\partial t} = 0$. Such a gauge is called a *Lorenz gauge* (where the name is to be distinguished from that of *Lorentz*). In this gauge the equations become

$$\nabla^2\mathbf{A} = \frac{1}{c^2}\frac{\partial^2 \mathbf{A}}{\partial t^2} - \frac{1}{\epsilon_0 c^2}\mathbf{J} \tag{7.41}$$

and

$$\nabla^2\phi = \frac{1}{c^2}\frac{\partial^2 \phi}{\partial t^2} - \frac{\rho}{\epsilon_0} \tag{7.42}$$

In this gauge, the potentials propagate as waves at speed c in empty space where there are no sources. Writing $A^\mu = (\phi, c\mathbf{A})$, we can write the gauge equation for the potentials as

$$\partial^\mu A_\mu = 0 \tag{7.43}$$

and the equations for the potentials and the sources as

$$\Box A^\mu = \frac{J^\mu}{\epsilon_0 c} \tag{7.44}$$

where the D'Alembertian $\Box \equiv -\nabla^2 + \frac{1}{c^2}\frac{\partial^2}{\partial t^2}$. $A^\mu = (\phi, c\mathbf{A})$ and $J^\mu = (c\rho, \mathbf{J})$.

7.5.2 Electromagnetic Field Tensor

The electromagnetic fields are derived from the potentials via

$$\mathbf{E} = -\nabla\phi - \frac{\partial \mathbf{A}}{\partial t}$$
$$\mathbf{B} = \nabla \times \mathbf{A} \tag{7.45}$$

Writing out the x components of these equations, we obtain

$$E_x = -\frac{\partial A_x}{\partial t} - \frac{\partial \phi}{\partial x} = \partial^0 A^1 - \partial^1 A^0$$

and

$$cB_x = c\left(\frac{\partial A_z}{\partial y} - \frac{\partial A_y}{\partial z}\right) = \partial^2 A^3 - \partial^3 A^2 \tag{7.46}$$

where $\partial^\alpha = (-\frac{\partial}{\partial x_0}, \nabla)$. Since ∂ and A are rank 1 tensors, it follows that the electric and magnetic fields are components of a second rank tensor.

We define an electromagnetic field-strength tensor as follows:

$$F^{\alpha\beta} = \partial^\alpha A^\beta - \partial^\beta A^\alpha \tag{7.47}$$

The components of this tensor are expressed in matrix form as

$$F^{\alpha\beta} = \begin{bmatrix} 0 & E_x & E_y & E_z \\ -E_x & 0 & cB_z & -cB_y \\ -E_y & -cB_z & 0 & cB_x \\ -E_z & cB_y & -cB_x & 0 \end{bmatrix} \tag{7.48}$$

Thus, $F^{\alpha\beta}$ is a skew symmetrical tensor, for which the diagonal elements are all 0. (At this point we need to emphasize that there is no unique form of the field-strength tensor. Some authors prefer the transpose of our matrix as their definition. Other authors suppress the speed of light c factor for the sake of simplicity. Other conventions define the Minkowski metric differently, so that $g_{00} = 1, g_{11} = -1, g_{22} = -1, g_{33} = -1$. Their field-strength tensor will be correspondingly different.)

We can apply the metric tensor twice to convert the contravariant indices to covariant indices:

$$F_{\alpha\beta} = g_{\alpha\gamma} F^{\gamma\delta} g_{\delta\beta} = \begin{bmatrix} -1 & 0 & 0 & 0 \\ 0 & 1 & 0 & 0 \\ 0 & 0 & 1 & 0 \\ 0 & 0 & 0 & 1 \end{bmatrix} \begin{bmatrix} 0 & E_x & E_y & E_z \\ -E_x & 0 & cB_z & -cB_y \\ -E_y & -cB_z & 0 & cB_x \\ -E_z & cB_y & -cB_x & 0 \end{bmatrix} \begin{bmatrix} -1 & 0 & 0 & 0 \\ 0 & 1 & 0 & 0 \\ 0 & 0 & 1 & 0 \\ 0 & 0 & 0 & 1 \end{bmatrix} \tag{7.49}$$

Hence

$$F_{\alpha\beta} = \begin{bmatrix} 0 & -E_x & -E_y & -E_z \\ E_x & 0 & cB_z & -cB_y \\ E_y & -cB_z & 0 & cB_x \\ E_z & cB_y & -cB_x & 0 \end{bmatrix} \tag{7.50}$$

We notice that in the covariant form the signs of the electric field components are flipped, but not those of the magnetic field components. This reflects the signs of the metric elements: $g_{00} = -1, g_{11} = g_{22} = g_{33} = 1$.

We may now write the four Maxwell equations in terms of the field tensor: $\nabla \cdot \mathbf{E} = \frac{\rho}{\epsilon_0}$ and $c^2 \nabla \times \mathbf{B} = \frac{\mathbf{J}}{\epsilon_0} + \frac{\partial \mathbf{E}}{\partial t}$ can be combined into a single tensor equation

$$\partial_\alpha F^{\beta\alpha} = \frac{J^\beta}{c\epsilon_0} \tag{7.51}$$

and the homogeneous equations $\nabla \cdot \mathbf{B} = 0$ and $\nabla \times \mathbf{E} + \frac{\partial \mathbf{B}}{\partial t} = 0$ can be written as

$$\partial_\alpha F^{\beta\gamma} + \partial_\beta F^{\gamma\alpha} + \partial_\gamma F^{\alpha\beta} = 0 \tag{7.52}$$

7.6 Lorentz Transformation of Tensors

Let us now examine the transformation rules for a (1, 1) tensor $G^\alpha{}_\beta$. We contract this tensor with a contravariant vector to obtain another contravariant vector:

$$G^\alpha{}_\beta D^\beta = A^\alpha \tag{7.53}$$

The equation can be expressed in matrix form as

$$GD = A \tag{7.54}$$

Suppose we measure the same components in a different inertial frame. We would get a corresponding equation

$$G'D' = A' \tag{7.55}$$

where the transformed vectors are related to the original vectors by the equations $D' = \Lambda D$ and $A' = \Lambda A$. In order for Eq. (7.55) to be valid in the new frame, the tensors G and G' should be related by

$$G' = \Lambda G \Lambda^{-1} \tag{7.56}$$

Now, Tr (AB) $=\sum_{i,j} A_{ij}B_{ji} = \sum_{j,i} B_{j,i}A_{i,j}$ = Tr (BA) for any two square matrices of the same order. So, Tr $(\Lambda G\Lambda^{-1})$ = Tr $(\Lambda^{-1}\Lambda G)$ = Tr G. Thus Tr G' = Tr G.

Let $G^{\alpha}{}_{\beta} = F^{\alpha\gamma}F_{\gamma\beta}$. The elements of this tensor are obtained by matrix multiplication:

$$G = \begin{bmatrix} 0 & E_x & E_y & E_z \\ -E_x & 0 & -cB_z & cB_y \\ -E_y & cB_z & 0 & -cB_x \\ -E_z & -cB_y & cB_x & 0 \end{bmatrix} \begin{bmatrix} 0 & -E_x & -E_y & -E_z \\ E_x & 0 & -cB_z & cB_y \\ E_y & cB_z & 0 & -cB_x \\ E_z & -cB_y & cB_x & 0 \end{bmatrix} =$$

$$\begin{bmatrix} E^2 & (cE_yB_z - cE_zB_y) & (-cE_xB_z + cE_zB_x) & (cE_xB_y - cE_yB_x) \\ (-cB_zE_y + cB_yE_z) & (E_x^2 - c^2B_z^2 - c^2B_y^2) & (E_xE_y + c^2B_yB_x) & (E_xE_z + c^2B_xB_z) \\ (cE_xB_z - cB_xE_z) & (E_xE_y + c^2B_xB_y) & (E_y^2 - c^2B_z^2 - c^2B_x^2) & (E_yE_z + c^2B_zB_y) \\ (-cE_xB_y + cB_xE_y) & (E_xE_z + c^2B_xB_z) & (E_zE_y + c^2B_zB_y) & (E_z^2 - c^2B_x^2 - c^2B_y^2) \end{bmatrix}$$

Tr $G = 2(E^2 - c^2B^2)$. This must therefore be invariant under a Lorentz transformation. Hence $E^2 - c^2B^2$ is a Lorentz invariant or a Lorentz scalar.

We will now obtain expressions for the Lorentz transformation of the electric and magnetic fields, via the electromagnetic field tensor, using the transformation equation

$$F'^{\alpha\beta} = \frac{\partial x'^{\alpha}}{\partial x^{\gamma}}\frac{\partial x'^{\beta}}{\partial x^{\delta}}F^{\gamma\delta} \tag{7.57}$$

If we are to write this equation in matrix form we must respect the order of the indices. Specifically, the sum $A^{\alpha}_{\beta}B^{\gamma\beta}$ cannot be written as the matrix product AB. The correct matrix order is expressed as follows:

$$A^{\alpha}_{\beta}B^{\gamma\beta} = B^{\gamma\beta}A^{\alpha}_{\beta} \equiv BA \tag{7.58}$$

Thus, it is preferable to write the transformation as

$$F'^{\alpha\beta} = \frac{\partial x'^{\alpha}}{\partial x^{\gamma}}F^{\gamma\delta}\frac{\partial x'^{\beta}}{\partial x^{\delta}} \tag{7.59}$$

which is expressed in matrix form as

$$F' = \Lambda F \Lambda \tag{7.60}$$

This is the transformation law for a (2, 0) tensor, which is different from that of a (1, 1) tensor, cf. Eq. (7.56).

Writing out the matrix elements explicitly, we obtain

$$\begin{bmatrix} \gamma & -\frac{\gamma u}{c} & 0 & 0 \\ -\frac{\gamma u}{c} & \gamma & 0 & 0 \\ 0 & 0 & 1 & 0 \\ 0 & 0 & 0 & 1 \end{bmatrix} \begin{bmatrix} 0 & E_x & E_y & E_z \\ -E_x & 0 & -cB_z & cB_y \\ -E_y & cB_z & 0 & -cB_x \\ -E_z & -cB_y & cB_x & 0 \end{bmatrix} \begin{bmatrix} \gamma & -\frac{\gamma u}{c} & 0 & 0 \\ -\frac{\gamma u}{c} & \gamma & 0 & 0 \\ 0 & 0 & 1 & 0 \\ 0 & 0 & 0 & 1 \end{bmatrix}$$

$$= \begin{bmatrix} 0 & E_x & \gamma(E_y + uB_z) & \gamma(E_z - uB_y) \\ -E_x & 0 & -\gamma(\frac{uE_y}{c} + cB_z) & \gamma(-\frac{uE_z}{c} + cB_y) \\ -\gamma(E_y + uB_z) & \gamma(\frac{uE_y}{c} + cB_z) & 0 & -cB_x \\ -\gamma(E_z - uB_y) & \gamma(\frac{uE_z}{c} - cB_y) & cB_x & 0 \end{bmatrix} \tag{7.61}$$

The transformation equations for the individual components become

$$E'_x = E_x \tag{7.62}$$

$$E'_y = \gamma(E_y + uB_z) \tag{7.63}$$

$$E'_z = \gamma(E_z - uB_y) \tag{7.64}$$

$$B'_x = B_x \tag{7.65}$$

$$B'_y = \gamma\left(B_y - \frac{uE_z}{c^2}\right) \tag{7.66}$$

$$B'_z = \gamma\left(B_z + \frac{uE_y}{c^2}\right) \tag{7.67}$$

The above equations list the Lorentz transformation equations for electric and magnetic fields between two inertial frames having a relative speed u in the x direction. A little calculation shows that our earlier prediction that $E^2 - c^2B^2$ is invariant under Lorentz transformations is justified, because when we plug in the components of the primed and the unprimed vectors we obtain

$$E'^2 - c^2B'^2 = E^2 - c^2B^2 \tag{7.68}$$

Further, it is seen that the dot product $\mathbf{E} \cdot \mathbf{B}$ is invariant under a Lorentz transformation,

$$\mathbf{E}' \cdot \mathbf{B}' = \mathbf{E} \cdot \mathbf{B} \tag{7.69}$$

There is an interesting complex vector called the Riemann-Silberstein vector which can be expressed as $\mathbf{R} = \mathbf{E}+ic\mathbf{B}$. One of the interesting properties of this vector is that its magnitude is invariant under Lorentz transformations:

$$\mathbf{R}\cdot\mathbf{R} = (\mathbf{E}+ic\mathbf{B})\cdot(\mathbf{E}+ic\mathbf{B}) = E^2 - c^2B^2 + 2ic(\mathbf{E}\cdot\mathbf{B}) \tag{7.70}$$

The electromagnetic energy density can be written in terms of the Riemann-Silberstein vector as

$$u = \frac{\epsilon_0}{2}(E^2 + c^2B^2) = \frac{\epsilon_0}{2}\mathbf{R}\cdot\mathbf{R}^* \tag{7.71}$$

Note that the electromagnetic energy density is not invariant under Lorentz transformations. Another interesting result is obtained by taking a cross product:

$$\mathbf{R}\times\mathbf{R}^* = -2ic\mathbf{E}\times\mathbf{B} = \frac{2}{i\epsilon_0 c}\mathbf{S} \tag{7.72}$$

where $\mathbf{S} = \epsilon_0 c^2\mathbf{E}\times\mathbf{B}$, the Poynting Vector.

The transformation equations listed above apply to a translation along the x direction. It is possible to obtain the transformation equations for a general relative velocity $\mathbf{v}$, though we will not derive these expressions. We define the dimensionless vector symbol $\boldsymbol{\beta} = \mathbf{v}/c$ and write the transformation equations as

$$\mathbf{E}' = \gamma(\mathbf{E} - c\boldsymbol{\beta}\times\mathbf{B}) - \frac{\gamma^2}{\gamma+1}\boldsymbol{\beta}(\boldsymbol{\beta}\cdot\mathbf{E}) \tag{7.73}$$

$$\mathbf{B}' = \gamma\left(\mathbf{B} + \frac{\boldsymbol{\beta}}{c}\times\mathbf{E}\right) - \frac{\gamma^2}{\gamma+1}\boldsymbol{\beta}(\boldsymbol{\beta}\cdot\mathbf{B}) \tag{7.74}$$

Exercises:

1. Use Eqs. (7.62) through (7.67) to show that $\mathbf{E}\cdot\mathbf{B}$ and $E^2 - c^2B^2$ are Lorentz invariant scalars.
2. Show that Eqs. (7.62) through (7.67) are special cases of the more general Eqs. (7.73) and (7.74).
3. Use Eqs. (7.73) and (7.74) to show that $\mathbf{E}\cdot\mathbf{B}$ and $E^2 - c^2B^2$ are Lorentz invariant scalars.

7.7 Doppler Effect

7.7.1 *Classical Formula*

Suppose there is a stationary source emitting a wave having speed v in the x direction, and an observer traveling in the negative x direction at speed

u. Suppose the frequency of the wave emitted by the source is f_s and the frequency of the same wave as measured by the observer is f_o. In this situation f_o will be greater than f_s and the difference is due to the Doppler effect.

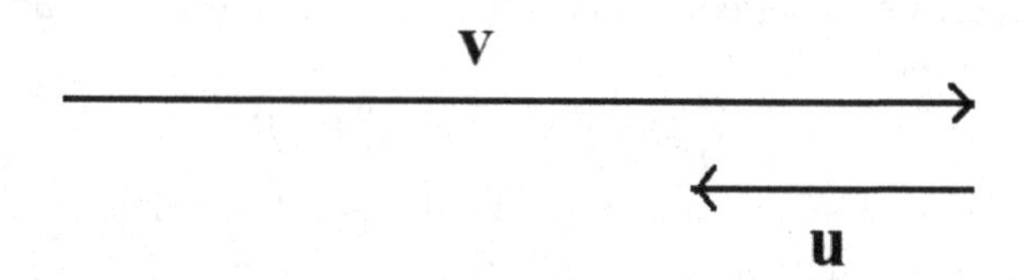

In 1 second the wave advances a distance v and the observer a distance u. In 1 second the source emits n_s waves. This is numerically equal to f_s the frequency of the source. Therefore there are n_s waves contained in a distance v along the path of the wave. If the observer were stationary, they would count n_s waves passing them every second. But because the observer is in motion towards the source, they would count an additional number of waves that are contained in the distance u advanced by the observer in one second. Since there are n_s waves in the distance v, the number of waves in the distance u equals $n_s(u/v)$. So, the number of waves measured by the observer in one second would be not n_s but $n_s + n_s(u/v) = n_s(v+u)/v$. So, the number of waves counted by the observer in one second would be

$$n_o = n_s \left(1 + \frac{u}{v}\right) \tag{7.75}$$

Thus, the frequency measured by the observer would be

$$f_o = f_s \left(1 + \frac{u}{v}\right) \tag{7.76}$$

Exercise:
Derive the Doppler formula for the frequency f_o measured by an observer moving away at speed u from a stationary source emitting sound of frequency f_s.

7.7.2 *Relativistic Formula*

In deriving the classical formula for the Doppler effect produced by an observer approaching a stationary source, we assumed that the elapsed time is the same for both the source and the observer that are in relative motion. But this assumption is not valid when we include the postulates of special relativity.

So, if Δt_{obs} is the time taken by the observer to count the waves, this will not be the same as the time Δt_{sou} taken by the source to count the waves. Δt_{sou} will be greater than Δt_{obs}. The relationship between these two time intervals

$$\Delta t_{obs} = \Delta t_{sou}\sqrt{1 - \frac{u^2}{c^2}} \tag{7.77}$$

The frequency of the wave as measured by the observer will be

$$f_o = \frac{n_o}{\Delta t_{obs}} \tag{7.78}$$

and the frequency of the wave as measured by the source will be

$$f_s = \frac{n_s}{\Delta t_{sou}} \tag{7.79}$$

To obtain the relativistic Doppler formula we substitute these results into Eq. (7.75) and obtain

$$f_o = \frac{f_s}{\sqrt{1 - \frac{u^2}{c^2}}}\left(1 + \frac{u}{v}\right) \tag{7.80}$$

The above equation is valid for any type of wave, be it a sound wave or a light wave. For the case of a light wave we replace v by c:

$$f_o = \frac{f_s}{\sqrt{1 - \frac{u^2}{c^2}}}\left(1 + \frac{u}{c}\right) \tag{7.81}$$

7.7.3 *Alternative Derivation of Relativistic Doppler Formula for Light*

The displacement of the medium due to a plane transverse wave propagating in the positive x direction and having angular frequency ω can be expressed as $y = \cos(kx - \omega t)$ or $y = \sin(kx - \omega t)$ where the wave number k is related to the angular frequency and the wave velocity c by the formula $\omega = kc$. It is also customary to write the solution of the wave equation as $y = \exp i(kx - \omega t)$, where the physical displacement is the real part of y. For any electromagnetic wave the displacement — which represents some component of the electric or the magnetic field — is always perpendicular to the direction of propagation. So, this displacement will be invariant under a Lorentz transformation or boost in the x direction. So, let the displacement due to the light wave as measured in the frame of the source be $\exp i(kx - \omega t)$ and the same displacement measured in a frame traveling in the negative x direction and therefore approaching the source with velocity $-u$ be $\exp i(k'x' - \omega' t')$. Since these displacements are equal, we can set the phases equal

$$k'x' - \omega' t' = kx - \omega t \tag{7.82}$$

(Equal displacements require that the phases can differ only by $2\pi n$ with n being an integer. But here $n = 0$, because the two phases have to become identical as $u \to 0$.) Now, the unprimed frame is moving with velocity u in the x direction as measured by the primed system. The transformation equations are therefore

$$x = (x' - ut')\gamma \tag{7.83}$$

$$t = (t' - ux'/c^2)\gamma \tag{7.84}$$

Substituting into Eq. (7.82), we obtain

$$k'x' - \omega' t' = k(x' - ut')\gamma - \omega(t' - ux'/c^2)\gamma \tag{7.85}$$

Now, $\omega'/k' = \omega/k = c$. So we can eliminate k and k' from both sides of the equation to obtain

$$\frac{\omega'}{c}x' - \omega' t' = \frac{\omega}{c}(x' - ut') - \omega(t' - ux'/c^2)\gamma \tag{7.86}$$

Equating the coefficients of x' on both sides we obtain

$$\omega' = \omega\gamma\left(1 + \frac{u}{c}\right) \tag{7.87}$$

$\omega' = 2\pi f_o$ and $\omega = 2\pi f_s$. Therefore

$$f_o = \frac{f_s}{\sqrt{1 - \frac{u^2}{c^2}}}\left(1 + \frac{u}{c}\right) \tag{7.88}$$

It is easy to show that for the case of an observer receding from a source emitting light of frequency f_s the observed frequency of the light will be given by the expression

$$f_o = \frac{f_s}{\sqrt{1 - \frac{u^2}{c^2}}}\left(1 - \frac{u}{c}\right) \tag{7.89}$$

Equations (7.88) and (7.89) enable us to calculate the relative speed between the earth and any star which is bright enough for us to study its emission spectrum. An observed increase in frequency of the light emitted by the star — evidenced in the shift of the recognizable spectral lines towards the blue end of the spectrum — is called a *blue shift*, and indicates an approaching star. An observed decrease in the frequency of the light — evidenced by the shift towards the red end of the spectrum — is called a *red shift*, and indicates a receding star. The relative speed between the star and the earth can then be calculated.

Galaxies have been observed to recede from the earth, and the rate of recession is proportional to the distance of the galaxy from the earth. This red shift of the galaxies — first observed by Hubble — is now interpreted as a feature of the expanding universe. In the general relativity model of the universe we do not picture all the other galaxies in a dynamic motion away from our own galaxy. We picture the galaxies as more or less statically embedded in a space that itself is expanding, causing a steady increase in the distance between the galaxies.

Earlier we found that an electromagnetic field propagates with momentum proportional to the Poynting Vector. Electromagnetic waves interact with material objects having mechanical mass and momentum. Indeed, every electric and magnetic field requires a source which is a charge. Charges are embedded in material objects, and even the electron which is considered as an elementary charge itself has mass and momentum. The conservation of global momentum is a consequence of a basic symmetry of space called *translational symmetry*, that the nature of the surrounding space will not change when we displace a particle in any direction. (Conservation of global *angular momentum* follows from the *rotational symmetry* of space.)

We will now see that the law of conservation of global momentum implies a definite relationship between the frequency and the momentum of radiation.

Suppose we have two spheres colliding along their line of centers. This collision is viewed from two different frames. We define motion to the right as positive. In the first frame sphere B is at rest, and sphere A has positive momentum. In the second frame B has positive momentum, and A has a greater positive momentum than in the first frame.

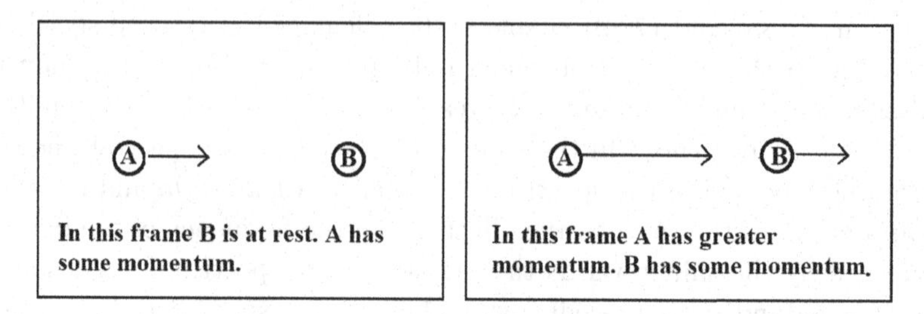

We now replace the sphere A by an incoming electromagnetic wave, and the sphere B by a solid object that absorbs the radiation. In the figure below a solid object with a rectangular cross-section is shown absorbing a beam of electromagnetic radiation, and in this reference frame the object is at rest.

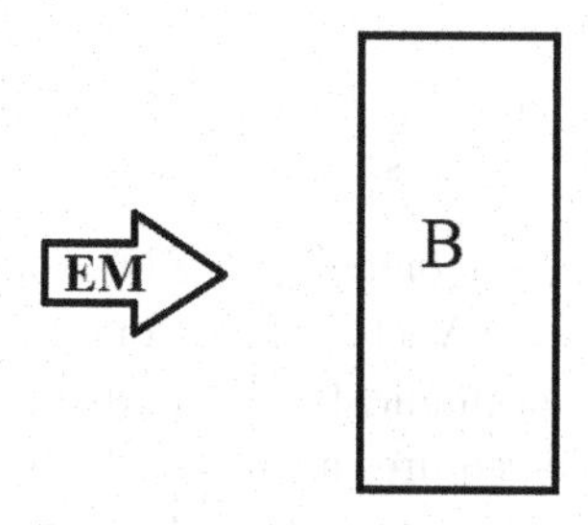

In this frame the object has zero momentum, and the radiation has positive momentum — directed to the right.

Next, suppose the same process is viewed from a different reference frame moving opposite to the direction of the electromagnetic radiation. In this frame the object has a velocity relative to the observer, and so will appear contracted according to the Lorentz contraction formula $\Delta L = \Delta L_0\sqrt{1-\frac{v^2}{c^2}}$. The radiation will have the same speed, but the frequency of the wave will be increased due to the Doppler effect.

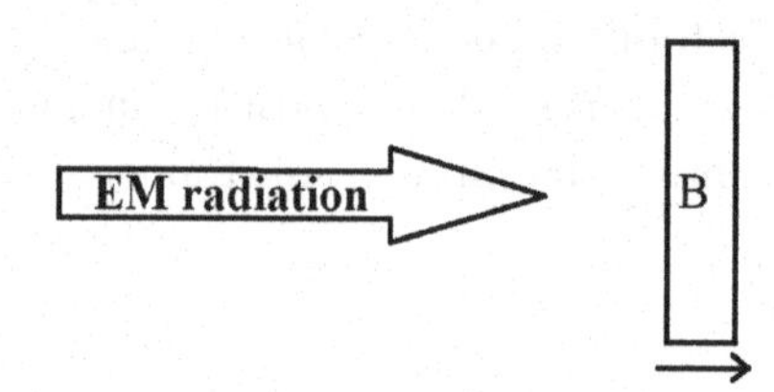

In this frame both the object and the radiation will have momentum directed to the right, and the analogy to the colliding spheres suggests that the momentum of the radiation will be greater in this frame than the previous one in which the object was at rest.

Now, the radiation has increased frequency due to Doppler effect. And it appears logical that the radiation has greater momentum in this frame. This suggests that an increase in frequency is accompanied by an increase in momentum. The inference is that there is a direct correlation between the frequency and the momentum of electromagnetic radiation — the greater the frequency, the greater the momentum.

We saw earlier that the momentum of electromagnetic radiation is proportional to the Poynting Vector $\epsilon_0 c^2 \mathbf{E} \times \mathbf{B}$, and hence the momentum is proportional to the product of the amplitudes of the electric and magnetic fields, and since the electric and magnetic fields are proportional to one another in a wave, the momentum of a wave is proportional to the square of the amplitude of the wave. Also, the energy carried by a wave is also proportional to the square of the amplitude of the wave. But the Doppler effect suggests that the momentum of a beam of electromagnetic radiation is also proportional to the frequency, a quantity associated not with matter but with waves. Thus electromagnetic waves have both material and undular (wavelike) properties.

This matter-wave duality of radiation bears an interesting relationship to the concept of *intensive* and *extensive* variables. Broadly speaking, an intensive variable is independent of quantities such as mass and volume, whereas extensive variables are dependent on such quantities. So mass and volume are both extensive, but density is not. Gaseous pressure is also an intensive variable — the pressure inside our lungs is equal to the pressure of the atmosphere. The internal energy of an object is an extensive variable, but its temperature is intensive. It is evident that the extensive-intensive dichotomy is a category that occurs only in the study of large aggregates of particles, such as a gas made up of molecules, a solid constituted of atoms or ions, etc.

The total momentum of electromagnetic radiation contained within some region of space is extensive, since it depends on the volume of the region. But now we see that the momentum of radiation also depends on its frequency. However, the frequency of the radiation is an intensive variable, independent of the volume.

That the distinction between extensive and intensive variables is a category that is also applicable to the momentum of electromagnetic radiation suggests that electromagnetic radiation itself has a corpuscular nature. The total momentum in some region is proportional to two quantities: (a) the number of corpuscles present in that region, which in turn is proportional to the volume of the region, and (b) the frequency of the radiation, which is independent of the volume of the region and hence also independent of the number of corpuscles in that region.

Everything that was said about the momentum of radiation is equally applicable to the energy of radiation. We saw earlier that the electromagnetic energy within a given volume of space is equal to the magnitude of the momentum of the field in that volume multiplied by c. The quantum theory of radiation first introduced by Planck and applied by Einstein to explain the photoelectric effect confirms the corpuscular nature of electromagnetic radiation. The total energy within a region of space is proportional on the one hand to the number of corpuscles of radiation within that region, and also to the frequency of the radiation, a quantity independent of the number of corpuscles. This last sentence implies that each corpuscle of radiation has an energy that is a function of the frequency of the radiation. As it turns out, the function is a simple linear dependence: the energy of a corpuscle of radiation is directly proportional to the frequency of the radiation. The constant of proportionality h is Planck's constant. We will discuss these ideas in more detail in the following chapters.

7.8 Relativistic Dynamics of Charged Particles

The Lorentz force acting on a charged particle is given by

$$\frac{d\mathbf{p}}{dt} = \mathbf{F} = q(\mathbf{E} + \mathbf{v} \times \mathbf{B}) \tag{7.90}$$

The power expended during this interaction between the field and the particle is therefore

$$\frac{dE}{dt} = \mathbf{F} \cdot \mathbf{v} = q\mathbf{E} \cdot \mathbf{v} \tag{7.91}$$

These equations can be checked for consistency with the energy-momentum four-vector equation. Consider a particle of rest mass m, having momentum

$\mathbf{p}$ and total energy E in some reference frame. These quantities are related by the energy-momentum equation

$$E^2 = \mathbf{p} \cdot \mathbf{p}c^2 + m^2c^4 \tag{7.92}$$

Taking the time derivative,

$$2E\frac{dE}{dt} = 2\mathbf{p} \cdot \frac{d\mathbf{p}}{dt}c^2 \tag{7.93}$$

Therefore

$$\frac{dE}{dt} = \frac{\mathbf{p}}{E} \cdot \frac{d\mathbf{p}}{dt}c^2 = \mathbf{v} \cdot \frac{d\mathbf{p}}{dt} \tag{7.94}$$

It is easily seen that this equation is consistent with Eqs. (7.90) and (7.91).

We can rewrite Eqs. (7.90) and (7.91) in covariant form by multiplying both sides by $\gamma = \frac{1}{\sqrt{1-\frac{v^2}{c^2}}}$.

The force equation becomes

$$\frac{d\mathbf{p}}{d\tau} = \gamma q(\mathbf{E} + \mathbf{v} \times \mathbf{B}) \tag{7.95}$$

The power equation becomes

$$\frac{dE}{d\tau} = \gamma q\mathbf{E} \cdot \mathbf{v} \tag{7.96}$$

The four-velocity U of a particle is defined with components $(\gamma c, \gamma \mathbf{v})$, where $\mathbf{v}$ is the measured velocity of the particle in some reference frame. Now $\mathbf{v} \cdot \mathbf{v} \equiv v^2$ is a scalar under rotations and translations, but not under Lorentz boosts. But the four-velocity is indeed a scalar under boosts, because $U^\alpha U_\alpha = \gamma^2(-c^2 + v^2) = -c^2$, which is an invariant or scalar.

The two Eqs. (7.95) and (7.96) can be combined into a single tensor equation by employing the four-velocity $U = (\gamma c, \gamma \mathbf{v})$.

Newton's Second Law takes on the covariant form

$$\frac{dp^\alpha}{d\tau} = m\frac{dU^\alpha}{d\tau} \tag{7.97}$$

Equations (7.95) and (7.96) can be written in compact form as

$$\frac{dp^\alpha}{d\tau} = \frac{q}{c}F^{\alpha\beta}U_\beta \tag{7.98}$$

Exercise:
Show that Eq. (7.98) is equivalent to Eqs. (7.95) and (7.96).

7.9 Lagrangian Dynamics of the Electromagnetic Field

7.9.1 *The Principle of Least Action*

In geometric ray optics one employs the laws of reflection (angle of incidence = angle of reflection) and the laws of refraction (Snell's law relating the angles of incidence and refraction to the speeds of light in the two different media). An alternative formulation of ray optics uses the principle of least time — the path taken by a ray of light between two points is the path of least time. The laws of geometric optics emerge as a consequence of the principle of least time (which is a consequence of Huygens' theory of wave propagation). This equivalence between two very different mathematical approaches finds a parallel in classical mechanics. Newtonian dynamics is based on the second law according to which the net force acting on a system is equal to the rate of change of momentum of the system:

$$\mathbf{F} = \frac{d\mathbf{P}}{dt} \tag{7.99}$$

However, in many situations it is easier to use an alternative mathematical formulation of the same physical principles. One such formulation is the Lagrangian formalism. The Lagrangian is defined as a functional (basically a function of functions) of the positions and the velocities of the particles of a system, and possibly the time as well: $L[q_i(t), \dot{q}_i(t), t]$ In Lagrangian dynamics the action is defined as the integral of the Lagrangian between two points in time:

$$A = \int_{t_1}^{t_2} L[q_i(t), \dot{q}_i(t), t]dt \tag{7.100}$$

These moments in time are analogous to the starting and ending positions in the least time principle of ray optics. The principle of Lagrangian dynamics states that the physical process undergone by the system between the initial and the final times will be that process for which the action is an extremum — i.e. a maximum or a minimum, expressible mathematically as $\delta A = 0$. The actual equations of motion will emerge as a consequence of this extremum condition. Stated as an integral, this condition is expressed as

$$\int_{t_1}^{t_2} \delta L[q_i(t), \dot{q}_i(t), t]dt = 0 \tag{7.101}$$

We can express the variation of the integrand in terms of a variation of each of the coordinates q_i:

$$\delta L = \sum_i \left(\frac{\partial L}{\partial q_i} \delta q_i + \frac{\partial L}{\partial \dot{q}_i} \delta \dot{q}_i \right) = \sum_i \left(\frac{\partial L}{\partial q_i} \delta q_i + \frac{\partial L}{\partial \dot{q}_i} \frac{d}{dt} \delta q_i \right)$$

The integral of the second term within brackets can be manipulated as follows:

$$\int_{t_1}^{t_2} \frac{\partial L}{\partial \dot{q}_i} \frac{d}{dt} \delta q_i dt = \int_{t_1}^{t_2} \frac{\partial L}{\partial \dot{q}_i} d\delta q_i = \left[\frac{\partial L}{\partial \dot{q}_i} \delta q_i \right]_{t_1}^{t_2} - \int_{t_1}^{t_2} \frac{d}{dt} \left(\frac{\partial L}{\partial \dot{q}_i} \right) \delta q_i dt$$

We allow each δq_i to vary independently only within the open interval (t_1, t_2), and so $\delta q_i = 0$ at both t_1 and t_2. And so, we obtain

$$\int_{t_1}^{t_2} \delta L[q_i(t), \dot{q}_i(t), t] dt = \int_{t_1}^{t_2} \sum_i \left[\frac{\partial L}{\partial q_i} - \frac{d}{dt} \left(\frac{\partial L}{\partial \dot{q}_i} \right) \right] \delta q_i dt = 0$$

Now, each δq_i is an arbitrary variation in q_i. Hence the vanishing of the integral between t_1 and t_2 implies the Euler-Lagrange equations

$$\frac{\partial L}{\partial q_i} - \frac{d}{dt} \left(\frac{\partial L}{\partial \dot{q}_i} \right) = 0$$

for each one of the generalized coordinates q_i.

7.9.2 *Examples*

7.9.2.1 *Harmonic oscillator*

The Lagrangian function of a system takes on the form $T - U$ in classical mechanics where $T = \sum_i \frac{1}{2} m_i v_i^2$ is the kinetic energy of the system, where the ith particle has mass m_i and speed v_i, and U is the potential energy of the system.

So, for a ball of mass m attached to a spring of constant k which is capable of extending without friction along the x direction, $T = \frac{1}{2} m v^2$ and $U = \frac{1}{2} k x^2$.

The Euler-Lagrange equation becomes

$$kx = -\frac{d}{dt}(mv) = -F \tag{7.102}$$

Thus, we obtain Hooke's Law for a spring $F = -kx$.

Moreover, from Eq. (7.102) we obtain the second order differential equation

$$\frac{d^2 x}{dt^2} = -\frac{k}{m} x \tag{7.103}$$

which is the simple harmonic motion equation having the general solution $x = A\cos(\omega t + \phi)$ where the amplitude A and the phase constant ϕ are constants of integration, and $\omega = \sqrt{\frac{k}{m}}$. Equivalently, the solution can also be expressed as

$$x = Ae^{i\omega t} + Be^{-i\omega t} \tag{7.104}$$

where A and B are in general complex numbers.

7.9.2.2 *Ring around a sphere*

A non-conducting sphere of radius R has a positive charge Q distributed uniformly over its surface. A narrow non-conducting solid ring of radius slightly larger than R, which carries a uniformly distributed charge $-|q|$, floats around the "equator" of the sphere:

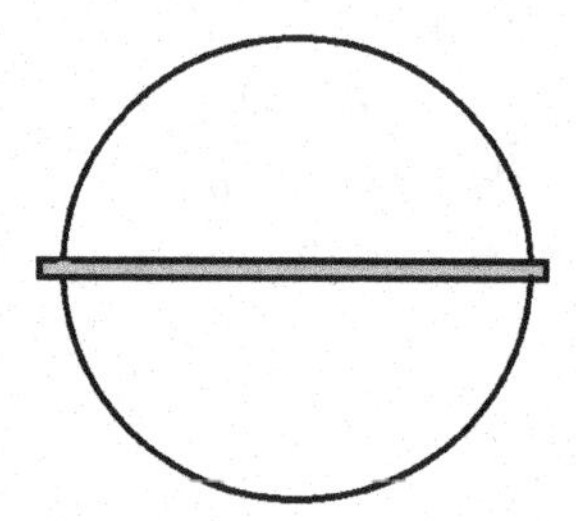

The mass of the ring is m. The system is in a weightless environment, such as a space ship. Electrostatic attraction between the ring and the sphere keeps the ring in place. If now the ring is given a slight displacement in the axial direction, the potential energy of the sphere-belt system becomes, for small θ,

$$V = -\frac{qQ\cos\theta}{4\pi\epsilon_0 R} = -\frac{qQ}{4\pi\epsilon_0 R}\left(1 - \frac{\theta^2}{2}\right)$$

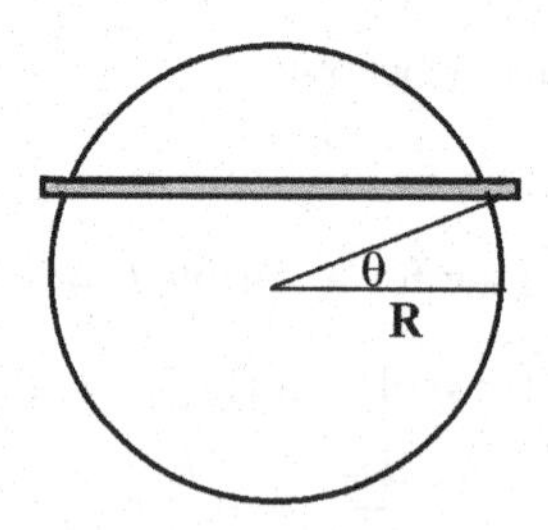

The speed of the ring is $\frac{d}{dt}(R\tan\theta)$ which for small θ becomes $\frac{d}{dt}(R\theta)$. The kinetic energy of the belt is therefore $T = \frac{1}{2}mR^2\dot{\theta}^2$. The Lagrangian for this system is

$$L = \frac{1}{2}mR^2\dot{\theta}^2 + \frac{qQ}{4\pi\epsilon_0 R}\left(1 - \frac{\theta^2}{2}\right)$$

The solution is a harmonic oscillation of angular frequency

$$\omega = \sqrt{\frac{qQ}{4\pi\epsilon_0 mR^3}}$$

Next, suppose the sphere is the earth of mass M and radius R, and the ring has radius slightly greater than R, and mass m. Such a system would be entirely analogous to the electrostatic system, and the angular frequency for small oscillations becomes

$$\omega' = \sqrt{\frac{GM}{R^3}}$$

It is interesting that in the electrostatic problem a measurement of the oscillation frequency can only give us the charge to mass ratio q/m of the ring, and in the gravitational problem the mass of the ring cannot be obtained.

7.9.3 *LC Circuit*

An idealized L-C circuit (with zero resistance) consists of a capacitor of capacitance C connected to a solenoid of self-inductance L. The electrical energy stored in the capacitor at any moment is given by $U_e = \frac{Q^2}{C}$ and the magnetic energy inside the solenoid at any moment is given by $\frac{1}{2}LI^2$. The current $I = \frac{dQ}{dt}$. We could set Q as a generalized coordinate q and I as the corresponding generalized velocity, which would identify the electrical energy stored in the capacitor as the potential energy and the magnetic energy in the solenoid as the kinetic energy. So, the Lagrangian for this circuit becomes

$$\mathcal{L} = \frac{1}{2}LI^2 - \frac{1}{2}\frac{Q^2}{C} \tag{7.105}$$

and the Euler-Lagrange equation becomes

$$\frac{\partial\mathcal{L}}{\partial Q} - \frac{d}{dt}\left(\frac{\partial\mathcal{L}}{\partial I}\right) = -\frac{Q}{C} - L\frac{dI}{dt} = -\frac{Q}{LC} - \frac{d^2Q}{dt^2} = 0 \tag{7.106}$$

This is also a harmonic oscillator equation whose general solution is expressible as

$$Q = Ae^{i\frac{t}{\tau}} + Be^{-i\frac{t}{\tau}} \tag{7.107}$$

where $\tau = \sqrt{LC}$ is the time constant.

7.9.4 Energy of an Electromagnetic Field

Now, the total energy in an electromagnetic field is given by

$$U = \frac{\epsilon_0}{2}\int (E^2 + c^2B^2)dV \tag{7.108}$$

Treating the electrical energy as potential and the magnetic energy as kinetic, the Lagrangian for an L-C circuit can be expressed as

$$\mathcal{L} = \frac{\epsilon_0}{2}\int (c^2B^2 - E^2)dV \tag{7.109}$$

We could model an electromagnetic wave as an oscillating LC circuit, with constantly changing electric and magnetic fields, where the electromagnetic field energy alternates between electric and magnetic. In terms of a harmonic oscillator model, we could picture the electromagnetic energy as oscillating between potential (electric) and kinetic (magnetic).

The discussion so far has been non-relativistic. A relativistic Lagrangian must satisfy other conditions. We have seen that a non-relativistic Lagrangian yields the action defined by

$$A = \int_{t_1}^{t_2} \mathcal{L}\left[q_i(t), \dot{q}_i(t), t\right] dt \tag{7.110}$$

To incorporate relativistic invariance into the Lagrangian dynamics of the electromagnetic field we can write the time element in terms of the γ factor $(\gamma = \frac{1}{\sqrt{1-\frac{v^2}{c^2}}})$ as follows:

$$dt = \gamma d\tau \tag{7.111}$$

where $d\tau$ is the proper time. Thus, we get the action integral

$$A = \int_{\tau_1}^{\tau_2} \gamma\mathcal{L}d\tau \tag{7.112}$$

Since the proper time interval $d\tau$ is relativistically invariant, the condition for A to be invariant is that $\gamma\mathcal{L}$ should be invariant.

Multiplying Eq. (7.109) by γ, we obtain

$$\gamma\mathcal{L} = \frac{\epsilon_0}{2}\int (c^2B^2 - E^2)\gamma dV \tag{7.113}$$

Now, γdV is invariant. (Consider a relative motion along the x axis. $\gamma dV = \gamma dxdydz$. Here dy and dz are invariant, and γdx is invariant.) Earlier we had shown that $c^2B^2 - E^2$ is Lorentz invariant. Thus $\mathcal{L} = \frac{\epsilon_0}{2}\int(c^2B^2 - E^2)dV$ is an appropriate Lagrangian for the field of an electromagnetic wave (in the absence of charges and currents).

Hence, the relativistic action of an electromagnetic wave is expressible as

$$A = \int \gamma L d\tau = \frac{\epsilon_0}{2}\int(c^2B^2 - E^2)d^4x \qquad (7.114)$$

where $d^4x = dVdt = \gamma dVd\tau$ is the invariant four-vector volume element.

7.9.5 *Electric and Magnetic Fields in a Propagating Wave*

If the force between any two particles of a many particle system results from a potential energy $V(r) = ar^n$ that is proportional to some power n of the inter-particle distance r, the *virial theorem* states that

$$2\langle T\rangle = n\langle V\rangle \qquad (7.115)$$

where $\langle T\rangle$ is the average kinetic energy, and $\langle V\rangle$ is the average total potential energy, of the system.

For a harmonic oscillator potential $n = 2$, and so we get the simple result

$$\langle T\rangle = \langle V\rangle \qquad (7.116)$$

Treating $\frac{1}{2}\epsilon_0 E^2$ as the average potential energy and $\frac{1}{2}\epsilon_0 c^2B^2$ as the average kinetic energy of the electromagnetic wave we obtain

$$\langle E\rangle = c\langle B\rangle \qquad (7.117)$$

In an electromagnetic wave, the direction of propagation is perpendicular to the directions of the electric and the magnetic fields, as shown by the Poynting Vector:

$$\mathbf{S} = \epsilon_0 c^2\mathbf{E}\times\mathbf{B} \qquad (7.118)$$

If the direction of propagation is the z axis, then $\mathbf{E}$ and $\mathbf{B}$ are both perpendicular to the z axis. Let us choose the x axis as the direction of $\mathbf{E}$. We will show that $\mathbf{B}$ must be directed along the y axis.

Since both $\mathbf{E}$ and $\mathbf{B}$ satisfy the equation for a wave propagating at speed c, the solutions can be written in complex form as

$$\mathbf{E} = \hat{\imath}E_0 e^{i(kz-\omega t)} \qquad (7.119)$$

(where the unit vector $\hat{\imath}$ is to be distinguished from the unit imaginary number i), and

$$\mathbf{B} = \mathbf{B}_0 e^{i(kz-\omega t)} \tag{7.120}$$

where the constants k and ω are related by the phase velocity $c = \omega/k$. Substituting these solutions into Maxwell's equation $\nabla \times \mathbf{E} = -\frac{\partial \mathbf{B}}{\partial t}$, we obtain

$$\frac{\partial E_x}{\partial z} = -\frac{\partial B_y}{\partial t} \tag{7.121}$$

and

$$\frac{\partial B_x}{\partial t} = 0 \tag{7.122}$$

Thus, if $\mathbf{E}$ is directed along the x axis, then $\mathbf{B}$ is directed along the y axis, and so the electric and magnetic fields are perpendicular to each other in an electromagnetic wave. Also, Eq. (7.121) implies that $E_0 = cB_0$, which is in agreement with the result from the virial theorem.

7.9.6 *Lagrangian for a Charge Interacting with a Field*

7.9.6.1 *Free particle*

The relativistically invariant action for a Lagrangian was shown earlier to have the form

$$A = \int_{\tau_1}^{\tau_2} \gamma \mathcal{L} d\tau \tag{7.123}$$

where $\gamma\mathcal{L}$ is Lorentz invariant.

We first construct the Lagrangian for a free particle, in the absence of fields. Since there are no forces, there can be no dependence on position coordinates. So, the Lagrangian must be a function of the velocity. Moreover, it must be Lorentz invariant, and so it must be a constant multiplied by $U_\alpha U^\alpha = -c^2$. So, in order for $\gamma\mathcal{L}$ to be invariant, $\mathcal{L}$ must be a constant multiple of $\frac{1}{\gamma}$. A function that works is

$$\mathcal{L} = -mc^2\sqrt{1 - \frac{u^2}{c^2}} \tag{7.124}$$

Substituting into the Euler-Lagrange equation

$$\frac{\partial \mathcal{L}}{\partial q} - \frac{d}{dt}\left(\frac{\partial \mathcal{L}}{\partial \dot{q}}\right) = 0 \tag{7.125}$$

we obtain the equation of a free particle, with constant relativistic momentum:

$$\frac{d}{dt}(\gamma m\mathbf{u}) = 0 \tag{7.126}$$

7.9.6.2 *Electron in an electromagnetic field*

We have seen that a relativistic Lagrangian must satisfy the condition that $\gamma\mathcal{L}$ is invariant. Thus, we seek a Lagrangian that is of the form $\frac{G}{\gamma}$ where G is an invariant function that includes the charge and the velocity of the electron as well as the scalar potential Φ and the vector potential $\mathbf{A}$. The simplest such function takes the form $eU_\alpha A^\alpha$ where U is the four-velocity and A the four-potential.

We will therefore try a Lagrangian of the form

$$\mathcal{L} = -mc^2\sqrt{1-\frac{u^2}{c^2}} - \frac{e}{\gamma c}U_\alpha A^\alpha \tag{7.127}$$

which written out explicitly becomes

$$\mathcal{L} = -mc^2\sqrt{1-\frac{u^2}{c^2}} - e\Phi + e\mathbf{u}\cdot\mathbf{A} \tag{7.128}$$

> **Exercise:**
> Show that Eq. (7.128) leads to the Lorentz force equation through an application of the Euler-Lagrange equation. The total time derivative is related to the partial derivative through $\frac{d}{dt} = \frac{\partial}{\partial t} + \mathbf{u}\cdot\nabla$.

7.9.6.3 *Hamiltonian formulation*

The Hamiltonian of a dynamical system is defined as

$$H = \sum_i P_i\dot{q}_i - \mathcal{L} \tag{7.129}$$

where P_i is the canonical momentum conjugate to the coordinate q_i, defined as

$$P_i = \frac{\partial\mathcal{L}}{\partial\dot{q}_i} \tag{7.130}$$

Taking the Lagrangian of Eq. (7.128), we obtain the canonical momentum coordinates as

$$P_i = \gamma m u_i + eA_i \tag{7.131}$$

Writing in vector notation, we express the relation between the canonical momentum $\mathbf{P}$ and the mechanical momentum (in terms of the mass of the particle) $\mathbf{p} = \gamma m\mathbf{u}$ as

$$\mathbf{P} = \mathbf{p} + e\mathbf{A} \tag{7.132}$$

The mechanical momentum $\mathbf{p}$ is related to the velocity $\mathbf{u}$ by the relativistic equation

$$\mathbf{p} = \frac{m\mathbf{u}}{\sqrt{1 - \frac{u^2}{c^2}}} \tag{7.133}$$

Inverting the variables we obtain

$$\mathbf{u} = \frac{\mathbf{p}c}{\sqrt{p^2 + m^2c^2}} \tag{7.134}$$

The Lagrangian formulation treated the position q_i and the corresponding velocity $\dot{q}_i$ as the independent variables. In Hamiltonian dynamics the variables are position and the conjugate canonical momentum. The velocity has to be expressed in terms of these other quantities. Since the conjugate variables are $\mathbf{P}$ and $\mathbf{r}$, we need to eliminate the velocity using the equation

$$\mathbf{u} = \frac{c(\mathbf{P} - e\mathbf{A})}{\sqrt{(\mathbf{P} - e\mathbf{A})^2 + m^2c^2}} \tag{7.135}$$

This will yield the relativistic Hamiltonian for a charged particle in a field:

$$H = \sqrt{c^2(\mathbf{P} - e\mathbf{A})^2 + m^2c^4} + e\Phi \tag{7.136}$$

Exercises:
1. Derive Eq. (7.136).
2. Using Eq. (7.136), and the Hamiltonian equations of motion ($\dot{P}_i = -\frac{\partial H}{\partial q_i}$, and $\dot{q}_i = \frac{\partial H}{\partial P_i}$), derive the Lorentz force equation.

7.9.6.4 *Gauge invariance*

The electric and magnetic fields are related to the potentials by the equations $\mathbf{E} = -\nabla\Phi - \frac{\partial \mathbf{A}}{\partial t}$ and $\mathbf{B} = \nabla \times \mathbf{A}$. These fields are invariant under a change of gauge of the potentials $\mathbf{A} \to \mathbf{A} + \nabla\chi$, and $\Phi \to \Phi - \frac{\partial \chi}{\partial t}$. A gauge transformation will affect the form of the Lagrangian of Eq. (7.128). However, it can be shown that a gauge transformation will not affect the action integral or the equations of motion.

Exercises:
1. Show that a gauge transformation does not affect the action integral of Eq. (7.123).
2. Show that a gauge transformation does not affect the equations of motion.

7.10 Abstract Covariant Formulation of the Lagrangian

For more general situations, it is useful to set up an abstract covariant formulation of the Lagrangian of the electromagnetic field.

The three-dimensional position vector **r** and the corresponding velocity vector **u** are replaced by the corresponding 4-vectors which we can express as x^α and the four-velocity U^α. The free-particle Hamiltonian of a particle of mass m $\left(= -mc^2\sqrt{1-\frac{u^2}{c^2}}\right)$ becomes, using the equality $U_\alpha U^\alpha = -c^2$

$$L_0 = -\frac{mc}{\gamma}\sqrt{-U_\alpha U^\alpha} \tag{7.137}$$

The action integral ($A = \int_{\tau_1}^{\tau_2} \gamma L d\tau$) becomes

$$A = -mc\int_{\tau_1}^{\tau_2} \sqrt{-U_\alpha U^\alpha} d\tau \tag{7.138}$$

The integrand can be rewritten as follows:

$$\sqrt{-U_\alpha U^\alpha} d\tau = \sqrt{\frac{dx_\alpha}{d\tau}\frac{dx^\alpha}{d\tau}} d\tau = \sqrt{g^{\alpha\beta} dx_\alpha dx_\beta}$$

Replacing the variable of integration τ by the parameter s that increases monotonically with τ (to ensure causality), the action integral can be expressed as

$$A = -mc\int_{s_1}^{s_2} \sqrt{g^{\alpha\beta}\frac{dx_\alpha}{ds}\frac{dx_\beta}{ds}}\, ds \tag{7.139}$$

The Lagrangian variables are now x^α and $\frac{dx^\alpha}{ds}$, where s is an arbitrary parameter. The subscripts and superscripts are significant for the time-component terms. So $x^0 = ct$, and $x_0 = -ct$, $\partial^0 = \frac{\partial}{\partial x_0} = -\frac{1}{c}\frac{\partial}{\partial t}$, and $\partial_0 = \frac{\partial}{\partial x^0} = \frac{1}{c}\frac{\partial}{\partial t}$.

Exercises:

1. Using the Lagrangian of Eq. (7.139) derive the Euler-Lagrange equation

$$mc\frac{d}{ds}\left[\frac{dx^\alpha/ds}{\left(\frac{dx_\beta}{ds}\frac{dx^\beta}{ds}\right)^{1/2}}\right] = 0$$

2. Show that the previous equation reduces to the free particle equation by setting the parameter $s = \tau$: $m\frac{d^2x^\alpha}{d\tau^2} = 0$.

We may next write the Lagrangian for the particle in an electromagnetic field as

$$\mathcal{L} = -mc\sqrt{g^{\alpha\beta}\frac{dx_\alpha}{d\tau}\frac{dx_\beta}{d\tau}} - \frac{e}{c}\frac{dx_\alpha}{d\tau}A^\alpha(x) \tag{7.140}$$

Exercises:
Use the Lagrangian of Eq. (7.140) for these two exercises:
1. The canonical momentum four-vector is defined as $P^\alpha = -\frac{\partial\mathcal{L}}{\partial(dx_\alpha/d\tau)}$. Show that $P^\alpha = mU^\alpha + \frac{e}{c}A^\alpha$.
2. Show that the Euler-Lagrange equations yield the covariant equations of motion $m\frac{d^2x^\alpha}{d\tau^2} = \frac{e}{c}\left(\partial^\alpha A^\beta - \partial^\beta A^\alpha\right)\frac{dx_\beta}{d\tau}$.

7.11 Maxwell's Equations from the Lagrangian

We will now consider the electromagnetic field itself as a system, and obtain a Lagrangian for this system. We want the action integral to be invariant under a Lorentz transformation, i.e. we want it to be a Lorentz scalar. And we want the Lagrangian itself to be also a Lorentz scalar. The reasons for this two-fold requirement will become clear presently.

The electromagnetic field is described by an electric and a magnetic field vector at every point in space. Space itself is a flat or Euclidean three-dimensional space that is not affected by the presence of charges or currents, unlike a gravitational field, which we will discuss in a later chapter.

We have seen that at every point in space there is a scalar potential Φ and a vector potential $\mathbf{A}$. The electric and magnetic fields can be written in terms of space and time derivatives of the potentials. This suggests that a Lagrangian of the electromagnetic field can be written as a functional of the functions $\phi(x)$ of the position, and the space and time derivatives of x which we shall express as $\partial^\alpha\phi(x)$. The latter is called a four-vector gradient of ϕ. Since the set of points in space is isomorphic to the set of real numbers, there is an uncountably infinite number of points in space. By dividing space into a large number of infinitesimal volume elements of size d^3x, we can divide space into a countably infinite number of volume elements. Each such volume element corresponds to a different degree of freedom of the system. One could also think of each such volume element as a particle belonging to the system. So, we define a Lagrangian density $\mathcal{L}$ such that the Lagrangian corresponding to each volume element is $\mathcal{L}d^3x$.

The total Lagrangian is obtained by adding these infinitesimal Lagrangians, i.e. by integrating over the entire region. Now, if there are multiple sources, there are also multiple potentials and corresponding fields. Considering these sources to be independent of one another, they would correspond to different dimensions, or different components. Thus, the most general Lagrangian is expressible as

$$L = \int \mathcal{L}(\phi_k, \partial^\alpha \phi_k) d^3x \tag{7.141}$$

The action integral becomes

$$A = \int \int \mathcal{L} d^3x dt = \int \mathcal{L} d^4x \tag{7.142}$$

The four-volume element d^4x is a Lorentz scalar, and so the Lorentz invariance of the action integral requires that the Lagrangian density $\mathcal{L}$ should also be relativistically invariant.

The time derivative is replaced by a four-vector derivative, with time and the three spatial coordinates as independent variables of differentiation. So the Euler-Lagrange equations become (using the abbreviation $\partial^\beta \equiv \frac{\partial}{\partial x_\beta}$)

$$\partial^\beta \frac{\partial \mathcal{L}}{\partial(\partial^\beta \phi_k)} = \frac{\partial \mathcal{L}}{\partial \phi_k} \tag{7.143}$$

In Section 7.9.4 we employed the analogy of the LC circuit to obtain an expression for the Lagrangian of an electromagnetic field in the absence of sources: $L = \frac{\epsilon_0}{2} \int (c^2B^2 - E^2)dV$. In Section 7.6 we derived the relationship

$$F_{\alpha\beta}F^{\alpha\beta} = 2(E^2 - c^2B^2)$$

Then, we may write the Lagrangian density of an electromagnetic field in the absence of sources as

$$\mathcal{L} = -\frac{\epsilon_0}{4} F_{\alpha\beta}F^{\alpha\beta} \tag{7.144}$$

Next, we include the sources. Here we can look to Eq. (7.127) of Section 7.9.6.2 for inspiration. There is a potential term of the form $-\frac{e}{\gamma c}U_\alpha A^\alpha$ which comes from the *effect* of the field on the electron. We can now invert this process and seek a *causal* term, one which expresses the effect of a source — i.e., a charge or a current — on the field. A logical procedure might be to replace the term eU_α/γ by the four-current J_α. (We do not bring in the γ factor, because our Lagrangian density is Lorentz invariant.)

Then, the additional term due to the sources becomes $-\frac{1}{c}J_\alpha A^\alpha$. And so, the Lagrangian density of the electromagnetic field becomes

$$\mathcal{L} = -\frac{\epsilon_0}{4}F_{\alpha\beta}F^{\alpha\beta} - \frac{1}{c}J_\alpha A^\alpha \tag{7.145}$$

The above equation can be written in terms of the potentials as

$$\mathcal{L} = -\frac{\epsilon_0}{4}g_{\lambda\mu}g_{\nu\sigma}(\partial^\mu A^\sigma - \partial^\sigma A^\mu)(\partial^\lambda A^\nu - \partial^\nu A^\lambda) - \frac{1}{c}J_\alpha A^\alpha \tag{7.146}$$

All the four Maxwell equations flow from this Lagrangian density of the electromagnetic field. First we note that it is not necessary to derive the homogeneous equations $\nabla \times \mathbf{E} = -\frac{\partial \mathbf{B}}{\partial t}$ and $\nabla \cdot \mathbf{B} = 0$ from the Lagrangian density. The homogeneous equations follow from the very definitions of the scalar and vector potentials $\mathbf{B} = -\nabla \times \mathbf{A}$ and $\mathbf{E} = -\nabla\Phi - \frac{\partial \mathbf{A}}{\partial t}$:

$$\nabla \times \mathbf{E} = \nabla \times \nabla\Phi - \nabla \times \frac{\partial \mathbf{A}}{\partial t} = -\frac{\partial \mathbf{B}}{\partial t} \quad \text{and} \quad \nabla \cdot \mathbf{B} = \nabla \cdot \nabla \times \mathbf{A} = 0$$

The Euler-Lagrange equations reduce to the equations of the electromagnetic field

$$\partial^\beta F_{\beta\alpha} = \frac{1}{\epsilon_0 c}J_\alpha \tag{7.147}$$

These are the inhomogeneous Maxwell equations $c^2\nabla \times \mathbf{B} = \frac{1}{\epsilon_0}\mathbf{J} + \frac{\partial \mathbf{E}}{\partial t}$ and $\nabla \cdot \mathbf{E} = \frac{\rho}{\epsilon_0}$.

Chapter 8

Microscopy of the Electromagnetic Field

8.1 Charges and Fields

The concept of an electromagnetic field arose in the context of understanding the mechanism of an interaction between two different charges placed at some spatial distance apart. Fields are generated by charges, and charges experience forces in fields.

The Maxwell equation $\nabla \cdot \mathbf{E} = \frac{\rho}{\epsilon_0}$ expresses the relationship between a charge source and the field generated by the charge. The Biot-Savart law expressible as $d\mathbf{B} = kI\frac{(d\mathbf{l}\times\mathbf{r})}{r^3}$ states the relationship between a current element and the magnetic field generated by it. Both these equations indicate that electromagnetic fields are generated by sources.

And these fields, in turn, are detected by the forces experienced by electric charges. The Lorentz force equation expresses the force applied by an electric and a magnetic field on a charge moving through the fields: $\mathbf{F} = q(\mathbf{E} + \mathbf{v} \times \mathbf{B})$.

Electric charges exist in nature as microscopic units, each unit being a positive or a negative value of magnitude $e = 1.602 \times 10^{-19}$ C. Since the electromagnetic field is generated by microscopic charges, and in turn the field itself acts upon microscopic charges, it is necessary to understand the microscopic nature of charges themselves, insofar as this is possible. In this chapter we will develop a conceptual picture of the microscopic realm of electromagnetism, and save the mathematical formalism — otherwise called quantum mechanics — for the next chapter.

8.2 Macroscopic and Microscopic Domains

The importance of the microscopic realm arose only in the 19th century, though the reality of the microscopic world had been recognized in antiquity. The Greek philosopher Democritus of Abdera had suggested around 430 BCE that matter is made of indivisible (*atomos* in Greek) particles. And because these particles are in motion in a gas such as air, the space in which they move must be empty. Thus the concept of a vacuum or empty space is an integral part of the atomic theory of matter. And as for these particles, today we call them *molecules*, and since the molecules can be split up into smaller chemically indivisible component particles, these component particles have come to be called *atoms* (*indivisible objects*) in modern science.

Experiments on gases at temperatures much higher than the condensation or boiling point and at pressures of the order of one atmosphere (about 10^5 Pa) enabled researchers to conclude that a gas at such ordinary pressures and temperatures — called an ideal or perfect gas — obeys an equation connecting pressure P, volume V and absolute temperature T:

$$PV = nRT \tag{8.1}$$

where R is the Universal Gas Constant, and n is the ratio of the mass of the gas expressed in grams to the molecular weight of the gas, and this ratio is also called the mole number of the gas. Avogadro hypothesized that one mole of any gas contains the same number of molecules. This number, which has come to be called Avogadro's number, written as N_A or N_0, expresses the connection between the macroscopic and the microscopic realms. Once this number was measured, temperature could be related to the mechanical energy of the molecules of the gas. Temperature and pressure belong to a class of quantities called intensive variables, because they are independent of the total mass of the object. Other intensive variables are density and specific heat. The opposite class of quantities, called extensive variables, include mass, number of moles, volume, and internal energy. The great size of Avogadro's number (6.02×10^{13}) enables the application of statistical mechanics to the treatment of matter in bulk.

The Universal Gas Constant R is a macroscopic quantity, and has the value 8.31 J/K mol^{-1}. The ratio $R/N_0 = k_B$ is called Boltzmann's constant, and being essentially an inverse of Avogadro's number it also provides a bridge between the macro and the micro worlds. Whereas Democritus had adduced evidence for the existence of invisible microscopic units of matter in the aromas that were carried by the wind from flowers to human nostrils, scientific proof of their existence could only come with direct observation and mathematical calculations. This meant that Avogadro's number (or Boltzmann's constant) had to be measured.

The discovery of the random zig-zag Brownian motion of particles such as pollen grains suspended in water provided irrefutable evidence that water consisted of microscopic particles in constant rapid, random motion. The erratic motion of the pollen grains was explained as the resultant motion due to the impulses imparted by collisions with the water molecules.

Using statistical mechanics, it can be shown that the expected displacement Δx of a spherical grain of pollen or other matter suspended in a liquid over a period of time t is given by

$$\Delta x^2 = 2Dt \tag{8.2}$$

where D is a constant called the Diffusion Coefficient of the particle.

Einstein obtained an expression for the diffusion coefficient in terms of the temperature T, the viscosity η of the medium, and the radius a of the suspended particle:

$$D = \frac{k_B T}{6\pi a \eta} \tag{8.3}$$

By measuring D and a microscopically, and by putting in the values of η and T, Boltzmann's constant could be determined. The present day value of k_B is 1.38×10^{-23} J/K. And thus Avogadro's number $N_A = 6.02 \times 10^{23}$ particles/mole.

Thus, the bridge between the macroscopic (pressure, volume, temperature, mass of a body) and the microscopic (mass of a molecule, average speed of a molecule) was achieved through the observation of the *mesoscopic* pollen grains which are considerably larger than molecules, and small enough to be invisible to the naked eye, but visible through a microscope. A parallel development occurred in the determination of the fundamental unit of electric charge e.

The charge to mass ratio e/m of an electron can be determined in a simple experiment. A stream of electrons produced by a source is sent through an electric field $\mathbf{E}$ perpendicular to their direction of motion. A magnetic field $\mathbf{B}$ perpendicular to both the electron velocity and the electric field

E generates a force on the electrons that is equal and opposite to that produced by the electric field, causing the electrons to move undeflected along a straight line. Thus

$$eE = evB \tag{8.4}$$

This yields $v = E/B$ and so the speed of the electron is determined. The electron beam then enters a region where the electric field is 0 and the magnetic field remains the same, **B**. Now the electron has a circular path of radius r, and experiences a centripetal force given by

$$\frac{mv^2}{r} = veB \tag{8.5}$$

Hence $v = erB/m$. Since $v = E/B$, we obtain

$$\frac{e}{m} = \frac{E}{rB^2} \tag{8.6}$$

There are also other methods of obtaining the charge to mass ratio of an electron. This ratio was first obtained by J. J. Thomson.

The actual charge or the actual mass could not be determined by such electromagnetic methods which employ Newtonian dynamics. But the charge of an electron was finally measured, again using a mesoscopic approach.

Avogadro's number — or the equivalent Boltzmann's constant — had been measured using the mesoscopic pollen grains, and now the electron charge was measured by R. A. Millikan using mesoscopic oil droplets suspended in air.

Millikan sprayed oil droplets into a space between two vertically separated flat plates at opposite electric potential, and used x-rays to ionize the air. Electrons that were freed by this ionization process attached themselves to the droplets, so that each droplet acquired a negative charge. Droplets could have charges of magnitude e, $2e$, $3e$, etc., depending on the number of electrons attached to the droplet. The electric field between the plates was directed downwards, and thus the charged droplets experienced an upward attraction which was opposite to the net downward force due to gravity minus buoyancy on each droplet. The strength of the electric field was adjusted until the droplets were suspended motionless.

The weight of a droplet of radius r and density $\rho_{oil} = \frac{4}{3}\pi r^3 \rho_{oil} g$.

The buoyant force on a droplet in air of density $\rho_{air} = \frac{4}{3}\pi r^3 \rho_{air} g$.

Electrostatic force on a droplet $= qE$.

Hence, applying the conditions for equilibrium, the charge on a droplet is obtained from the equation

$$q = \frac{4}{3E}\pi r^3 g(\rho_{oil} - \rho_{air}) \tag{8.7}$$

The radius of the droplet cannot be measured directly with a microscope. A different measurement is needed.

Stokes' equation for the force acting on a sphere moving slowly through a liquid or gas with speed v is given by

$$F = 6\pi\eta r v \tag{8.8}$$

where η is the viscosity of the fluid, and r the radius of the sphere. Now, if the sphere reaches terminal speed, that is, it moves downward with *constant* speed v, the net force on the sphere is zero. Then the resistive force due to viscosity is balanced by gravity and buoyancy:

$$6\pi\eta r v = \frac{4}{3}\pi r^3 \rho_{oil} g - \frac{4}{3}\pi r^3 \rho_{air} g \tag{8.9}$$

By measuring v using a microscope, the radius of the sphere can be obtained, and substituted into Eq. (8.7). Now, a droplet could acquire one, two or more electrons. When this experiment is repeated multiple times, one gets the following values for the magnitude of the charge of a droplet: $q = e, 2e, 3e, \ldots.$ Thus the electron charge e is determined. Knowing the experimentally established value of e/m, the electron mass is obtained. The electron constants are: charge $= 1.60 \times 10^{-19}$ C; mass $= 9.11 \times 10^{-31}$ kg.

8.3 Frequency and Energy of an Electromagnetic Wave

We now review some results we obtained in Chapter 6 (Energy and Momentum of Fields). Electromagnetic energy is propagated through space via electromagnetic waves, at speed c. Light is a form of electromagnetic radiation. The rate at which energy crosses unit area perpendicular to the radiation is given by the Poynting Vector $\mathbf{S} = \epsilon_0 c^2 \mathbf{E} \times \mathbf{B}$. The amount of energy in a cylinder of length c and unit cross-section is $\epsilon_0 c^2 |\mathbf{E} \times \mathbf{B}|$ and the momentum carried by this quantity of energy has magnitude $\epsilon_0 c |\mathbf{E} \times \mathbf{B}|$. So, if E is the energy of the field within a given volume, and p the magnitude of the momentum of that amount of energy, $E = pc$, and so $E^2 = p^2 c^2$.

Comparing this to the relativistic energy-momentum equation for a material object $E^2 = p^2c^2 + m^2c^4$, we can think of a traveling electromagnetic field as a substance constituted of units or bodies having energy E and momentum p, but with rest mass 0. And since the rest mass of a volume of electromagnetic energy is zero, we should **not** think of the energy density of a "static" electromagnetic field $\frac{1}{2}\epsilon_0(E^2 + c^2B^2)$ as a sort of rest energy analogous to the rest energy mc^2 of a material object.

In the foregoing paragraph we discussed the concept of energy per unit volume or energy density of a field, in analogy with the mass density of a body. But the concept of density is macroscopic, and breaks down at the atomic level. Matter is composed of molecules, and charge is composed of electrons and protons. Neither mass nor charge is truly continuous. A question can be raised concerning electromagnetic radiation. Is electromagnetic energy continuously divisible, or is there a graininess or atomicity to this energy? We shall discuss this question in what follows.

In Chapter 7 (Special Relativity and Electromagnetism) we derived the equations for the Doppler effect whereby the frequency of light measured by an observer is different from the frequency emitted by the source, when there is relative motion between the source and observer. So, for a source receding from the observer we got the equation:

$$f_o = \frac{f_s}{\sqrt{1 - \frac{u^2}{c^2}}}\left(1 - \frac{u}{c}\right) \tag{8.10}$$

In the discussion that followed we argued that the momentum, and hence the energy, of a region of space carrying electromagnetic radiation, is proportional to the frequency of the radiation.

In classical electromagnetism the energy density of electromagnetic radiation is proportional to the product of the electric and the magnetic fields of the radiation. Thus the energy density is proportional to the square of the amplitude of the electric field, and hence it is proportional to the intensity of the radiation. Yet here we see that the energy density is apparently also proportional to the *frequency* of the radiation, a quantity independent of the field strength. We suggested therefore that electromagnetic radiation may be granular in nature, since its energy is dependent on both an extensive property — its intensity, which can be raised by increasing the power supplied to the generating source — and an intensive property, which is the frequency. Such duality is the property of matter consisting of a large number of units whose aggregate behavior is described by statistical mechanics.

These properties of electromagnetic energy enable us to postulate that electromagnetic radiation exists as units of energy, with each unit being proportional to the frequency of the radiation. And the total amount of energy within a given volume of space is proportional to the number of such units within that space, where this number is proportional to the strengths of the electric and the magnetic field (i.e. intensity). Writing this hypothesis in algebraic notation, if E is the energy of a unit of radiation, then $E = hf$ where f is the frequency and h is a constant of proportionality. The total energy within a volume is therefore nhf where the number of energy units n is proportional to the intensity of the field.

If the energy of a unit is $E = hf$, then the momentum of a unit is $p = hf/c$. We will now show that the Doppler formula of Eq. (8.10) is indeed the Lorentz transformation equation for the energy of a body as measured in two different inertial frames of reference with a relative speed u along their x axis. Multiplying both sides by h, and making the identification $E = hf$, and using the symbol $\gamma = \frac{1}{\sqrt{1-\frac{u^2}{c^2}}}$ we obtain

$$E' = (E - up)\gamma \tag{8.11}$$

And multiplying both sides of Eq. (8.10) by h/c, we obtain

$$p' = (p - uE/c^2)\gamma \tag{8.12}$$

Equations (8.11) and (8.12) describe the Lorentz transformation of an object having energy E and momentum p. It can be readily verified that $E'^2 - c^2p'^2 = E^2 - c^2p^2$.

Thus the packets or units of electromagnetic energy obey the Lorentz transformation equations. And so we may treat these units of energy as material particles having energy hf where f is the frequency of the radiation and h a constant.

8.4 Determination of the Value of *h*

The constant h is called Planck's Constant, because this number was first introduced and calculated by Planck in his solution to a thermodynamic problem involving standing electromagnetic waves inside a hollow container. This is an important historical development, and we will take it up presently, but for now we continue along the conceptual path of treating the units of electromagnetic energy as particles having energy and momentum.

If the units of electromagnetic energy behave like particles with energy hf and momentum hf/c, then they could exchange momentum and energy with other particles through collisions. This means that a unit of radiation could collide with an electron and the two particles could exchange energy and momentum in that process.

We will now replace the phrase "unit of electromagnetic energy" by a more economical term. A convenient word is "photon" which means something like "a light particle" though it is used of a unit of any sort of electromagnetic radiation, not limited to the visible spectrum. Experimental evidence for the existence of photons carrying momentum and energy was provided by the phenomenon called the Compton Effect.

Suppose a photon of frequency f and wavelength λ_i were to collide with a stationary electron of mass m. Let us say this is a "glancing" collision, so that the photon is scattered at an angle ϕ relative to its initial direction, and the electron travels at some speed in a different direction as shown below:

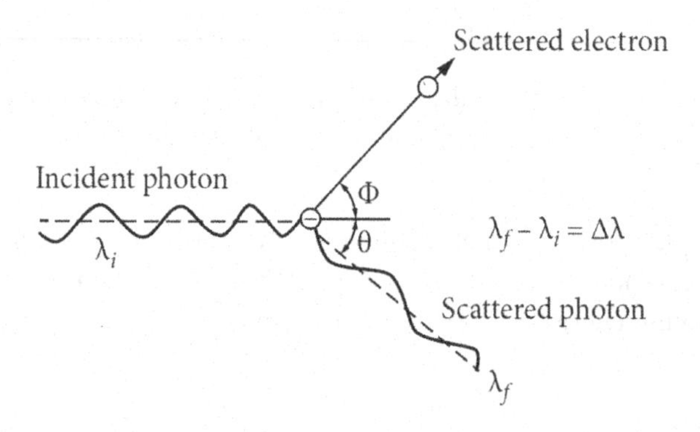

Let θ be the angle between the direction of the scattered photon and that of the incident photon. λ_i is the wavelength of the incident photon, and let λ_f be the wavelength of the scattered photon. The energy and momentum of the incident photon are hc/λ_i and h/λ_i respectively, and those of the scattered photon are hc/λ_f and h/λ_f.

Let P represent the energy-momentum four-vector of the incident photon, and Q the initial energy-momentum four-vector of the electron. Let R represent the energy-momentum four-vector of the scattered photon, and let S represent the final energy-momentum four-vector of the electron.

Conservation of energy and momentum in the collision is represented by the four-vector equation

$$P_\mu + Q_\mu = R_\mu + S_\mu \tag{8.13}$$

For the electron, $Q_\mu Q^\mu = S_\mu S^\mu = -E^2/c^2 + p^2 = -m^2c^2$, which is an invariant or Lorentz scalar. And for the photon $P_\mu P^\mu = R_\mu R^\mu = 0$. One can also form invariant inner products. So $P_\mu Q^\mu = R_\mu S^\mu$.

We now take the inner product of both sides of Eq. (8.13) with R^μ:

$$P_\mu R^\mu + Q_\mu R^\mu = R_\mu R^\mu + S_\mu R^\mu \tag{8.14}$$

Now, the inner product of two energy-momentum four vectors labeled 1 and 2 is $-E_1E_2/c^2 + p_1p_2\cos\theta$ where θ is the angle between the momenta. So,

$$P_\mu R^\mu = -\frac{h}{\lambda_i}\frac{h}{\lambda_f} + \frac{h}{\lambda_i}\frac{h}{\lambda_f}\cos\theta$$

$Q_\mu R^\mu = -hmc/\lambda_f$, $R_\mu R^\mu = 0$, and $S_\mu R^\mu = Q_\mu P^\mu = -mhc/\lambda_i$.

Substituting into Eq. (8.14), we obtain

$$-\frac{h^2}{\lambda_i\lambda_f}(1-\cos\theta) - \frac{hmc}{\lambda_f} = 0 - \frac{hmc}{\lambda_i}$$

Hence, we obtain the equation

$$\lambda_f - \lambda_i = \frac{h}{mc}(1-\cos\theta) \tag{8.15}$$

The wavelength of a photon is the wavelength of the radiation of which the photons are the individual units. This wavelength can be measured by spectroscopy. The angle between the initial and final directions of the radiation can also be measured. Hence h can be measured. $h = 6.63 \times 10^{-34}$ J.s.

Historically, the Compton Effect described above was discovered after the value of h had been determined by other means, and it served to confirm that electromagnetic radiation is composed of individual units which came to be called photons, with energy hf and momentum $hf/c = h/\lambda$.

The notion that electromagnetic radiation exists in discrete units first arose in the context of statistical mechanics, which was created to provide microscopic explanations for the laws of thermodynamics.

8.5 Thermodynamics

The study of thermodynamics is classified under four laws, which are briefly explained below.

Zeroth Law:

The 0th law states that if two bodies are separately in thermal equilibrium with a third body, then they must be in thermal equilibrium with one another. By thermal equilibrium we mean simply that no heat flows between the two bodies when they are placed in contact with each other. The zeroth law enables the definition of temperature. Two bodies that are in thermal equilibrium with one another have the same temperature. So if body A and body B have the same temperature, and body A and body C have the same temperature, then by the zeroth law body B and body C must have the same temperature. This means we can define temperature as an absolute quantity, regardless of the nature of the body that has the temperature. Temperature is also an intensive variable, since it is independent of mass or volume.

First Law:

The 1st law is an expression of the conservation of energy. It states that when heat energy is given to a body, part of it goes to raise the temperature, and hence the internal energy of the body increases, and the rest goes to do external work, which is done by the expansion of the body. Since this is a law of conservation of energy, it can also be applied to a case when work is done on a gas by compressing it. In this case the gas gets heated, and may yield some of its heat to the surroundings. No energy is created, and no energy is destroyed.

If the amount of heat energy supplied to a body is ΔQ, the rise of internal energy of the body ΔU, and the external work done by the body ΔW, then the first law can be expressed as

$$\Delta Q = \Delta U + \Delta W$$

The internal energy U of a body is an extensive variable.

Second Law:

The 2nd law is a statement of irreversibility. It can be stated in many different ways. The simplest statement is that heat always flows naturally from a hotter to a cooler body.

Both the first and the second law deal with the conversion of heat into work and work into heat. The first law tells us that heat and work are different forms of energy and that one can be converted to the other. The second law places restrictions on the conversion of heat energy into work. Heat and work (mechanical energy) are not fully reversible. Whereas mechanical energy can be converted entirely into heat energy, the reverse cannot take place. When brakes are applied to a moving car, the kinetic energy of the car is converted to heat energy in the wheels and the road. But the heat energy that is so generated cannot be converted back to obtain the kinetic energy that was lost by the car.

Third Law:

The 3rd law states that it is impossible to cool a body right down to absolute zero (0 K) in any finite number of steps, even though it is theoretically possible to come closer to this temperature with each step.

8.6 Statistical Mechanics

All the laws of thermodynamics can be explained by the atomic or molecular theory of matter. Heat is a *manifestation* of the internal energy of a body — the sum of the kinetic and potential energies of all the molecules. Here the forces that give rise to the potential energy are due to intermolecular attraction and repulsion. The kinetic energy is due to the motion of the molecules. The phenomenon of Brownian motion showed that the motion of the molecules in a liquid is erratic and random. The molecules move in all possible directions with a range of velocities that change in magnitude and direction each time a molecule collides with another or with the walls of the container — or more precisely with the atoms or molecules of the solid container. Thus it is futile to try and follow the movements of any one molecule. The best we can do is to investigate the overall aggregate or statistical behavior of these molecules. The study of the behavior of matter in terms of the collective motion of the molecules is therefore called *statistical mechanics.*

The temperature of a body is a measure of the average kinetic energy of the molecules of the body. If two bodies are in thermal contact with each other, there will be a transfer of kinetic energy from the molecules of one body to the molecules of the other body, until both bodies have the same average kinetic energy of their molecules. This means they will have the same temperature. This is the explanation for the zeroth law.

Consider a gas contained in a cylinder enclosed by a piston. If some heat is supplied to the gas, its temperature will increase, and so the molecules will have greater kinetic energy. This greater kinetic energy will mean that the molecules will pound on the piston with greater force, causing the piston to move outwards. So the gas expands and the force of this expanding gas does work on the piston. This is an illustration of the first law.

The second law merits an extensive and detailed statement. We will take up a discussion of the second law in the following section.

The third law can be explained by an analogy. Suppose a moving sphere A collides head-on with an identical stationary sphere B. By the laws of conservation of energy and momentum, the first sphere will stop, and the second sphere will move with the same velocity possessed by the first sphere before the collision. It is important to note that the first sphere will not stop if the second had *any velocity at all.* Suppose A is a molecule of a gas that we are trying to cool to absolute zero. At absolute zero the kinetic energy of the molecules is zero according to classical physics. So, if we want to reduce the temperature of a gas to zero, we must place it in contact with a gas that is *already at absolute zero, with stationary molecules.* Unless we can find such a gas somewhere in the universe — which is impossible, considering that the universe is constantly cooling down from a very hot initial state — it is impossible to reduce the temperature of any gas to absolute zero. (Our argument here is classical. The argument becomes stronger in quantum mechanics where it is impossible for the molecules of a gas to have zero energy no matter how low is the temperature of the gas.)

8.6.1 *One-dimensional Gas*

A helium molecule, which is a single atom, can be treated as a rigid sphere. Diatomic molecules such as hydrogen and oxygen have a dumbbell shape. Molecules with three or more atoms have more complex shapes. We now limit our discussion to the simplest kind of gas — one consisting of identical monatomic molecules such as Helium, Argon or Neon. Each of these molecules can be modeled as a tiny rigid sphere.

Suppose all these rigid spheres are lined up along their line of centers, i.e. like a one-dimensional array of identical billiard balls.

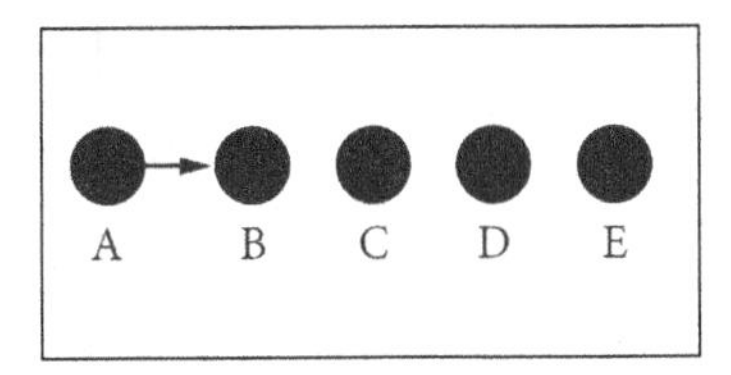

And let us say that this array of balls is suspended within a zero gravity box. If the sphere A at the far left were set in motion towards the right, it would collide with the next one, which in turn would collide with the sphere next to it, and so on till the last sphere E moves forward, hits the wall of the container, bounces back, hits the previous ball D, which in turn hits the ball C behind it, and so on until the ball A moves to the left wall, bounces back, hits the ball B, and the process continues indefinitely. As long as all the collisions are elastic, i.e. with no loss of kinetic energy at each collision, the process will be repeated forever. Moreover, again assuming the collisions to be elastic, the process is also time reversible. If we were to record the motion of the balls for a period of time and play the film backwards it would be impossible to find any essential difference between the forward time and backward time sequences of motion. What we have just described is a model of a one-dimensional gas, and a one-dimensional gas in a gravity-free environment is a time reversible system.

A *degree of freedom* is a particular way in which a molecule is free to move. And because a molecule in this scenario can execute only one kind of motion, which is to move along a straight line, such a molecule has a single degree of freedom.

8.6.2 *Equipartition of Energy*

Next we consider a two-dimensional gas. Again, we consider a container in a gravity-free environment. Here the balls are floating at different points but their centers are all in the same plane. This time, the spheres are not all aligned in straight lines. Now, if one ball were given a push in any direction, it would hit another, which would hit another, and so on, but these collisions would not necessarily be head on. As a result, the directions of the motion of the spheres will change, as well as their speeds.

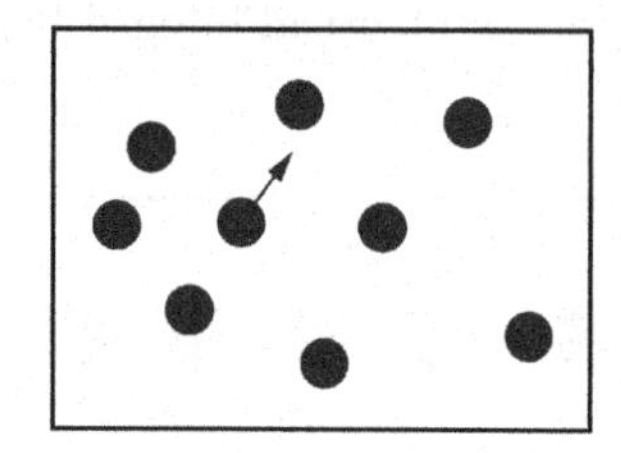

These collisions would be random. Eventually, the balls would be moving haphazardly in all directions, while remaining in the same two-dimensional plane. But the kinetic energy of the balls has now been distributed evenly along two dimensions. The random statistical nature of the motions ensures that the average kinetic energy due to the motion in any one direction equals the average kinetic energy due to motion in any other direction. Every two-dimensional motion can be resolved into motion in two mutually perpendicular directions, and we call each such perpendicular direction a *degree of freedom.* So a monatomic molecule that is capable of moving in two dimensions has 2 degrees of freedom. Each degree of freedom has the same average kinetic energy. The average energy per molecule has been divided equally between its two degrees of freedom. Clearly, the dynamics of this two-dimensional gas are not time reversible. In forward time the energy gets distributed evenly between the two degrees of freedom. One does not observe the reverse happening in nature. Thus a two-dimensional gas is an irreversible system.

The dynamics of a three-dimensional gas are not much different from those of a two-dimensional one. In an actual physical gas the molecules move rapidly like tiny bullets, and so gravity does not play a perceptible rôle in their motion. As they move within the space of the container — which we shall take to be a cube — they collide with the walls of the cube and change direction with each collision. They also collide with one another. With each collision the molecules abruptly change velocity and exchange kinetic energy. Since the collisions are haphazard, at every collision each molecule undergoes a random change of momentum and energy. With about 10^{24} molecules to deal with, we can only use statistics to analyze the motions of the molecules comprising the gas. When the molecules are moving in the most random fashion, the kinetic energy gets distributed equally among the three degrees of freedom of the three-dimensional gas, which is an example of the Principle of Equipartition of Energy.

8.6.3 *Second Law of Thermodynamics*

Because heat cannot flow from a cold body to a hot body by itself, the Second Law of Thermodynamics provides a unique arrow of time. Suppose an ice cube were placed in a glass of warm water. A video recording will show the ice melting as it receives heat from the water. If the video were played backwards it would show a tiny piece of ice gradually becoming bigger until it acquired the shape of a cube floating on the warm water. It is evident that this sort of time reversal cannot occur in nature. The flow of time is like the flow of heat. It cannot be reversed. If we have a large number of microscopic particles whose motions are random, no matter how orderly they are arranged in the beginning, once the system is set in motion, the random collisions will create a disorder from which the original order can never be retrieved. This has some important consequences.

One consequence is the diminishing of available energy. An array of molecules all moving together can apply a concerted force which can therefore do a lot of work on an object and thereby impart a corresponding energy to the object. But if the molecules are moving haphazardly, the force they can exert together is considerably less, and so the amount of energy that can be provided is less. Thus, in an irreversible process there is a decrease in the amount of available energy. So the Second Law can also be stated as: *Natural processes always occur in such a way that the amount of available energy decreases.*

Another consequence is the collapse of orderliness. An array of molecules all moving with the same velocity parallel to each other is a highly orderly system. But as the system is left to itself, the degree of orderliness will gradually diminish until there is total randomness. So the 2nd Law can also be stated thus: *Natural processes will always take place in such a way that there is a loss of order.*

Disorderliness is also called *entropy.* So another formulation of the 2nd Law: *Natural processes occur in such a way that the overall entropy increases.*

Yet another consequence is the loss of information. We could create different arrays of molecules which are all orderly, but not identical with each other. Let us say we have two boxes with the same number of molecules all moving parallel to each other. In one box we divide the molecules into two parallel arrays with a gap between them. In the second box we have the same number of molecules, all parallel to each other, but without a gap. We could label the first box **0** and the second box **1**. The distinction between the two boxes allows us to store *information.* The simplest information is binary — yes or no. We could agree that **0** means yes and **1** means no, or vice versa. Now, suppose we allow both the boxes to stand for a while. After some time all the molecules in both boxes will be moving at random, and the gap between the molecules in the first box will vanish. And so the distinction between the two boxes has disappeared. We can no longer tell which is **0** and which is **1**. The information is lost. So the 2nd Law can be stated thus: *Natural processes tend to destroy information.*

The example of the two boxes in the preceding paragraph offers an illustration of the principle of thermodynamic equilibrium. The two boxes were initially in thermodynamically unstable states. But as time passed, each of them underwent a development towards greater disorder and eventually reached thermodynamic equilibrium. This equilibrium is reached when the energy of the molecules is distributed evenly among all the different degrees of freedom. An enclosed system will reach equilibrium when its molecules achieve maximum randomness, so that the degree of disorderliness has reached its highest possible value. When this happens, energy is distributed equally among all the particles and degrees of freedom. Of course, due to the randomness of the motion of the molecules, an individual molecule will undergo a constant change of energy, but over a period of time its average kinetic energy will be the same as the average kinetic energy of every other molecule in the system. Now, if the molecules are diatomic (and rigid), in addition to translation, they will also be able to rotate, and since there are two mutually perpendicular axes about which the rotations can be resolved, each molecule has two rotational degrees of freedom. (A diatomic molecule does not have energy of rotation about an axis passing through the component atoms.)

8.7 Entropy and Temperature

The 0th and the 2nd laws of thermodynamics enable us to define temperature as a variable. If two bodies are in thermal equilibrium, they have the same temperature. And heat flows spontaneously from a body at higher temperature to a body at lower temperature when these bodies are placed in thermal contact. Temperature can also be defined in terms of entropy.

A body reaches thermodynamic equilibrium when its entropy reaches the highest possible value, when the energy is distributed over the maximum number of microscopic components. Suppose we label the number of microscopic components which share the internal energy of the body as Ω. The entropy S is clearly a monotonically increasing function of Ω. This relationship between entropy and the number of microscopic components enables us to define temperature in a different way.

Consider a gas consisting of a large number of molecules. For simplicity we shall consider a gravity-free environment, such as a laboratory in outer space. Let us say initially all the molecules are absolutely stationary, floating at equidistant fixed points in a three-dimensional lattice or grid. Let us say there is no force between the molecules. So, in this state of total suspended animation the internal energy of the gas is zero. The number of ways in which energy is distributed among the molecules, which is a measure of Ω, the number of different components of energy, is 1. Now, suppose some energy were given to the gas. This energy would set the molecules in motion. For a small input of energy only certain modes of the motion would be activated. For example, suppose the gas is contained in a cubical box, one wall of which is a movable piston. Suppose now a small amount of energy is given to the gas by compressing the piston a little bit, thereby imparting momentum to the molecules in the layer closest to the piston. These molecules would move forward and share some of their energy with the next layer of molecules, and this process would continue, but the effect would weaken with each successive layer. In this process a large number of molecules would be set in motion, but there would be even more molecules that remain at rest, because the energy has been dissipated before reaching them. So, a small input of energy sets in motion a small number of components or modes of motion. As more energy is inputted, more components of motion are activated. Thus, the number of components of motion Ω is a function of the internal energy E of the gas.

Suppose we have two boxes A and B having volumes V_1 and V_2, with numbers of particles N_1 and N_2, and internal energies E_1 and E_2. If we keep the volume and the number of particles constant for both boxes, the number of components Ω in each box will be a function of the only variable, which is the energy E. So, the number of components in A can be written as $\Omega_1(E_1)$ and that in B as $\Omega_2(E_2)$. Now, suppose the two boxes are placed in thermal contact with each other. Heat will flow from one to the other, until both reach the same temperature. Equilibrium is reached when the compound $A-B$ system has the maximum number of microscopic components. Let $\Omega^{(0)}$ denote the number of components of the AB system, which we enumerate as follows: For each component of A there are Ω_2 components of B, and for each component of B there are Ω_1 components of A, and so, the total number of components of the $A-B$ system is the product of the components of A and B. Hence, we can write

$$\Omega^{(0)}(E_1, E_2) = \Omega_1(E_1)\Omega_2(E_2) \tag{8.16}$$

Now, E_1 and E_2 will vary as long as heat flows between the bodies, but the total $E = E_1 + E_2$ will remain constant. Since E_1 and E_2 are not independent, we can write $\Omega^{(0)}$ as a function of E_1 alone. So

$$\Omega^{(0)}(E_1) = \Omega_1(E_1)\Omega_2(E_2) \tag{8.17}$$

Heat will flow from one body to the other until equilibrium is reached. Let the equilibrium heat of each body be $\bar{E}_1$ and $\bar{E}_2$, so that $\bar{E}_1 + \bar{E}_2 = E$. At equilibrium, the total number of components of the $A-B$ system will be a maximum, and so we can write

$$\frac{\partial\Omega^{(0)}(E_1)}{\partial E_1} = \frac{\partial\Omega_1(E_1)}{\partial E_1}\Omega_2(E_2) + \frac{\partial\Omega_2(E_2)}{\partial E_2}\Omega_1(E_1)\frac{\partial E_2}{\partial E_1} = 0 \tag{8.18}$$

Now, $E_1 + E_2 = E$, a constant, and so $\frac{\partial E_2}{\partial E_1} = -1$. Hence

$$\frac{1}{\Omega_1(E_1)}\frac{\partial\Omega_1(E_1)}{\partial E_1}\bigg|_{E_1=\bar{E}_1} = \frac{1}{\Omega_2(E_2)}\frac{\partial\Omega_2(E_2)}{\partial E_2}\bigg|_{E_2=\hat{E}_2} \tag{8.19}$$

This equality holds at thermal equilibrium between the two bodies, when both bodies have the same temperature. Thus, the quantity on both sides cannot depend on the extensive variables $\bar{E}_1$ and $\bar{E}_2$. It is a function solely of the final equilibrium temperature, an intensive variable. So we write

$$\frac{d\Omega}{\Omega} = f(T)dE \tag{8.20}$$

Or,

$$d(\ln\Omega) = f(T)dE \tag{8.21}$$

In thermodynamics the entropy S of a body is defined by

$$dS = \frac{dE}{T} \tag{8.22}$$

Here temperature is a macroscopic thermodynamic quantity, and dS is the infinitesimal increase of entropy of a body at temperature T which receives an infinitesimal amount of heat energy dE. A comparison of the two equations enables us to define entropy in terms of the number of microscopic components via $dS = k_B d(\ln\Omega)$, and temperature in terms of Ω and E, by writing $\frac{1}{f(T)} = kT$. Here k_B is Boltzmann's constant, also written as k_0 or as k, that relates the macroscopic temperature to the microscopic variable f. And so, we write

$$d(\ln\Omega) = \frac{dE}{k_B T} \tag{8.23}$$

Integrating, we get

$$\ln\Omega = \frac{E}{k_B T} + c \tag{8.24}$$

$$\Omega = Ae^{\frac{E}{k_B T}} \tag{8.25}$$

Ω is the number of different possible microscopic ways that the body can exist at any point in time. This number is also called the number of microstates. The probability of finding the body in any one of the states is inversely proportional to $\frac{1}{\Omega}$. Thus, the probability of finding a body in one of these states is proportional to $e^{-\frac{E}{k_B T}}$. So the probability of occurrence of a particular state is related to the energy of that state. If there are a total of N_0 possible states, the number N_i of states having a particular energy E_i is given by

$$N_i = N_0 e^{-\frac{E_i}{k_B T}} \tag{8.26}$$

Another relation we obtained is

$$dS = k_B d(\ln \Omega) \tag{8.27}$$

which, upon integration, yields $S = k_B \ln \Omega + c$.

Now, entropy increases with disorder. If there is only one state, there is no disorder, and the entropy is 0. Setting $\Omega = 1$, and $S = 0$, we find that $c = 0$. So

$$S = k_B \ln \Omega \tag{8.28}$$

We will next apply the basic formula of thermodynamics:

$$dE = TdS - PdV + \mu dN \tag{8.29}$$

where μ is the chemical potential, and the other quantities have their usual meanings. Taking the partial derivative,

$$\left(\frac{\partial S}{\partial V}\right)_{N,E} = \frac{P}{T} \tag{8.30}$$

So

$$\frac{k_B}{\Omega}\frac{d\Omega}{dV}\bigg|_{N.E} = \frac{P}{T} \tag{8.31}$$

We will now apply this equation to an ideal gas, consisting of non-interacting distinct particles. Both the volume V of the gas and the number of particles N within the volume are extensive variables. At thermodynamic equilibrium the number of microstates per unit volume is constant throughout the gas. So, given any molecule, the total number of distinct states Ω is proportional to V. And for N molecules the number of distinct microstates is proportional to V^N.

So, we obtain

$$PV = Nk_BT \tag{8.32}$$

This is the perfect or ideal gas equation for a gas containing N molecules. From the kinetic theory of gases, we obtain the average kinetic energy of a molecule in a monatomic ideal gas as $\frac{3}{2}k_BT$ and so the average energy per degree of freedom is $\frac{1}{2}k_BT$. For diatomic molecules there are more degrees of freedom. By the principle of equipartition of energy each degree of freedom has an average energy of $\frac{1}{2}k_BT$.

8.8 Electromagnetic Radiation Gas

When light is scattered by matter, the process can be described as an absorption of the light followed by an emission. When it gets reflected off a mirror or other shiny metal surface the direction of the waves changes, but the coherence of the absorbed waves is maintained in the emitted waves. It is a different case with a surface such as soil, wood, or other solid material. There is no coherence between the absorbed and the emitted waves. Suppose we have a hollow cavity made of walls that have the property that they absorb electromagnetic radiation of all frequencies, and also emit radiation of all frequencies in an incoherent manner. So, if there is some radiation introduced into the cavity, this radiation will undergo absorption by the walls, and each absorption will be followed by an emission of radiation at possibly a different frequency in a direction that is uncorrelated to the absorbed radiation.

A surface that absorbs radiation of all frequencies — and emits at all frequencies — is called a *black-body.* So, if some radiation is introduced into such a cavity containing black-body walls, this radiation will exchange energy with the molecules of the walls. Some radiation will be absorbed by the walls, and some radiation emitted by the walls. Such a system is akin to a gas, except that unlike the molecules of a gas, where the molecules collide with one another as well as with the walls, in the case of a radiation gas the waves do not interact with one another, but only with the walls.

If allowed to stand for some time, there will be thermodynamic equilibrium between the radiation and the walls. The radiation inside the cavity can then be called stationary waves or standing waves. Such radiation is called *black-body radiation.*

At equilibrium, there will be a range of frequencies of the radiation inside the cavity. Just as in a standing wave in a stretched string the energy of the different modes is different, so that the energy is unequally distributed among the different harmonics, so, too, the energy of the radiation gas is not uniformly distributed among all the frequencies.

Let $u(\omega)d\omega$ be the amount of energy possessed by radiation of angular frequencies ranging from ω to $\omega + d\omega$ within the cavity. The function $u(\omega)$ is called the equilibrium radiation density. According to Kirchhoff's First Law of Radiation, this function is independent of the nature of the material objects inside the cavity, and depends only on the temperature of the cavity. So we could write $u = u(\omega, T)$. The total energy within the cavity would then be $U = \int_0^\infty u(\omega, T) d\omega$.

We can obtain an expression for this energy in terms of the dimensions of the cavity, by treating the equilibrium radiation as standing waves. For simplifying the analysis we first assume that the cavity is shaped like a cube of side L, with each side parallel to one of the Cartesian coordinate axes.

Considering a standing wave parallel to one of the sides, the condition for standing waves is that $2L = n\lambda = \frac{2\pi n}{k}$ where k is the wave number. We could write a similar condition for each one of the three mutually perpendicular directions. For each direction we would get an independent number n, so that for integers n_x, n_y and n_z we could write

$$k_x L = \pi n_x, \; k_y L = \pi n_y, \; k_z L = \pi n_z \tag{8.33}$$

The number dN of waves whose wave numbers lie between k_x and $k_x + dk_x$, k_y and $k_y + dk_y$, k_z and $k_z + dk_z$ is equal to the number of integers in the interval $(n_x, n_x + dn_x), (n_y, n_y + dn_y)$, and $(n_z, n_z + dn_z)$. Now, electromagnetic waves, unlike sound waves, are transverse, and have two degrees of freedom as regards their polarization. Thus, the total number of different waves has to be doubled. Hence

$$dN = 2 dn_x dn_y dn_z = 2(L/\pi)^3 dk_x dk_y dk_z \tag{8.34}$$

This analysis can be simplified by transforming to an abstract phase space in which k_x, k_y and k_z are treated as Cartesian coordinates, with (k, θ, φ) as the corresponding spherical polar coordinates. But k_x, k_y, k_z are positive, and so the volume element in spherical coordinates is the spherical shell having inner radius k and outer radius $k + dk$ confined to the positive octant, hence having volume $\frac{4\pi k^2 dk}{8}$. The actual number of waves therefore becomes

$$dN = 2(L/\pi)^3 (1/8) 4\pi k^2 dk = (L/\pi)^3 \pi k^2 dk \tag{8.35}$$

The wave number k of a wave is related to its frequency ω by $k = \omega/c$.[1] And so, the number of waves per unit volume in the cavity is

$$dN/L^3 = \frac{\omega^2}{\pi^2 c^3} d\omega \tag{8.36}$$

[1]The symbol ω is called the angular frequency, and is related to the frequency f by $\omega = 2\pi f$. We will occasionally drop the adjective "angular" and refer to ω simply as the frequency.

Since we have replaced the wave number k by the frequency ω, the symbol k has been freed up, so that we can use it for Boltzmann's constant, and drop the subscript from k_B and write k instead.

Now, when the radiation is in equilibrium with the walls, the fraction of the waves having a particular energy drops off exponentially with the energy according to the rule (cf. Eq. (8.26))

$$\Delta N = Ne^{-\frac{\mathcal{E}}{kT}} \tag{8.37}$$

So, the average energy per oscillation mode having frequency ω becomes

$$\langle \mathcal{E} \rangle = \mathcal{E}(\omega)\Delta N/N = \mathcal{E}(\omega)e^{-\mathcal{E}/kT} \tag{8.38}$$

where $\mathcal{E}(\omega)$ is the energy of a mode of frequency ω.

8.8.1 *Transition to Quantum Theory*

The fundamental axioms of the quantum theory of radiation are the following:

1. The energy of radiation is proportional to the frequency of the radiation.

2. The energy of radiation is proportional to the number of individual units or quanta of radiation of that frequency.

These two axioms can be combined into one equation: $\mathcal{E} = h\nu = \hbar\omega$ where h is Planck's constant (or the Planck constant), and $\hbar = \frac{h}{2\pi}$ is known as Dirac's constant (or the Dirac constant). In this equation $\mathcal{E}$ is the energy of a single quantum of radiation.

The classical equivalent of the second axiom is that the energy is proportional to the intensity of the radiation. The classical and the quantum pictures can be reconciled by interpreting intensity as a measure of the number of photons present in unit volume of the radiation.

Here the classical and the quantum pictures lead to different functions for the energy density within a black-body cavity. We outline the quantum calculation below.

For any given frequency, there can be one, two, three, or more quanta within the cavity. The probability for the presence of a particular number of quanta depends on the total energy of these photons for that frequency.

This means, that if the energy of a quantum of radiation is $\mathcal{E} = \hbar\omega$, the total number of modes of a particular energy is not simply $\Delta N = Ne^{\frac{-\hbar\omega}{kT}}$, but rather

$\Delta N = \Delta N_1 + \Delta N_2 + \Delta N_3 + ...$ where ΔN_n is the number of modes having n quanta. Clearly, $\Delta N_n = Ne^{-\frac{n\hbar\omega}{kT}}$, and so

$$\Delta N = N(e^{-\frac{\hbar\omega}{kT}} + e^{-\frac{2\hbar\omega}{kT}} + e^{-\frac{3\hbar\omega}{kT}} + ...)$$

So, the average energy of a wave of frequency ω is $\hbar\omega(e^{-\frac{\hbar\omega}{kT}} + e^{-\frac{2\hbar\omega}{kT}} + e^{-\frac{3\hbar\omega}{kT}} + ...)$.

The number of waves per unit volume is given by Eq. (8.36). The energy per unit volume for waves between ω and $\omega + d\omega$ is therefore

$$\frac{du}{V} \equiv dW = \frac{\omega^2}{\pi^2 c^3}\hbar\omega(e^{-\frac{\hbar\omega}{kT}} + e^{-\frac{2\hbar\omega}{kT}} + e^{-\frac{3\hbar\omega}{kT}} + ...)d\omega \tag{8.39}$$

Unlike material particles, light waves can pass through each other without destroying each other. And this principle of superposition is independent of the intensity of the light waves. From the quantum perspective, this means that the number of quanta can be increased indefinitely. So, the sum in Eq. (8.39) extends indefinitely. Such an infinite series can be summed easily.

Exercise:
Show that

$$\sum_{n=1}^{\infty} e^{\frac{-n\hbar\omega}{kT}} = \frac{1}{e^{\frac{\hbar\omega}{kT}} - 1}$$

Writing $W_\omega(T) = \frac{dW}{d\omega}$, we get Planck's formula for the energy density of black-body radiation of frequency ω :

$$W_\omega = \frac{\hbar\omega^3}{\pi^2 c^3}\frac{1}{e^{\hbar\omega/(kT)} - 1} \tag{8.40}$$

8.8.2 *Correspondence with Classical Formulas*

In 1896 Wien used general thermodynamics to infer that the energy of a mode with frequency ω is proportional to ω. Using $\hbar$ as the constant of proportionality, and applying the Boltzmann distribution $\Delta N = Ne^{-\mathcal{E}/kT}$, Wien obtained this formula for the energy density for high frequencies:

$$W_\omega = \hbar\omega^3 e^{-\hbar\omega/(kT)}/(\pi^2 c^3) \tag{8.41}$$

> **Exercise:**
> Show that Planck's formula (Eq. (8.40)) agrees with Wien's formula (Eq. (8.41)) in the limit as $\omega \to \infty$.

Rayleigh and later Jeans developed an equation for the energy density of standing waves using Eq. (8.36). Using the principle of equipartition of energy, whereby each degree of freedom has energy $\frac{1}{2}kT$, they assigned two degrees of freedom to each mode of radiation, with each degree corresponding to the polarizations, and obtained this formula:

$$W_\omega(T) = \omega^2 kT/(\pi^2 c^3) \tag{8.42}$$

We see that W_ω diverges for large ω, which does not agree with experimental observation. Such a divergence for larger frequencies was called the *ultraviolet catastrophe.*

> **Exercise:**
> Show that Planck's law (Eq. (8.40)) agrees with the formula obtained by Rayleigh and Jeans (Eq. (8.42)) in the limit as $\omega \to 0$.

8.8.3 *Stefan-Boltzmann Law*

In Eq. (8.40), $W_\omega = \frac{dW}{d\omega}$, where W is the total energy density of all wavelengths of radiation. We may integrate this equation to obtain the total energy density:

$$W = \frac{\hbar}{\pi^2 c^3} \frac{\omega^3 d\omega}{e^{\hbar\omega/kT} - 1} \tag{8.43}$$

By a suitable change of variables ($\xi = \frac{\hbar\omega}{kT}$), this integral becomes

$$W = \frac{k^4 T^4}{\pi^2 c^3 \hbar^3} \int_0^\infty \frac{\xi^3}{e^\xi - 1} d\xi = \frac{k^4 \pi^2 T^4}{15 c^3 \hbar^3} \tag{8.44}$$

Thus the total energy contained within the cavity is proportional to the fourth power of the absolute temperature.

According to the Stefan-Boltzmann Law, a body at absolute temperature T radiates energy at the rate

$$M = \sigma T^4 \tag{8.45}$$

where $\sigma = 5.67 \times 10^{-8}$ W. m^{-2} K^{-4} is the Stefan-Boltzmann Constant. Planck's equation provides a quantum theoretical justification for the Stefan-Boltzmann Law.

Exercises:
1. Prove Eq. (8.44). (Assume the value $\sum_1^\infty \frac{1}{n^4} = \frac{\pi^4}{90}$.)
2. Two bodies A and B, in thermal contact with each other, are initially at temperatures T_A and T_B where T_A is slightly greater than T_B, so that $\Delta T = T_A - T_B$ is small compared to T_A or T_B. Show that the rate at which heat flows from A to B is proportional to ΔT. This is Newton's law of cooling.

8.9 The Photoelectric Effect

Ordinarily a strong electric field is required to entice free electrons to jump out of the body of a conductor. But there is a phenomenon in which electrons can be released from a conductor without the use of a strong field. This is the photoelectric effect. When light waves fall upon certain metallic conductors electrons are released from the surface of the conductor.

The following is a simplified description of an apparatus that illustrates the photoelectric effect. Two conducting plates are kept a small distance apart in a vacuum. One plate is connected to the negative terminal of a battery or some other DC power supply and the other plate is connected to the positive terminal. Thus there is an electric field set up between the positive and the negative plates which attracts the electrons that are on the negatively charged metal plate. But they cannot leave the surface of the plate because of the cohesive forces that bind them to the body of the conducting plate. However, if electromagnetic radiation such as light is shone on the plate, this radiation can communicate energy to the electrons, thereby enabling them to overcome the cohesive forces and leap through the vacuum to the positive plate. Once inside the positive plate they are attracted to the positive terminal of the power supply and flow through the ammeter which records the electric current — which is proportional to the rate at which electrons flow through the device.

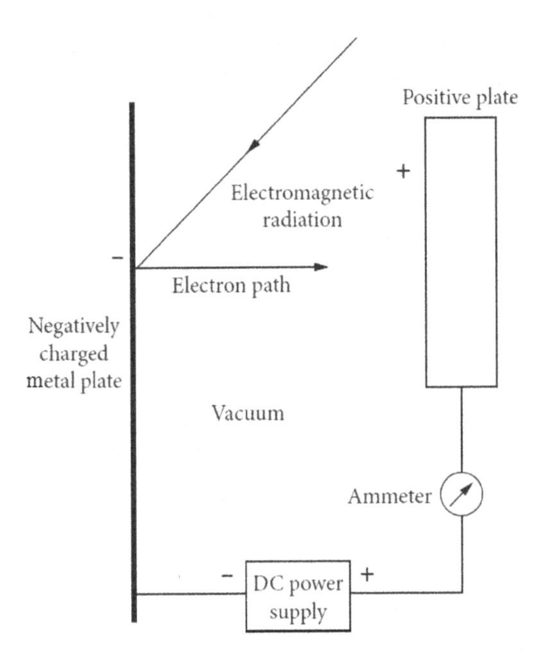

The results of the experiments can be summed up as follows:

1. When the negatively charged plate was bombarded with radiation of low frequency, there was no discharge of electrons, no matter how great was the intensity of the radiation. Thus even a large amount of energy given to the conductor did not serve to release any electrons from the conductor.

2. Then the frequency of the radiation was increased steadily. When a particular frequency was reached, the electrons began to be emitted. Next, keeping the frequency of the radiation constant, the intensity of radiation was raised. As the intensity increased, more electrons were released. The current measured by the ammeter increased in proportion to the intensity of the radiation.

3. The frequency of the radiation was now raised, without changing the intensity. It was now observed that the electrons were leaving the plate with greater kinetic energy. The change in kinetic energy could be measured by changing the potential difference between the plates. But the number of electrons — as measured by the current strength — did not increase.

As a particular example, the metal potassium did not emit any electrons when red light (low frequency) fell upon it. When green light (higher frequency) was shone on the metal it began to emit electrons. The energy of the electrons increased as the frequency of the light was further raised to blue and violet.

The results were given a simple interpretation:

1. A minimum frequency of radiation is necessary to release the electrons, i.e. to help them overcome the cohesive force that binds them to the conductor.

2. Once this minimum frequency was reached, the number of electrons released is proportional to the intensity of the radiation.

3. As the frequency of the radiation is further increased, the energy of the electrons also rises.

8.9.1 *Einstein's Explanation*

Einstein provided a coherent explanation for this photoelectric effect. He suggested that electromagnetic energy travels in the form of packets of energy, with each packet having energy proportional to the frequency of the wave, and on the basis of this suggestion he was able to explain the photoelectric effect.

Planck had suggested that radiation was emitted or absorbed in units or quanta of energy $\epsilon = h\nu$ where ν is the frequency of the radiation. Einstein's postulate carries Planck's hypothesis to the next logical step — that radiation is not only absorbed or emitted in quanta, but radiation *exists* only as quanta of energy $h\nu$.

Einstein's explanation for the photoelectric effect can be summed up as follows:

Each quantum of electromagnetic radiation has energy $h\nu$. An electron in the metallic conductor receives this energy from the bombarding radiation. If this energy is low, the electron cannot escape from the metal surface. As the frequency ν is steadily raised, a point is reached when the energy of each quantum of radiation is exactly equal to the energy needed for the electron to burst free of the cohesive force. If the intensity of the radiation is increased, more quanta are present. Each such quantum gives its energy to an electron, and releases the electron. Thus, the number of released electrons — which is measured as the current by the ammeter — is proportional to the intensity of the radiation, since the number of emitted electrons is equal to the number of quanta incident upon the metal surface. As the frequency of the beam is increased, the energy of each quantum increases according to $\epsilon = h\nu$. And so each electromagnetic quantum is able to give more energy to each electron.

Exercise:
Light of frequency 7.00×10^{14} Hz was shone on a metal surface. Electrons of kinetic energy up to 1.30×10^{-19} J were emitted. The frequency of the light was raised to 8.00×10^{14} Hz. This time electrons of kinetic energy up to 1.96×10^{-19} J were emitted. The relationship between the energy of the light ($\epsilon = h\nu$) and the maximum kinetic energy T of an emitted electron is the following:

$$\epsilon = T + \Phi$$

where Φ is a constant number equal to the work done by the electron in escaping from the metal surface. Use the data given in this exercise to calculate the value of h.

8.10 Uncertainty Principle

The photoelectric effect and the Compton effect have shown that electromagnetic energy is propagated as quanta or photons with energy $\epsilon = h\nu$, and momentum $p = \frac{h}{\lambda} = h\nu/c$. These properties of a photon are important for understanding the peculiarities of measuring the position or the momentum of an electron.

Suppose we want to measure the position of an electron. We know that an electron is subject to an electromagnetic field and a gravitational field. Since the electron mass is very small, its direct interaction with gravity is very hard to measure. The most direct way of measuring an electron's position is by measuring its interaction with an electromagnetic field.

Now, the electromagnetic field is quantized, and exists in units of energy $h\nu$. A single electron is so small that in order to measure its position, we need to employ a weak electromagnetic field. A strong field — carrying several photons — would disturb the electron significantly, and so the measurement would not be accurate. The weakest possible field would contain just a single photon. Let us do a thought experiment in which a photon tries to "measure" the position of an electron.

The interaction of the photon with the electron can be thought of as two successive events: first, the electron absorbs a photon, and next, the electron emits a photon.

As the electron absorbs the photon, it undergoes a change of momentum, as the momentum of the photon is transferred to the electron. If the photon has momentum p, then it imparts a momentum p to the electron. Thus, in the process of measuring the position of the electron, the momentum of the electron is not known accurately. There is an uncertainty in the momentum of the electron, an uncertainty which is of the order of the momentum of the measuring photon. So, if Δp_x is the uncertainty in the x-component of the momentum of the electron,

$$\Delta p_x \sim p \tag{8.46}$$

The photon has wavelength λ, which means that in the course of the absorption of the photon, the electron is displaced through a distance which is of the order of the wavelength of the photon. So, an error Δx has come into the measurement of the position of the electron. And $\Delta x \sim \lambda$, where λ is the wavelength of the photon. Now, $\lambda = \frac{h}{p}$, where p is the momentum of the photon. So

$$\Delta x \sim \frac{h}{p} \sim \frac{h}{\Delta p_x} \tag{8.47}$$

Thus, we obtain this relationship between the measured uncertainty in momentum and the corresponding uncertainty in position of the electron:

$$\Delta p_x \Delta x \sim h \tag{8.48}$$

This is a simple form of the uncertainty principle. A rigorous derivation of the uncertainty principle, employing a definition of uncertainty in terms of the standard deviation of outcomes of independent repeated measurements, yields the result

$$\Delta x \Delta p_x \gtrsim \frac{\hbar}{2} \tag{8.49}$$

The uncertainty principle implies that the process of measuring an electron causes it to undergo a displacement of the order of Δx. Suppose the temporal duration of this measurement is Δt. We could therefore write

$$\Delta t \frac{\Delta p_x}{\Delta t} \Delta x \gtrsim \frac{\hbar}{2} \tag{8.50}$$

In the course of the interaction between the photon and the electron a force F_x is exchanged between the two particles, which is of the order of $\frac{\Delta p_x}{\Delta t}$. During this process a certain amount of energy is transferred from one particle to the other, of the order of $F_x \Delta x$. Thus, in making any measurement on the electron, there is a minimum time Δt involved, and a corresponding minimum uncertainty in energy ΔE of the electron, and these are related by

$$\Delta E \Delta t \gtrsim \frac{\hbar}{2} \tag{8.51}$$

The quantization of electromagnetic energy implies that no measurement can be done in an arbitrarily short period of time.

8.10.1 *Wave Particle Duality*

Newton had suggested that light is a stream of particles. This seemed to be a reasonable theory. The reflection of light from smooth surfaces was similar to the kinematics of tiny spheres bouncing off a hard floor. And because these particles travel very fast the effect of gravity on their motion is unnoticeable. But other physicists such as Huygens and later Fresnel proposed the wave theory of light, which was eventually confirmed by Young's double slit experiment.

So, prior to the twentieth century it was known that light has wave-like properties — it is capable of interference and diffraction. But in the early twentieth century it was discovered that also has particle-like properties. However, this corpuscular nature of light is different from the model proposed by Newton. The wave-particle duality of light can be expressed in simple terms in two principles, the first we call the wavelike or undular principle, and the other the particle-like or corpuscular principle:

1. **Undular principle:** Photons *propagate as waves* — they undergo interference and diffraction according to the properties of waves.

2. **Corpuscular principle:** Photons are *detected as particles.* When they are "caught," they are found to have mass, momentum, and other properties traditionally associated with particles.

In the following we show how an updated version of Young's experiment illustrates these two principles.

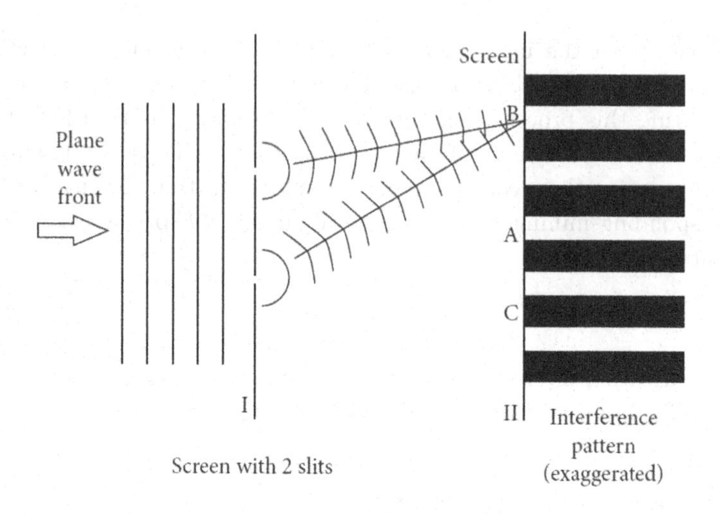

In the figure above a source of light is kept far to the left of screen I. So the light reaching screen I has a plane wave front. This wave is blocked by the screen except for a small amount that is allowed to pass through the two slits (cut perpendicular to the plane of the diagram). These two slits behave like sources producing waves having cylindrical wave fronts, which appear in the diagram as circular arcs. These two waves proceeding from the two slits will undergo *interference.* Consider a point A which receives light from both the slits. If the distance of A to the upper slit is equal to its distance to the lower slit, then both the waves will reach A in phase. There will be a large amplitude at A, and the intensity of light will be great at A. So there will be a bright spot at A. Suppose now there is a point B such that the difference in the distances of the two slits from B is equal to a whole number of wavelengths. Then the two beams will arrive at the point B in phase. There will therefore be a bright spot also at B. Next consider a point C such that the difference in the distances between C and the two slits is an odd multiple of half a wavelength, so that a crest of one wave and a trough of the other wave reach C simultaneously. The two waves will be exactly out of phase when they arrive at C, and so there will be destructive interference and hence a dark spot at C. Of course, since the openings on screen I are not tiny holes but narrow slits, the wave fronts between screens I and II are cylindrical, and so the pattern we find on screen II will not be dots of light but alternate bright and dark bands perpendicular to the plane of the figure. This interference pattern demonstrates that light travels in the form of waves.

The double slit experiment was originally carried out by Thomas Young who proved that light is a form of wave motion. In the following section we shall see how the double slit experiment can be refined to illustrate the corpuscular or quantum nature of light.

8.10.2 *Quantum Theory of Light*

Suppose we were to reduce the brightness of the source until it is so feeble that it cannot be seen with the human eye. If we now replace the screen II by a photosensitive plate of very high resolution, and mount a powerful microscope on the other side of the plate, we would observe an interesting thing.

Initially, the photosensitive plate would be dark. Then, a single bright spot would appear in a region where a bright band used to be, and later another bright spot where another bright band used to be when the source had full power. A little later a bright spot might appear in line with the first bright spot — within the same band as one of the original bright bands observed earlier — and another bright spot where another bright band used to be, and so on. After a long time, there will be millions of spots which will eventually merge to form the interference pattern that we saw when a bright source was used.

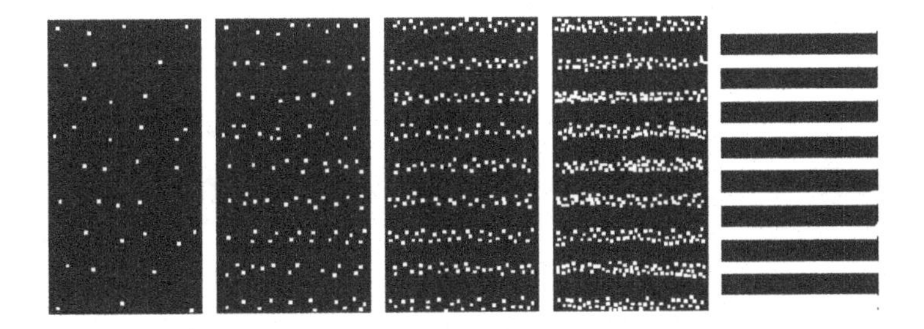

Each bright spot represents a unit of light energy — a quantum, or a photon. This experiment shows that light energy is not emitted or absorbed continuously, but only in drops or units. What we have done in this experiment is to lower the intensity of the source so much that only about 1 quantum of light energy — or 1 photon — is emitted per second. This experiment proves that light energy is quantized, that it exists only in discrete multiples of a basic unit of energy.

This then raises a question. If only one photon were traveling from the source to the screen II, through which slit did it travel across screen I — the upper or the lower slit? The surprising answer is that the photon went through both slits, or, more precisely, electromagnetic energy traveled through both slits, and this energy was detected somewhere along screen II as a single photon.

The electromagnetic wave went through both slits, and became two waves on the other side of screen I. The interference between these two waves determined where the photon would be found on a detector such as screen II. But we saw that there is a certain randomness to the process. Each individual photon did not seem to follow any set procedure for falling on the screen, except that it landed only where the bright bands appeared when the source had full intensity.

Suppose we were to keep track of all the light quanta by following the appearances of all the spots of light at different points on the screen. So we could draw numbers indicating the order in which the spots of light appeared at different places on the screen, 1 for the first spot that appears, 2 for the second spot, etc. A typical pattern with numbered spots may have 1 at the mid point or center, 2 at a point some distance below 1, 3 above 1, 4 close to 1, etc. After a long time, the spots will be clustered so closely that they are no longer discernible as individual spots but merge into the interference pattern we saw earlier.

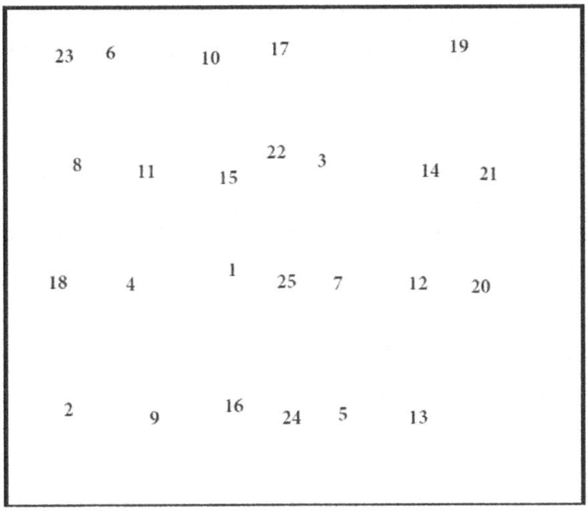

23 6 10 17 19

8 11 15 22 3 14 21

18 4 1 25 7 12 20

2 9 16 24 5 13

Now we switch off the source, and replace the photographic plate. We switch on the source as before. The order in which the spots arrive will not be the same as the former. But after millions of spots have appeared, the same interference pattern will emerge. This shows that while the wave determines all the possible locations where the light energy could hit the screen, it does not tell any one quantum where to go. There is a randomness in the way individual light quanta "decided" where to land on the screen.

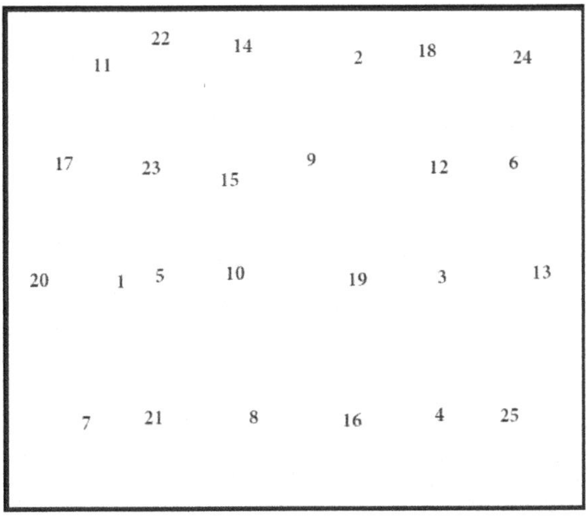

11 22 14 2 18 24

17 23 15 9 12 6

20 1 5 10 19 3 13

7 21 8 16 4 25

The regular pattern did not appear with one or two photons, but with billions of them. So the interference pattern is a statistical effect. The apparently well defined image is formed by billions of spots appearing at random but falling in preferred locations. The probability for the appearance of a spot is determined by the geometry of the apparatus, but an individual photon is free to choose any location that is permitted by the probability. Over the long run the number of photons arriving at any one location is proportional to the probability for a single photon to arrive at that location.

Now, the refraction of light through a lens is ultimately due to wave interference, even though elementary textbooks treat this subject as "geometric optics." What this means is that the image produced on our retina is due to the impression created by billions of photons arriving at random following the laws of probability as they pass through the lens in our eye. Our perception of reality is therefore a statistical effect. Maxwell had accurately predicted that "the true logic of this world is in the calculus of probabilities."

8.11 Electron Waves

Atoms emit and absorb infrared, visible, and ultraviolet radiations. Rutherford had shown that the atom consists of a positively charged central core or nucleus and negative electrons at some distance from the nucleus. The exact motion of these electrons was not understood. We know that electrons absorb and emit photons. Thus it made sense to think of these atomic electrons as the sources or the targets of radiation. Experimental physicists discovered interesting and puzzling relationships between the frequencies of the radiation emitted (or absorbed) by atoms. The radiation appears only with certain frequencies, and these have a mathematical relationship which cannot be explained by classical physics:

$$\text{Frequency of the radiation} = \text{Constant} \times \left(\frac{1}{n_1^2} - \frac{1}{n_2^2}\right)$$

where n_1 and n_2 are natural numbers 1, 2, 3, ...etc. with $n_2 > n_1$.

As we explained in the first chapter, if the result of a measurement suggests an exact rational number in the formula, there must be a theoretical explanation. Some new principle or law is being discovered. An explanation for this mathematical relationship was provided by Louis de Broglie, who postulated that it is not just light quanta that have a dual wave-particle nature, but that this duality is a property of every known particle, including electrons, protons, neutrons, and even composite objects such as nuclei, atoms and molecules. De Broglie proposed that every object having momentum p also has a wavelength λ associated with it, given by the formula

$$\lambda = \frac{h}{p} \tag{8.52}$$

This wavelength is called the de Broglie wavelength of the object. For macroscopic objects such as baseballs the de Broglie wavelength is too small to be measurable. But for electrons the wave properties are significant.

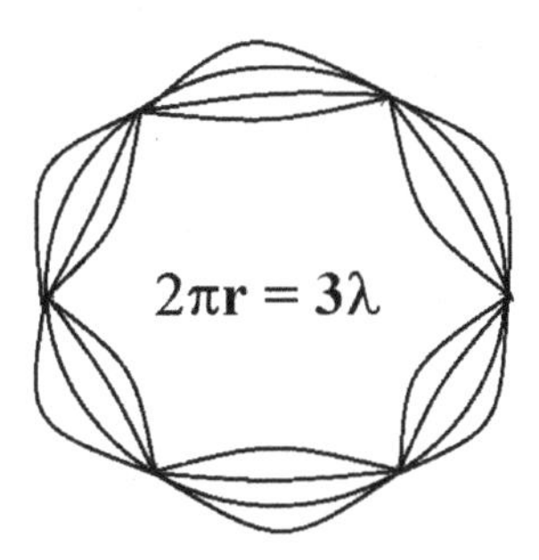

Fig. 8.1 Atomic orbitals as standing de Broglie waves. Here 3 wavelengths are contained in 1 orbit, corresponding to $n = 3$. There are 6 nodes (zero displacement) and 6 antinodes (maximum displacement).

De Broglie suggested that the permitted electron orbits were the circular standing waves of the electron. His explanation was that when an electron orbits the nucleus the circumference of the orbit should be equal to an integer multiple of the wavelength of the electron. In other words, the paths of the orbiting electrons should be thought of as stationary waves, similar to the standing waves in a vibrating string.

Using this principle, if the radius of the nth orbit is r, we would have

$$2\pi r = n\lambda = \frac{nh}{mv} \tag{8.53}$$

So not all radii are permitted, but only those that have a specific relationship to the speed v of the electron:

$$vr = \frac{nh}{2\pi m} \tag{8.54}$$

This equation may be rewritten as

$$mvr = \frac{nh}{2\pi} = n\hbar \tag{8.55}$$

The quantity on the far left is the angular momentum of the orbiting electron. This angular momentum is seen to be a whole multiple of $\hbar$.

Now, an electron in orbit round the nucleus is kept in a circular orbit by the centripetal force which is the electrical force of attraction between the electron and the proton (which constitutes the nucleus of a hydrogen atom). According to the laws of electrostatics this force is equal to

$$\frac{e^2}{4\pi\epsilon_0 r^2}$$

This centripetal force generates a centripetal acceleration. And the centripetal acceleration of a particle moving with speed v along a circle of radius r is given by the formula $\frac{v^2}{r}$. By Newton's second law (mass × acceleration = force),

$$\frac{mv^2}{r} = \frac{e^2}{4\pi\epsilon_0 r^2} \tag{8.56}$$

By combining Eqs. (8.54) and (8.56) we obtain an expression for the radius

$$r = \frac{n^2 h^2 \epsilon_0}{\pi m e^2} \tag{8.57}$$

The total energy of an atom = kinetic + potential. The kinetic energy of the atom is entirely due to the orbital motion of the electron, since the nucleus can be taken to be stationary in comparison with the fast motion of the electron round the nucleus. The potential energy of the atom is the potential energy due to the electrical force between the electron and the nucleus. So the total energy

$$E = \frac{1}{2}mv^2 - \frac{e^2}{4\pi\epsilon_0 r} \tag{8.58}$$

The potential energy is negative because the force between the electron and nucleus is attractive. Substituting expressions for v and r obtained earlier, and using some simple algebra, we get

$$E = -\frac{e^4 m}{8\epsilon_0^2 h^2 n^2} \tag{8.59}$$

We find that the total energy is negative. This means that the electron cannot escape from the nucleus unless it receives some energy from outside.

If this positive energy given to the electron is greater in magnitude than the negative energy of the atom that would enable the electron to break free from the nucleus. But if it receives energy somewhat less than this amount then it may be possible for the electron to jump to a higher orbit with a higher value of n. Normally the electron occupies the lowest possible energy state, for which $n = 1$. If it receives some energy, it could jump to an orbit where the value of $n = 2$, 3, 4, etc.

The lowest possible energy state of an atom is called its *ground state.* For a hydrogen atom the ground state is the energy state with $n = 1$. States with higher values of n are called *excited states.* When an atom is in an excited state it would normally return to the ground state by emitting a light particle or photon. The energy of this photon would be equal to the difference in energy between the excited state and the ground state of the atom. This energy is

$$E_2 - E_1 = \frac{e^4 m}{8\epsilon_0^2 h^2}\left(\frac{1}{n_1^2} - \frac{1}{n_2^2}\right) \tag{8.60}$$

If λ is the wavelength and ν the frequency of the emitted photon, we can write

$$E_2 - E_1 = h\nu = \frac{hc}{\lambda} = \frac{e^4 m}{8\epsilon_0^2 h^2}\left(\frac{1}{n_1^2} - \frac{1}{n_2^2}\right) \tag{8.61}$$

The experimental physicists used a quantity called *wave number*, which is the reciprocal of the wavelength:

$$\frac{1}{\lambda} = \frac{e^4 m}{8\epsilon_0^2 h^3 c}\left(\frac{1}{n_1^2} - \frac{1}{n_2^2}\right) = R\left(\frac{1}{n_1^2} - \frac{1}{n_2^2}\right) \tag{8.62}$$

where R is called the Rydberg constant, which was originally obtained from experiment. The value of R may be calculated theoretically from the above equation. It works out to be 10 973 732 m^{-1} and agrees with the experimental value.

In the calculation of the Rydberg constant we used both classical electrodynamics and classical mechanics, with two notable exceptions. We assumed that the electrons can only assume certain orbital paths, which are determined by the de Broglie wavelength, and that when an electron moves along such a path it does not emit electromagnetic radiation, as required by an accelerating charge. An atomic electron gives off electromagnetic radiation only when it drops from a higher to a lower energy state. The difference in energy is equal to the energy of the photon emitted by the atom. This model of the atom we have outlined above was originally postulated by Neils

Bohr, who was the first to suggest that the electrons in an atom occupy stable orbits in which they do not radiate energy, and only do so as they transition to orbits of lower energy. But it was de Broglie who provided the correct explanation for why these orbits were stable. Bohr himself had put forward an ad hoc hypothesis which relates the stability of the orbits to their angular momentum.

8.11.1 *Photon Angular Momentum*

Equation (8.55) shows that the angular momentum of an electron in a stable orbit is a whole multiple of $\hbar$. So, if an electron drops from the 3rd orbit to the 2nd orbit the orbital angular moment of the electron — which is essentially the angular momentum of the atom — decreases by an amount $\hbar$. This suggests that the emitted photon must have an angular momentum of magnitude $\hbar$, in order that the total angular momentum of the system may be conserved.

Whereas the Bohr-de Broglie atom model has been considerably refined by developments in quantum theory, the conclusion of the last paragraph remains valid. A photon does have intrinsic angular momentum of magnitude $\hbar$, regardless of its wavelength.

Chapter 9

The Quantum Mechanics of the Field

9.1 Feynman Graphs

The Compton effect can be depicted in terms of the world lines of the photon and the electron as follows:

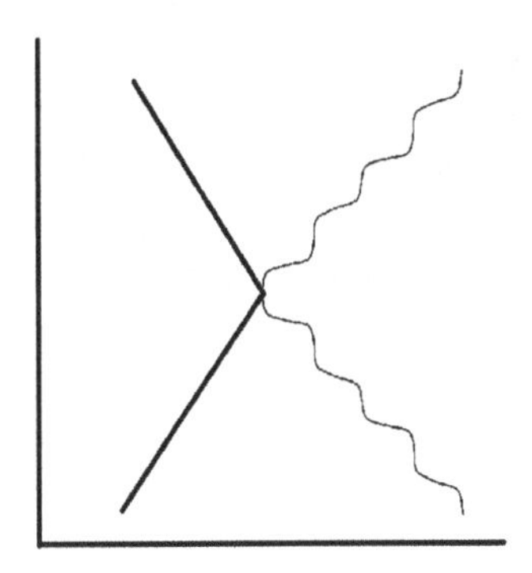

The thick line is the world line of the electron and the wavy line is the world line of the photon. The horizontal axis represents spatial displacement and the vertical axis represents the flow of time, and these coordinates are measured in some inertial frame of reference. This graph depicts a single spatial coordinate, which we have taken to be the direction of motion of the particles. So in this situation there is a head on collision between the electron and the photon.

The laws of physics are invariant in all inertial frames. Two frames that are in uniform relative motion with respect to each other are shown as rotated relative to each other. So relative motion is depicted as a rotation in Minkowski space.

What if we were to rotate the axes through ninety degrees? The graph of the world lines would appear thus:

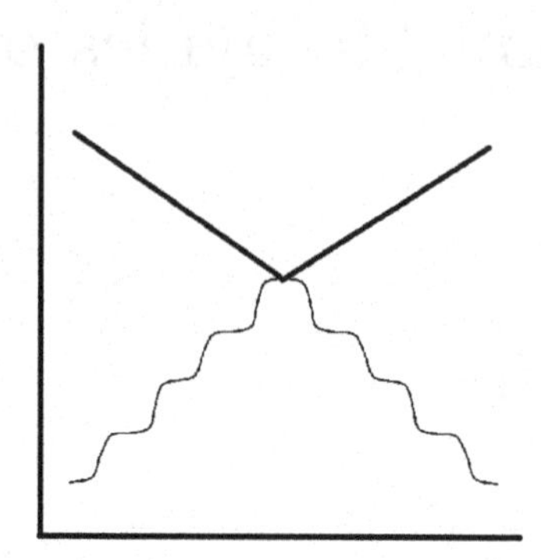

This is a physically different process, but by the laws of special relativity this graph also represents a physically valid process. Here we have two photons approaching each other, and then they collide and disappear, and two particles emerge and move in opposite directions. Now, electric charge cannot be created or destroyed, and therefore the two particles cannot have the same charge. They could be neutral, or they could have opposite charges. As it turns out, this diagram represents the process called *electron-positron pair production.* The collision of the photons occurs in a strong electromagnetic field, and two charged particles of equal mass and opposite charge — a negative electron and a positive positron — are created.

We could further rotate the diagram and obtain the following graph:

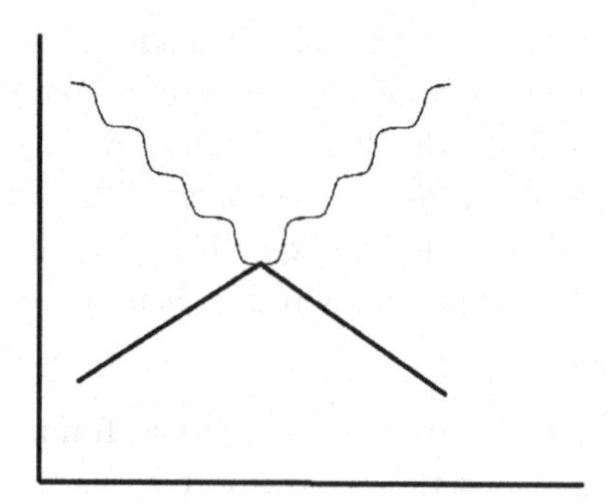

Here two particles approach each other, collide and disappear, and two photons emerge and move in opposite directions. Such a process is also physically possible. This figure represents *electron-positron pair annihilation.* When a negatively charged electron approaches a positively charged positron they fall into each other and annihilate each other, and the result is two photons that move in opposite directions.

In all three processes the total momentum, the total energy, and the total charge of all the particles is conserved before and after the interaction.

Such graphs are helpful for understanding interactions of particles, and are very useful for calculations. They were invented by Feynman, and are called Feynman graphs or Feynman diagrams. (We will see further down that the above diagrams need to be refined in order to be physically correct.)

9.2 Quantum Interactions

Classical electromagnetism deals with the laws of electricity and magnetism as they were known prior to the twentieth century. According to these laws charges can receive or emit electromagnetic energy in continuous streams, and not in discrete units or quanta. But Planck showed at the beginning of the twentieth century that energy can be absorbed or emitted only in discrete quanta of magnitude $h\nu$. This idea is the genesis of quantum theory and remains one of its central tenets.

Einstein took Planck's hypothesis one step further and showed that the reason why light is absorbed or emitted in packets or quanta is that light *exists* as packets or quanta. And these quanta of light are particles in their own right, with energy, mass, momentum and even angular momentum. They collide with electrons in the phenomenon called the Compton Effect. Light propagates as a wave but is detected as a particle.

De Broglie removed the basic distinction between light and matter when he showed that material particles such as electrons too have a wavelength. They too propagate as waves but are detected as particles.

The study of the energy, motion, angular momentum and other properties of a physical system is called *mechanics.* Quantum mechanics is the study of microscopic objects such as photons, electrons, protons, neutrons, and small collections of particles such as atoms and molecules.

We now come to a profound difference between classical and quantum mechanics. In any problem in classical mechanics we can concentrate on one aspect or one facet of the object and ignore the others. As an example, suppose we are interested solely in studying the forces that lift up a flying airplane. For most purposes it is sufficient to study the air flow above and below the wings to understand how the aircraft overcomes gravity and remains aloft. The remainder of the airplane — including the interior — is largely irrelevant for the investigation of the lift.

By contrast, there are no parts or components of a fundamental particle such as an electron or a photon. When we observe or detect an electron we do not detect its top part or its side. We detect the entire particle. Likewise for a photon.

Quantum theory as it is understood today states that in order to make any sort of measurement on a particle, the particle must be absorbed first, and then re-emitted. And the absorption and re-emission involve the whole particle, not any one portion of it.

Let us illustrate this in the case of the scattering of an electron in the Compton Effect. The schematic Feynman diagram illustrating this phenomenon is the following:

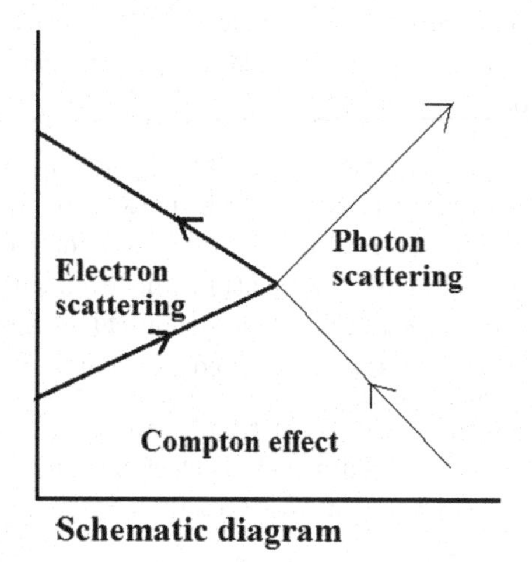

However, this is not the correct Feynman diagram. A correct Feynman diagram cannot have a vertex with two electron lines and two photon lines meeting at a point. A more accurate Feynman diagram has two electron lines and a photon line meeting at a point. So there are two such vertices in the complete Feynman diagram:

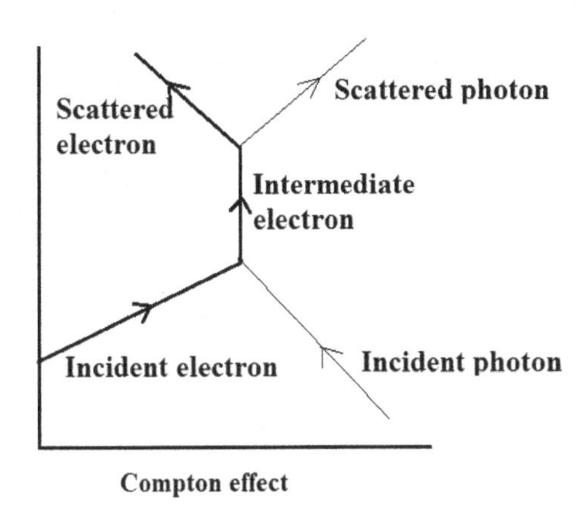

Compton effect

There are 5 *distinct* particles in this diagram: the incident electron, the incident photon, the intermediate electron, the scattered electron and the scattered photon. The incident photon disappears at the first (lower) vertex, and the "scattered" photon appears at the second (upper) vertex. It is common in quantum theory to use the terms "annihilated" and "created" for the disappearing and the appearing photons. These terms serve to indicate that the incident photon comes to the end of its career when it is absorbed by the electron and the "scattered" photon begins its own career when it is emitted by the electron. So they are two different particles.

What is perhaps really surprising is that the incident electron also ends its career when it absorbs a photon. It is annihilated the moment it absorbs the photon. The intermediate electron is a new particle that exists only for a short time before it is annihilated and a new electron is created which appears as the "scattered" electron.

It is easier to think of the electron as a continuously existing particle because it has charge, and charge cannot be destroyed according to Maxwell's theory of electromagnetism. But quantum theory permits a charge to be destroyed and recreated immediately. This is one of the surprising consequences of the Uncertainty Principle.

In the historical development of quantum mechanics particles such as electrons were treated as though they existed indefinitely, even as they interacted with other charges and with photons. The notion of electrons being created and destroyed as they interact with other particles is a later development, and this quantum mechanical approach was called *second quantization* or *quantum field theory*. But nowadays it is becoming increasingly recognized that this so-called field theoretical approach expresses a better understanding of the behavior of fundamental particles such as electrons and photons. And so we will pursue this approach and then we shall see that this path leads us in a more logical way to the same conclusions that were arrived at by the more ad hoc methods of earlier quantum theory.

Absorption or emission of a quantum has absolutely no place in classical physics. We therefore need to invent a new language to describe quantum theory. In this chapter we shall develop this new language in the context of photons. After learning the basics of the vocabulary and the grammar of this new language we shall apply it to other particles such as electrons.

9.3 Complex Numbers in Quantum Mechanics

Both absorption and emission are processes *in time*. It takes time to absorb a photon, and it takes time to emit a photon. But each of them is an elementary process, involving a photon and an elementary particle such as an electron. And such elementary processes — as distinct from processes involving a large number of particles — are time reversible. So an absorption is simply a time reversed emission, and vice versa. This has serious implications for the mathematical description of the two different processes.

Let us describe the process of absorption by a mathematical symbol a. Right now we do not know anything about a except that it describes the absorption or annihilation of a photon. Since this entity describes an action it is called an *operator*. And since it describes the action of annihilation it is called the *annihilation operator*.

In special relativity time is treated as an imaginary spatial dimension. So, imaginary numbers were given a physical meaning. But we learned that it is possible to camouflage the imaginary numbers by using the metric with positive and negative elements. As it turns out, quantum mechanics also involves imaginary numbers, but here it is not possible to camouflage them. And so, the mathematical descriptions of operators frequently contain imaginary numbers. And this would be true of creation and annihilation operators.

Notice that time reversal — changing ict to $-ict$ — can be accomplished *either* by changing t to $-t$ *or* by changing i to $-i$. This suggests a rule in quantum mechanics. The creation of a photon is the time reversal of its absorption or annihilation. So, wherever the imaginary number i appears in the operator expressing annihilation, the corresponding operator expressing creation would have $-i$. The process of changing i to $-i$ in any mathematical expression is called *complex conjugation*. We need to carry out a complex conjugation when converting the operator for annihilation into an operator for creation.

The operator representing emission or creation is written as $a^\dagger$ and is called the *creation operator*. So a and $a^\dagger$ are related to each other. Wherever i appears in a, the complex conjugate $-i$ appears in $a^\dagger$ (and vice versa).

9.4 States and Operators

Let us return to the black body cavity. We now consider these two distinct processes:

Process I: Wall A emits a photon.

Process II: Wall B absorbs a photon.

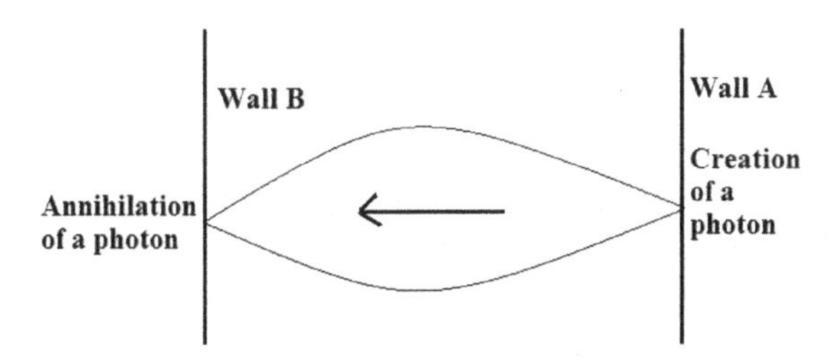

In this process a photon is first created and then annihilated. In order to represent this process mathematically, we require mathematical symbols for creation and annihilation. First, the creation. Suppose there were no photons in the cavity prior to the creation of this photon. We express such a state by the symbol $|0\rangle$. This symbol is called a *state function*, because it expresses the *state* of the cavity. The symbol is also called a *state vector*, for reasons that will become clear subsequently. This symbol expresses the fact that there are no photons in the cavity.

The *state* — (the word is related to the adjective *static*) — does not describe a process. A process is described by an *operator* (Latin for someone who does work). The process of emission of a photon is described by a *creation operator* written as $a^\dagger$. An operator works on a state. So the full description of the emission of a photon is expressed in quantum mechanical symbolism as $a^\dagger|0\rangle$.

The result of this process is that the cavity now contains one photon of light. This photon is subsequently absorbed by wall B and so the cavity then returns to its empty state. The process of absorption is the exact time reverse of emission. The quantum mechanical representation of absorption is by means of an absorption or annihilation operator which we write as a. So the entire process can be written as $aa^\dagger|0\rangle$. This symbol actually represents the state of the cavity after the emission and absorption have been completed. And since we now have 0 photons in the cavity, this state should be the same as $|0\rangle$. So, we can write a mathematical equation

$$aa^\dagger|0\rangle = |0\rangle \tag{9.1}$$

9.5 Physical Meaning of Symbols

9.5.1 *Creation and Annihilation of Photons*

But how we know that $aa^\dagger|0\rangle$ is identical with $|0\rangle$? That is, how do we know for sure that after the emission and absorption are completed we have returned to the original state? We can do a check on the state $aa^\dagger|0\rangle$ to see if it indeed has no photon. There are physical ways of doing this, and such a physical check is represented mathematically by the symbol $\langle 0|$. The significance of this symbol is that we are seeking an answer to the question: "Is it true that there is no photon in the cavity?"

Thus, we have carried out an experiment. First, we created a photon, then we annihilated the photon, and finally we checked the field to make sure there was no photon left. This sequence of actions is represented as a mathematical chain of symbols, to be read from right to left:

$$\langle 0|aa^\dagger|0\rangle$$

The earliest stage is at the extreme right, and the latest on the extreme left. So this symbol is shorthand for saying that initially there were no photons in the cavity ($|0\rangle$), then a photon was created ($a^\dagger$), then a photon was annihilated (a), and finally we checked that there were no photons in the cavity ($\langle 0|$).

The expression $\langle 0|aa^\dagger|0\rangle$ may be read as a *report* of the entire process. While the symbolic expression does have a narrative appearance, it also has a numerical value. This numerical value is in general a complex number. If the process is certain or inevitable, we give the report a value of 1. If the process is impossible, we give it a value of 0. If there is some possibility of its occurring even though it is not entirely certain, we give it a number whose absolute value is intermediate between 0 and 1. The absolute value of a complex number $a+ib$ is defined as $|a+ib| = \sqrt{(a+ib)(a-ib)} = \sqrt{a^2+b^2}$, keeping in mind that the square root symbol always indicates a positive number.

Thus the report of the entire process is a (complex) number. But the intermediate stages of the process are not numbers. The initial state $|0\rangle$ is not a number but a state function or a state vector. The process of the emission of a photon is not represented by a number but by an operator ($a^\dagger$) which is not a number. When the creation operator acts on the state with no photons, it changes the state into a state with 1 photon. This process is written as

$$a^\dagger|0\rangle = |1\rangle \tag{9.2}$$

An annihilation operator a performs the reverse operation as the creation operator. So if a were to operate on $|1\rangle$ it would give us $|0\rangle$:

$$a|1\rangle = |0\rangle \tag{9.3}$$

What do Eqs. (9.2) and (9.3) represent physically? The precise meaning will depend on the exact physical conditions. As an example, they could represent the interaction between an atom and a photon. The first expresses an atom emitting a photon and the second represents an atom absorbing a photon. So quantum mechanics offers us an elegant mathematical language for expressing the foundational tenet of quantum theory, that energy is absorbed or emitted not continuously, but in units or quanta.

If we know the actual physical conditions of the system in which a process occurs, then we can calculate the numerical value of the report of this process, a value which in general is a complex number whose absolute value lies between 0 and 1. What is the purpose of calculating this number? It turns out that every prediction that we can make in a particular situation depends on the calculation of this number for that situation.

Suppose the cavity is divided into two chambers by a wall with 2 openings labeled I and II. Suppose a photon is emitted at A and absorbed at B. How do we express this process quantum mechanically? We first write an expression for the creation of a photon in the right-hand chamber: $a_r^\dagger$. So we now have one photon in the right-hand chamber, and we can express this state as

$$a_r^\dagger|0\rangle_r$$

where the subscript r signifies that we are considering the right-hand chamber.

9.5.2 *Propagation of a Photon*

Now, the photon can pass through either of the two openings I or II. Let us see what happens if the photon passes through I. The passage of the photon can be thought of as two sequential actions — entering the opening from the right chamber, and exiting the opening into the left. Each of these is a separate action and needs a separate operator to describe the process. We represent the photon's entering the opening I by the operator $U(I)$. So the process so far reads:

$$U(I)a_r^\dagger|0\rangle_r$$

In the next step, the photon exits the opening I. Since entering the opening was described by the operator $U(I)$, we shall represent the exit of the photon from the opening by the operator $U^\dagger(I)$, where the dagger symbol indicates an operator representing a reverse process, similar to the creation and annihilation operators. So the process now reads:

$$U^\dagger(I)U(I)a_r^\dagger|0\rangle_r$$

Finally, the photon is absorbed in the left chamber and disappears. This step is represented by the operator a_ℓ. The state is then measured and no photon is found in the left chamber. This final step is represented by the measurement $\langle 0|_\ell$. The report of the entire process is given by:

$$\langle 0|_\ell a_\ell U^\dagger(I)U(I)a_r^\dagger|0\rangle_r$$

This report is a number, but we cannot give it the value 1. The reason is that this is not the only possible way for an electron originating in the right chamber to be absorbed in the left chamber. There is the possibility that the photon might have gone through opening II. So let us say

$$\langle 0|_\ell a_\ell U^\dagger(I)U(I)a_r^\dagger|0\rangle_r = \psi_1$$

where ψ_1 is a complex number and $|\psi_1|$ lies between 0 and 1.

Next, we consider the possibility that the photon went through the opening II. We would get a similar number for this path:

$$\langle 0|_\ell a_\ell U^\dagger(II)U(II)a_r^\dagger|0\rangle_r = \psi_2$$

where ψ_2 is another complex number and $|\psi_2|$ also lies between 0 and 1.

9.5.3 *Probability Amplitudes*

The numbers ψ_1 and ψ_2 are called *probability amplitudes* or simply *amplitudes.* So ψ_1 is the amplitude for the photon to go through I and ψ_2 the amplitude for the photon to go through II.

We saw above that a photon (or an electron or a proton, etc.) propagates as a wave but is detected or absorbed as a particle. When it is passing through the wall separating the two chambers *the photon is not detected.* So we must consider the kinematics of the photon's movement from one chamber to the other to be the kinematics of a wave. Now, one of the properties of a wave is interference. So the photon wave — or light wave — undergoes interference as it moves between the two chambers. So there is interference between the wave that flows through opening I and the one that flows through opening II. In quantum theory, interference is expressed by adding the amplitudes. So the complete description of the photon being emitted at wall A and absorbed at wall B is given by

$$\psi_1 + \psi_2 = \langle 0|_\ell a_\ell U^\dagger(I)U(I)a_r^\dagger|0\rangle_r + \langle 0|_\ell a_\ell U^\dagger(II)U(II)a_r^\dagger|0\rangle_r$$

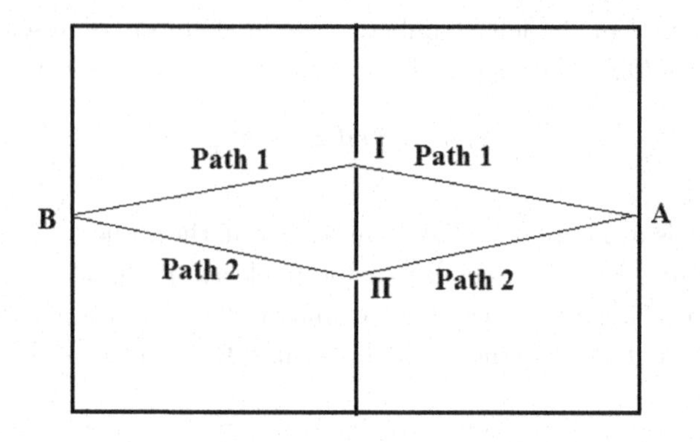

If we assume that a photon emitted at wall A *must* cross the partition and be absorbed at wall B, then the above equation tells the whole story. This means that $\psi_1 + \psi_2 = 1$. We notice that the expression for each amplitude summarizes the path. The expression begins on the right with the state function and ends on the left with the state function in reverse form, signifying closure of the process through a measurement. In between we have a string of operators.

9.5.4 *Addition of Paths*

Thus each amplitude is a product of possibly several operators between two state functions at the two ends. This amplitude is a complex number. And we add up the amplitudes for all the possible paths.

If we know for certain that the process is bound to happen — i.e. a photon emitted at A is bound to cross the partition through either I or II and be absorbed at B, then the sum of the amplitudes is 1. But if the process is not one hundred percent certain — say there is a possibility that the photon could be absorbed by the partition itself, then $\psi_1 + \psi_2$ will not be equal to 1, but $|\psi_1 + \psi_2|$ would be less than 1. In this case the probability of this entire process occurring is given by the square of the absolute value of the amplitude:

$$\text{Probability of I or II occurring} = |\psi_1 + \psi_2|^2 \tag{9.4}$$

So the rule is as follows:

1. Multiply all the operators representing a particular path, making sure that the order in which the operators are placed — right to left — follows the path correctly.

2. Multiply this operator product on the right by the initial state and on the left by the final state (written in reverse form). This product will be a complex number, and is called the amplitude for this path.

3. Add up the amplitudes for all the paths.

4. The square of the absolute value of this sum of the amplitudes is the probability for the entire process to occur. This could be anywhere between 0 and 1.

9.5.5 *Classical and Quantum Probabilities*

Suppose we close up one of the two openings, say opening II. Then the photon would be able to go through I only. The amplitude for this process is simply ψ_1. The probability for this process to occur is given by $|\psi_1|^2 = \psi_1^*\psi_1$.

If now we close up opening I and open II the probability for the photon to go through this opening is $|\psi_2|^2 = \psi_2^*\psi_2$.

Now we come to one of the biggest differences between classical and modern physics.

Probability for the photon to go through I and NOT through II $= |\psi_1|^2$.

Probability for the photon to go through II and NOT through I $= |\psi_2|^2$.

If now both I and II are open, according to classical physics, the probability that the photon goes through either I or II is simply the sum of the probabilities:

$|\psi_1|^2 + |\psi_2|^2 = \mathrm{P}$ (either I or II) $= \mathrm{P}_c$.

But quantum physics makes a very different prediction. If both I and II are open, the probability that the photon goes through these openings is

$|\psi_1 + \psi_2|^2 = \text{P (both I and II)} = \text{P}_q$.

$\text{P}_q = |\psi_1|^2 + |\psi_2|^2 + \psi_1^*\psi_2 + \psi_1\psi_2^* = \text{P}_c + \psi_1^*\psi_2 + \psi_1\psi_2^*$.

According to quantum theory the photon is detected as a particle, but it propagates as a wave. And so when both I and II are open the photon travels through these openings and there is *interference* between these two portions of the wave. This interference term is the difference between the two probabilities, which equals $\psi_1^*\psi_2 + \psi_1\psi_2^*$. This term can be positive or negative.

9.5.6 *Constructive Interference*

The highest positive value for the expression $\psi_1^*\psi_2 + \psi_1\psi_2^*$ is when $\psi_1 = \psi_2$, in which case this expression has the value $2|\psi_1|^2$ (or $2|\psi_2|^2$). So then we would have

$\text{P}_q = 4|\psi_1|^2$ and $\text{P}_c = 2|\psi_1|^2$ so that the quantum mechanical probability P_q is twice the classical probability P_c.

9.5.7 *Destructive Interference*

When $\psi_1 = -\psi_2$ the expression $\psi_1^*\psi_2 + \psi_1\psi_2^*$ has its lowest negative value equal to $-|\psi_1|^2 = -|\psi_2|^2$. In this case $\psi_1 + \psi_2 = 0$ and $\text{P}_q = 0$. But P_c remains the same at $2|\psi_1|^2$. So if we think of the photon as a classical particle, it has a probability of going through one or the other opening, but as a quantum mechanical wave particle, the photon cannot cross over through either opening. The quantum paradox is therefore that if only one of the openings P or Q is open, the photon has a chance of crossing to the other side, but if both are open, the photon cannot cross.

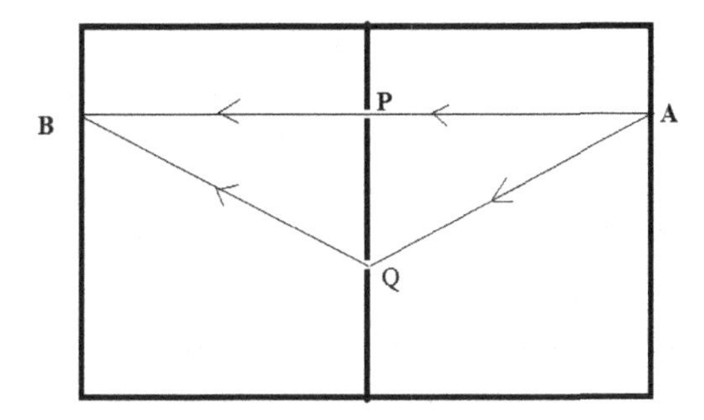

The above diagram illustrates how the quantum mechanical probability could become zero. A photon is created at A. A detector is placed at B to observe this photon. If we think of the photon as a classical particle, this photon has some probability of going through P and then to B, and another probability of going through Q and then to B. So the probability that the photon is detected at B is the sum of these two probabilities.

The resolution of the paradox is that the quantum mechanical probability is calculated by considering the propagation of the photon as a wave. Because it is a wave, it undergoes interference between the two paths. One wave travels from A to P to B, and another wave travels from A to Q to B. Suppose the difference between these two paths is exactly an odd multiple of half a wavelength:

$$AQB - APB = (2n+1)\frac{\lambda}{2}$$

where n is any whole number 0, 1, 2, 3, etc. Then there is destructive interference between the two waves at the point B, with the result that the intensity at B is zero. Hence there is zero probability of finding the photon at B.

9.6 Matrix Representation

9.6.1 *States, Operators and Matrices*

A state with no photons is written $|0\rangle$. The question "Does this state have zero photons?" is expressed as $\langle 0|$. The answer is yes, and so we write $\langle 0|0\rangle = 1$. A state with 1 photon is written $|1\rangle$. If now we ask "Does this state have one photon?" the answer is yes, and we write $\langle 1|1\rangle = 1$.

But if the given state is $|1\rangle$ and the question is "Does this state have zero photons?" the answer is no, and we write $\langle 0|1\rangle = 0$. Likewise $\langle 1|0\rangle = 0$.

So far we have been treating the states $|0\rangle$ and $|1\rangle$ (as well as $\langle 0|$ and $\langle 1|$ as abstract entities. We shall now make them concrete in terms of familiar mathematical structures.

One very important mode of representing states and operators is matrix notation. We shall see that it is possible to represent the relationships between the states $|0\rangle$, $|1\rangle$, and the operators a and $a^\dagger$ by drawing a "numerical picture" of each one of these entities by an array of numbers. Specifically, we shall seek to represent each one of these abstract quantum theoretical objects by a rectangular array of numbers — in general complex numbers.

The matrix representation of quantum mechanical objects is helpful because these matrices follow universal algebraic rules of addition, subtraction, multiplication, etc. It is remarkable that these rules can be validly employed in the quantum mechanical calculation of physical observable and measurable quantities.

9.6.2 *A Highly Simplified Special Case*

We will now consider a highly simplified and artificial situation. A cavity has just enough energy to produce one photon. So at any time there may be no photon in the cavity, or exactly one photon. So the state of the cavity is *binary*, because it can take one of two values: $|0\rangle$, or $|1\rangle$.

Of course, a real cavity can have billions of photons, since as long as there is sufficient available energy there is no reason in principle why more and more photons cannot be emitted by the walls. But our cavity is idealized and artificial and serves merely to illustrate how quantum mechanical states and operators can be represented by matrices or arrays of numbers.

Let us represent the state $|0\rangle$ by the column matrix or column vector $\begin{bmatrix} \frac{1}{\sqrt{2}} \\ \frac{i}{\sqrt{2}} \end{bmatrix}$.
What is the explanation for this particular array of numbers? Right now we are offering no explanation. We are just suggesting that this rather arbitrary pair of numbers will help to make somewhat more concrete the abstract equations we have written so far.

Since $\langle 0|0\rangle = 1$ we see that it makes sense to set $\langle 0| = \begin{bmatrix} \frac{1}{\sqrt{2}} & -\frac{i}{\sqrt{2}} \end{bmatrix}$ so that the number $\langle 0|0\rangle$ is equal to the matrix product

$$\begin{bmatrix} \frac{1}{\sqrt{2}} & -\frac{i}{\sqrt{2}} \end{bmatrix} \begin{bmatrix} \frac{1}{\sqrt{2}} \\ \frac{i}{\sqrt{2}} \end{bmatrix} = 1 \tag{9.5}$$

Note that the elements of $\langle 0|$ are the complex conjugates of the elements of $|0\rangle$.

We could also set up a matrix or vector for the $|1\rangle$ state as $\begin{bmatrix} \frac{i}{\sqrt{2}} \\ \frac{1}{\sqrt{2}} \end{bmatrix}$. The corresponding $\langle 1|$ state vector becomes $\begin{bmatrix} -\frac{i}{\sqrt{2}} & \frac{1}{\sqrt{2}} \end{bmatrix}$.

Exercise:
Taking $|0\rangle = \begin{bmatrix} \frac{1}{\sqrt{2}} \\ \frac{i}{\sqrt{2}} \end{bmatrix}$, $|1\rangle = \begin{bmatrix} \frac{i}{\sqrt{2}} \\ \frac{1}{\sqrt{2}} \end{bmatrix}$, $\langle 0| = \begin{bmatrix} \frac{1}{\sqrt{2}} & \frac{-i}{\sqrt{2}} \end{bmatrix}$ and $\langle 1| = \begin{bmatrix} \frac{-i}{\sqrt{2}} & \frac{1}{\sqrt{2}} \end{bmatrix}$ use the rules of matrix multiplication to show that
(a) $\langle 0|0\rangle = 1$ (b) $\langle 1|1\rangle = 1$ (c) $\langle 0|1\rangle = 0$ (d) $\langle 1|0\rangle = 0$.

$|n\rangle$ is called a *ket* or a ket vector and $\langle n|$ is called a *bra* or bra vector, where these words are derived from *bracket*. So a bra vector can be written as a horizontal row matrix, and a ket vector as a vertical column matrix. We notice that when a bra vector or row matrix is converted into a ket vector or column matrix or vice versa the imaginary numbers changes sign. In other words, the elements of the bra form of a vector are the complex conjugates of the corresponding elements of the ket form of the same vector.

If after transposing a matrix X — i.e. switching columns and rows — we change each element into its complex conjugate, then the resulting matrix is called the adjoint of the original matrix, written as $X^\dagger$. The converse is also true. So if $X = \begin{bmatrix} 2i & -4i \end{bmatrix}$ then $X^T = \begin{bmatrix} 2i \\ -4i \end{bmatrix}$ and $X^\dagger = \begin{bmatrix} -2i \\ 4i \end{bmatrix}$. Thus $|0\rangle^\dagger = \langle 0|$ and $\langle 0|^\dagger = |0\rangle$.

9.6.3 *Operators and Square Matrices*

If a square matrix A has this property that $A^\dagger = A$, then we say that A is Hermitian. So the matrices $\begin{bmatrix} 0 & -i \\ i & 0 \end{bmatrix}$ and $\begin{bmatrix} 1 & 0 \\ 0 & -1 \end{bmatrix}$ are Hermitian matrices.

Suppose A and B are both $n \times n$ matrices. Let $C = AB$. It can be shown easily that $C^T = B^T A^T$ and that $C^\dagger = B^\dagger A^\dagger$.

Next we shall set up matrices for the operators a and $a^\dagger$. For the operator a let us try the matrix $\frac{1}{2}\begin{bmatrix} -i & 1 \\ 1 & i \end{bmatrix} = \begin{bmatrix} -\frac{i}{2} & \frac{1}{2} \\ \frac{1}{2} & \frac{i}{2} \end{bmatrix}$. Here we have followed the rule that multiplying a matrix by a number means multiplying each element of the matrix by that number.

Exercises:

1. Taking $a = \frac{1}{2}\begin{bmatrix} -i & 1 \\ 1 & i \end{bmatrix}$ and $|1\rangle = \begin{bmatrix} \frac{i}{\sqrt{2}} \\ \frac{1}{\sqrt{2}} \end{bmatrix}$ and $|0\rangle = \begin{bmatrix} \frac{1}{\sqrt{2}} \\ \frac{i}{\sqrt{2}} \end{bmatrix}$ show that $a|1\rangle = |0\rangle$ by matrix multiplication.
2. Given $A = \begin{bmatrix} 1 & -i \\ 2 & -3i \end{bmatrix}$ and $B = \begin{bmatrix} 2i & 0 \\ -i & 1 \end{bmatrix}$ show that $(AB)^\dagger = B^\dagger A^\dagger$.

We next need a corresponding matrix for the operator $a^\dagger$. It seems plausible that $a^\dagger$ is the adjoint of a. Let us recall that wherever a has the number i, we expect $a^\dagger$ to have the number $-i$. So let us try $a^\dagger = \frac{1}{2}\begin{bmatrix} i & 1 \\ 1 & -i \end{bmatrix}$.

Exercise:

Taking $a^\dagger = \frac{1}{2}\begin{bmatrix} i & 1 \\ 1 & -i \end{bmatrix}$ and $|1\rangle = \begin{bmatrix} \frac{i}{\sqrt{2}} \\ \frac{1}{\sqrt{2}} \end{bmatrix}$ and $|0\rangle = \begin{bmatrix} \frac{1}{\sqrt{2}} \\ \frac{i}{\sqrt{2}} \end{bmatrix}$ show that $a^\dagger|0\rangle -$ $|1\rangle$ by matrix multiplication.

Thus we have obtained matrices for $|0\rangle, |1\rangle, a$ and $a^\dagger$. We find that these matrices obey the equations $\langle 0|0\rangle = 1, \langle 1|1\rangle = 1, \langle 0|1\rangle = 0, \langle 1|0\rangle = 0, a|1\rangle = |0\rangle$ and $a^\dagger|0\rangle = |1\rangle$.

Are these matrices purely ad hoc, meaning they have been cooked up only to satisfy the equations we want them to satisfy? Or can they actually be used to make some predictions?

Suppose we were to apply the annihilation operator to a state with no photons. This cannot be done, since it is impossible to annihilate a photon when there is no photon present. Quantum mechanically, we represent this fact in an equation as follows:

$$a|0\rangle = 0$$

The product of a 2×2 square matrix and a 2×1 column matrix is another 2×1 column matrix. So the number 0 on the right-hand side of the above equation is actually a 2×1 column matrix with both elements equal to 0, i.e. $\begin{bmatrix} 0 \\ 0 \end{bmatrix}$. Let us check if we do get this matrix as a result of multiplying a and $|0\rangle$. We obtain the result:

$$\frac{1}{2}\begin{bmatrix} -i & 1 \\ 1 & i \end{bmatrix}\begin{bmatrix} \frac{1}{\sqrt{2}} \\ \frac{i}{\sqrt{2}} \end{bmatrix} = \begin{bmatrix} 0 \\ 0 \end{bmatrix} \tag{9.6}$$

So we have some degree of confidence in our choice of matrix elements to describe the annihilation operator a.

Thus the matrix representation of these states and operators has served to illustrate how matrices work in quantum mechanics. But now we want to reiterate something we emphasized at the beginning of this discussion, that these particular matrices that we created are artificial and work only in the limited situation where there is at most one photon in the cavity. Actual physical states distinguished by the number of photons are not binary, because the number of photons in a real cavity is not limited to 0 or 1. Because of the limitations imposed on our physical setup, we find that the matrices representing $a^\dagger$ and $|1\rangle$ obey this equation $a^\dagger|1\rangle = 0$, which means that it is impossible to add another photon to the cavity which already has 1 photon. Real cavities do not behave this way, and so we cannot use *these* matrices in real situations.

What do the matrices look like for a real cavity? To answer that question, let us first ask what sort of matrices would describe a cavity that can contain a maximum of 2 photons. Now there are 3 possible states — no photon, one photon or two photons. It is a good guess that the state vectors would have 3 elements and the creation and annihilation operators would be 3×3 matrices. That would be correct. We could continue this way. A cavity capable of holding a maximum of n photons would require an $n+1$ state vector and $(n+1) \times (n+1)$ operators. Since there is no limit to the number of photons a real cavity can hold, the state vectors and operators would have infinite numbers of elements. We will not attempt to write down these infinite matrices.

9.7 Orthonormal Vectors

We see that quantum mechanical entities such as state functions and operators can all be expressed in matrix form. A state function is expressible as a column matrix, which is also called a column vector. The rules of vector algebra basically carry over to the algebra of state functions in quantum mechanics. Because of the close analogy between vectors and state functions, the latter are also called state vectors. The analogy is purely mathematical, not physical. A quantum mechanical state cannot be pictured as a physical quantity with magnitude and direction. Vectors such as force, displacement, velocity, etc., can be measured in physical space. Quantum mechanical state functions are vectors in an abstract space called *Hilbert space.* An important property of Hilbert space is that the components of a state vector are in general complex numbers. So, there is no one-to-one correspondence between Hilbert space and the space of the electromagnetic field defined by Maxwell's equations, since the coordinates of the latter are all real numbers.

Suppose we have a vector $\mathbf{r} = \hat{\imath}x + \hat{\jmath}y$ in two dimensions. The length or magnitude of this vector $r = \sqrt{x^2 + y^2}$. If we divide a vector by its magnitude we do not change the direction of the vector but we alter its magnitude to unity. This procedure is called *normalizing* the vector. So any vector $\vec{r}$ can be normalized to $\hat{r}$ by dividing $\vec{r}$ by its magnitude r: $\hat{r} = \vec{r}/r$.

So if $\vec{r} = [x\ y]$ (the vector expressed in matrix form) then the unit vector $\hat{r}$ has the components

$$\frac{x}{\sqrt{x^2 + y^2}} \text{ and } \frac{y}{\sqrt{x^2 + y^2}}$$

We saw above that $\langle 0|0\rangle = 1$. So that tells us that the state vector $|0\rangle$ has been normalized. In quantum mechanics we want $\langle x|x\rangle = 1$ for every state function. Normalization is important for vector state functions. So, henceforth, when we write the state functions $|0\rangle$,$|1\rangle$, $|2\rangle$, etc. it will be understood that these functions are all normalized.

What is the value of the quantity $\langle 0|1\rangle$? This expression means that a state having a single photon is measured to check if it is a state having no photon. These two states are wholly disjoint. According to quantum theory there can be no overlap between these two states. If we measure one state by another that is wholly disjoint then the result is 0. So we write $\langle 0|1\rangle = 0$. Another way of stating this result is that a state having one photon cannot simultaneously be a state having no photon. More precisely, the amplitude for a state having one photon to be also a state with no photon is zero.

In mathematical language we say that the states $|0\rangle$ and $|1\rangle$ are *orthogonal* to each other. If two state vectors are normalized and are also orthogonal to each other we call them *orthonormal* vectors. $|x\rangle$ and $|y\rangle$ are orthonormal if $\langle x|y\rangle = \langle y|x\rangle = 0, \langle x|x\rangle = 1, \text{and} \langle y|y\rangle = 1$. Clearly, $|0\rangle$ and $|1\rangle$ are orthonormal vectors. For n dimensions one can find n vectors with every pair of the set orthogonal to each other. In a space with a maximum of n photons, $|k\rangle$ and $|m\rangle$ are orthonormal if $\langle k|m\rangle = \delta_{k,m}$.

9.8 Operators and State Vectors

9.8.1 *Changing a Vector by an Operator*

We now eschew the artificial matrix formulations of the creation and annihilation operators that we had set up earlier (Section 9.6) for illustrative purposes. The actual matrix forms of these operators are complicated by the fact that there is no limit to the number of photons we can add to the cavity. And so we will not consider the actual matrix structures of these operators in the ensuing discussions.

An operator acting on a vector would in general produce a different vector. So $a^\dagger$ changes $|0\rangle$ to $|1\rangle$. From the standpoint of physics $a^\dagger|0\rangle = |1\rangle$ means the creation operator adds a photon to a field with no photons and $a|1\rangle = |0\rangle$ means the annihilation operator removes a photon from a field with one photon. We would therefore expect that $a^\dagger|1\rangle = |2\rangle$ but this would not be quite right because we do not know if $a^\dagger|1\rangle$ is a normalized state. Some calculations show that

$$a^\dagger|n\rangle = \sqrt{n+1}|n+1\rangle \tag{9.7}$$

We shall next explain how this formula can be derived.

9.8.1.1 *The Number Operator*

Suppose we initially have a state with 1 photon: $|1\rangle$. Now, let us annihilate this photon: $a|1\rangle$. This yields a state with no photons: $a|1\rangle = |0\rangle$. We then re-create a photon and thus restore the original state: $a^\dagger|0\rangle = |1\rangle$. Now, $\langle 1|1\rangle = 1$. Therefore, $\langle 1|a^\dagger a|1\rangle = 1$.

Next, let us begin with a state having two photons: $|2\rangle$. We annihilate one photon: $a|2\rangle$. We cannot assume that this is equal to $|1\rangle$, since we do not know that $a|2\rangle$ is a normalized state. We next create a photon: $a^\dagger a|2\rangle$.

When this state is measured by a state with 2 photons the result is $\langle 2|a^\dagger a|2\rangle$. We cannot assume that this is equal to 1. In fact, it is not. The reason is that photons are *indistinguishable* particles. Since there are 2 photons in the field, which we shall label "photon A" and " photon B", in annihilating one photon and creating a photon we could have done it in two different ways. We could have annihilated photon A and then created a photon in its place, or we could have annihilated photon B and then created a photon in its place.

These two alternative ways of doing the job represent two different "paths" that we could have taken from the initial state to the final state. As we saw earlier, when there are two different paths for a process, and we do not make an observation of which path is taken in the process, we add up the amplitudes for the different paths. The example given earlier dealt with an actual physical path. Here we are talking about different *choices.* But the same quantum mechanical principle applies:

If a process can take place in two different ways, and we cannot tell which way the system actually "chose", then we find out the amplitudes for each of the two ways, and add up the amplitudes to obtain the amplitude for the completed process.

So this is how we proceed in this situation. We focus on one photon, which we shall call A. We consider the state containing this one photon, which is $|1\rangle_A$. We then obtain the amplitude for annihilating and creating this photon: $\langle 1|_A a^\dagger a|1\rangle_A = 1$.

Next we focus on photon B, and obtain the amplitude for annihilating and creating this photon: $\langle 1|_B a^\dagger a|1\rangle_B = 1$.

Since the two photons are indistinguishable, each of the two processes described just now represents an alternative "path" that we can take in the process of removing and then adding a photon to the field containing two photons. Thus the total amplitude that we want is

$$\langle 2|a^\dagger a|2\rangle = \langle 1|_A a^\dagger a|1\rangle_A + \langle 1|_B a^\dagger a|1\rangle_B = 2.$$

If there are 3 photons, there are 3 different ways of annihilating and creating a photon in this field, and so

$$\langle 3|a^{\dagger}a|3\rangle = 3.$$

Proceeding this way, it follows that we can write a general formula as

$$\langle n|a^{\dagger}a|n\rangle = n. \tag{9.8}$$

When the annihilation operator acts on $|n\rangle$, it reduces the number of photons in the state from n to $n-1$. So, $a|n\rangle = \lambda_1|n-1\rangle$, where λ_1 is some (complex) number. Next, we operate on this state by the creation operator $a^{\dagger}$, which adds a photon to the state and brings the number back to n: $a^{\dagger}\lambda_1|n-1\rangle = \lambda_2\lambda_1|n\rangle$, where λ_2 is another complex number. So we have operated on the state $|n\rangle$ by two operators in succession, first a and then $a^{\dagger}$. This process is written as $a^{\dagger}a|n\rangle$. And so, we have found that

$$a^{\dagger}a|n\rangle = \lambda_2\lambda_1|n\rangle \tag{9.9}$$

Operating on the left by $\langle n|$,

$$\langle n|a^{\dagger}a|n\rangle = \langle n|\lambda_2\lambda_1|n\rangle = \lambda_2\lambda_1\langle n|n\rangle \tag{9.10}$$

Comparison with Eq. (9.8) shows that $\lambda_2\lambda_1 = n$. Hence Eq. (9.9) can be written as

$$a^{\dagger}a|n\rangle = n|n\rangle \tag{9.11}$$

$a^{\dagger}a$ is called the *number operator*, because when it operates on a state with n photons, it multiplies the state by the number of photons in the state.

Exercise:
Go back to the special case of the cavity with a maximum of 1 photon. Write down the operators $a^{\dagger}$ and a in 2×2 matrix form and the state vectors $|0\rangle$ and $|1\rangle$ as 2×1 column matrices. Obtain the matrix product $a^{\dagger}a$ and show that $a^{\dagger}a|0\rangle = 0$ and $a^{\dagger}a|1\rangle = |1\rangle$.

9.8.2 *Eigenfunctions and Eigenvalues*

In mathematics, when an operator acts upon a function and leaves the function unchanged except perhaps for a constant factor, we say that the function is an *eigenfunction* of the operator, and the constant factor is called the corresponding *eigenvalue*. So $|n\rangle$ is an eigenfunction of the operator $a^{\dagger}a$ with eigenvalue n.

But a and $a^\dagger$ do not have eigenfunctions. Since the annihilation operator by its very definition represents the absorption of a photon, it cannot possibly have an eigenfunction, since it changes the state which it operates upon (e.g. from one photon to no photon). Similarly the creation operator cannot have an eigenfunction.

The states $|n\rangle$ and $|n-1\rangle$ are each normalized. And we know that $a|n\rangle$ yields the state $|n-1\rangle$ but we cannot simply write $a|n\rangle = |n-1\rangle$ because there may be a factor that we do not know. Earlier we called this factor λ_1. Now we will call this factor k_n, where the subscript indicates that this number may vary with the value of n. In general k_n could be a complex number.

So

$$a|n\rangle = k_n|n-1\rangle. \tag{9.12}$$

Earlier we saw that for two matrices A and B the following relationship holds good: $(AB)^\dagger = B^\dagger A^\dagger$. Taking the adjuncts of both sides of Eq. (9.12) we obtain $\langle n|a^\dagger = k_n^*\langle n-1|$. Combining these two equations we obtain

$$\langle n|a^\dagger a|n\rangle = k_n^* k_n\langle n-1|n-1\rangle = |k_n|^2.$$

So $|k_n|^2 = n$. Can k_n be a real number? If so, then $k_n = \sqrt{n}$. Let us check if this would work. Since $a|1\rangle = |0\rangle$, we see that $k_1 = 1$. So k_n is real for $n = 1$ and so we may take k_n as a real number for all n. So $k_n^2 = n$ and therefore $k_n = \sqrt{n}$.

And so we have $a|n\rangle = \sqrt{n}|n-1\rangle$. Multiplying on the left by $a^\dagger$ we obtain

$$a^\dagger a|n\rangle = \sqrt{n}a^\dagger|n-1\rangle$$

Since the left-hand side equals $n|n\rangle$, we obtain the relation $\sqrt{n}a^\dagger|n-1\rangle = n|n\rangle$. Thus

$$a^\dagger|n-1\rangle = \sqrt{n}|n\rangle$$

So $a^\dagger|0\rangle = |1\rangle$, $a^\dagger|1\rangle = \sqrt{2}\,|2\rangle, a^\dagger|2\rangle = \sqrt{3}\,|3\rangle, a^\dagger|3\rangle = \sqrt{4}\,|4\rangle$, etc. And from Eq. (9.12) we get $a|4\rangle = \sqrt{4}|3\rangle$, $a|3\rangle = \sqrt{3}\,|2\rangle, a|2\rangle = \sqrt{2}\,|1\rangle, a|1\rangle = |0\rangle$, and $a|0\rangle = 0$. The last equation implies that it is impossible to annihilate a photon from a state having zero photons.

The number operator $a^\dagger a$ has as eigenvalue the number of photons in the field. We can therefore write $a^\dagger a = \hat{N}$. $\hat{N}|n\rangle = n|n\rangle$.

There is no upper limit to the number of photons in any cavity, and so $\hat{N}$ has many eigenfunctions $|n\rangle$, theoretically infinite in number, for all possible integer values of n, and correspondingly an infinite number of eigenvalues n. Therefore the operator $aa^{\dagger}$ has an infinite number of rows and columns, and consequently $a^{\dagger}$ and a are infinite-dimensional matrices.

What about the operator $aa^{\dagger}$? Does this operator have an eigenvalue and a corresponding eigenfunction? Let us apply this operator to the state function $|n\rangle$. We obtain

$$aa^{\dagger}|n\rangle = a\sqrt{n+1}|n+1\rangle = \sqrt{n+1}\sqrt{n+1}|n\rangle = (n+1)|n\rangle \qquad (9.13)$$

So $aa^{\dagger}$ also has an infinite number of eigenfunctions $|n\rangle$ with an infinite number of corresponding eigenvalues $n+1$.

It is also evident that $a^{\dagger}a \neq aa^{\dagger}$. Operators in quantum mechanics do not in general follow the commutative law of multiplication. The operators $a^{\dagger}$ and a do not commute. But the operators $a^{\dagger}a$ and $aa^{\dagger}$ do commute, and we can show easily that $a^{\dagger}aaa^{\dagger}|n\rangle = aa^{\dagger}a^{\dagger}a|n\rangle$. Commutation and non-commutation between operators is a central theme in quantum mechanics. In fact, one could say it is this very property that defines quantum mechanics and distinguishes it from classical mechanics. We shall say more about this in a subsequent section of this chapter.

9.9 Energy of a Photon Field

Suppose we were to try to measure the energy of a photon field inside a cavity. In order to do that we need to interact with the field. In quantum theory we cannot possibly make a measurement on a system without disturbing the system in some way. So the act of measurement is itself a physical process that follows the rules of quantum theory.

Without going into specific physical details we present the summary of the process. Photons are being created and annihilated inside the cavity. Suppose this cavity contains only photons of a particular wavelength and therefore of a particular frequency ν. So this cavity is not a black body. We could think of it as a box with reflecting inner walls. And let us say that at any time there are n photons in this cavity. How do we measure this number? These photons are constantly bouncing off the walls and thereby imparting momentum to the walls. But when we say a photon bounces off a wall what we actually mean is that a photon is absorbed by a wall and another photon is emitted by the wall. We could therefore count the number of emissions or the number of absorptions. According to classical physics these two numbers should be the same, but we cannot make that assumption in quantum physics. If we count both of these separately, and find they are not equal, we shall take the average in order to obtain the measured energy of the photons inside the container, which we shall obtain by multiplying this number by the energy of a single photon $h\nu$.

There are two ways of counting the photons: either count the number of ways a photon could be emitted first, and then absorbed, or count the number of ways a photon could be absorbed first, and then emitted. If it turns out that these two numbers are different, we take the average of the two numbers.

Number of ways a photon could be emitted first and then absorbed:

$$\langle n|aa^\dagger|n\rangle = n + 1 \tag{9.14}$$

Number of ways a photon could be absorbed first and then emitted:

$$\langle n|a^\dagger a|n\rangle = n \tag{9.15}$$

The two numbers are not the same. So the required number is

$$\frac{1}{2}(n + 1 + n) = n + \frac{1}{2}.$$

Thus, the measured energy of the field is given by

$$E = h\nu\left(n + \frac{1}{2}\right) = \hbar\omega\left(n + \frac{1}{2}\right) \tag{9.16}$$

It is interesting that even when the number of photons $n = 0$ the energy is not zero, but is $\frac{1}{2}\hbar\omega$. This energy is called the *zero point energy* and is present even in the absence of any photon. This is a purely quantum phenomenon, and is related to the uncertainty principle. The uncertainty in energy comes from the uncertainty in the number of counted photons.

In a real field one has to consider the contributions due to photons of all frequencies. So even in the absence of any photons, there is a zero point energy contribution from every possible frequency. If we were to add up all these contributions, we would get infinity. Thus we obtain this apparently unphysical result that a vacuum contains an infinite amount of energy. However, this is not really a problem, because what is actually measurable is not the energy stored in the field, but the work done by this energy in displacing an object through a force. So if we bring two metal plates very close to each other, there is some vacuum energy between them. This energy is a potential energy. Every system tends to move in such a way as to reduce its potential energy. And this tendency is experienced as a force. So the plates will experience a force when they are brought very close to each other. It turns out that for the case of two conducting plates there is a force of attraction, and this force is finite, even though the vacuum energy is infinite. (But, as we shall see in a later chapter, this infinite positive energy could be canceled by the infinite negative energy of electrons according to the Dirac theory.)

The force between two conducting plates due to the vacuum energy is called the *Casimir force*, which was predicted by Casimir in 1948 and measured experimentally about 50 years later.

9.10 Hermitian Operators

9.10.1 *Eigenvalues of Hermitian Operators*

$\hat{N}$ is a Hermitian operator. When it operates on the state function $|n\rangle$ it multiplies the state function by the eigenvalue n.

It can be proved mathematically that for any Hermitian operator it is possible to find an eigenfunction such that when the operator acts on the eigenfunction, it multiplies the eigenfunction by a constant which is the eigenvalue. We shall not go into this proof, but accept it as true.

An important theorem states that the eigenvalues of a Hermitian operator must be real numbers. This proof is very simple, and we provide it below:

Let λ be an eigenvalue of a Hermitian operator A. Let $|\psi\rangle$ be the corresponding eigenfunction. So

$$A|\psi\rangle = \lambda|\psi\rangle \tag{9.17}$$

Let us take the adjoint of both sides of the equation:

$$\langle\psi|A^{\dagger} = \langle\psi|\lambda^{*} \tag{9.18}$$

Now we multiply on the right by the state function $|\psi\rangle$:

$$\langle\psi|A^{\dagger}|\psi\rangle = \lambda^{*}\langle\psi|\psi\rangle = \lambda^{*} \tag{9.19}$$

Since A is Hermitian, $A^\dagger = A$. And so,

$$\langle\psi|A|\psi\rangle = \lambda^* \tag{9.20}$$

Let us next multiply on the left of Eq. (9.17) by $\langle\psi|$:

$$\langle\psi|A|\psi\rangle = \lambda \tag{9.21}$$

Comparing Eqs. (9.20) and (9.21) we see that $\lambda^* = \lambda$ and so λ is real.

Since every eigenvalue of a Hermitian operator is a real number, it follows that a Hermitian operator is a suitable mathematical tool to describe a physical quantity that is the result of a measurement, such as position, momentum, energy, angular momentum, number of particles, electric charge, magnetic moment, etc.

Indeed, one of the fundamental principles of quantum theory is that every physically observable or measurable quantity can be represented mathematically by a Hermitian operator, and that the eigenvalues of this operator represent the possible measured values of the physical quantity. Thus the energy operator of a hydrogen atom has as its eigenvalues all the possible energy levels of the atom.

To summarize:

a and $a\dagger$ are not Hermitian. $a \neq a^\dagger$. The operators $\hat{N} = a^\dagger a$ and $\hat{M} = aa^\dagger$ are Hermitian. $(a^\dagger a)^\dagger = a^\dagger a$ and $(aa^\dagger)^\dagger = aa^\dagger$. The state function $|n\rangle$ is an eigenfunction of both $a^\dagger a$ and $aa^\dagger$ with eigenvalues n and $n+1$ respectively.

9.11 Other Operators in Quantum Mechanics

We are now familiar with the creation operator which creates a photon in a field, and the annihilation operator which removes a photon from the field. We are also familiar with combinations of such operators.

Suppose we are interested in measuring some physical attribute of a particle such as its energy, or its momentum. Such measurements are also represented by operators. These operators act upon different types of Hilbert spaces. An important type is the spatial wave function.

9.12 The Spatial Wave Function

9.12.1 *Probability Density*

A photographic plate records the spot at which a photon hit the plate. But this spot is not a geometric point. If it were, it would not be visible to the human eye, and we could never detect the photon. The spot is actually a smudge, smeared over a small area. The spot on the photographic plate also has a small depth. So we have really localized the photon to within a small volume, and not a geometrical point.

We have discussed photons within cavities or chambers. Now we make the chamber infinitesimal in size, and ask: What is the probability of finding the photon within a small volume ΔV? Since this volume is very small, we shall call it a *volume element.* We first define a quantity that we call a probability density ρ which has the meaning that the product $\rho\Delta V$ is the probability of finding the photon within the volume element ΔV. Let us choose a point at the center of this volume element and label this point by its coordinates as $P(x, y, z)$. Since we expect that the probability density should vary from point to point, it is in general a function of the coordinates of the point: $\rho(x, y, z)$. ρ is a real scalar field or *scalar point function.*

Suppose we know for a fact that the photon is somewhere inside a macroscopic cavity. Let us divide up the cavity into a large number of microscopic volume elements ΔV_i. The probability of finding the photon within the element ΔV_i is equal to $\rho(x_i, y_i, z_i)\Delta V_i$. Here i can have a value from 1 to some very large number equal to the number of volume elements into which we have divided up the cavity.

9.12.2 *Amplitude and Probability*

We saw earlier that a probability P for an event to occur — such as detecting a photon on one of the walls of the cavity — is related to the amplitude α for the event to occur by the equation

$$P = \alpha^*\alpha = |\alpha|^2$$

We make a slightly different definition when we come to spatial probability, or the probability of finding the particle in some small volume located at a point.

We define an amplitude ψ, which we call the *spatial wave function*, which is related to the probability density of finding the particle at a point i by the equation

$$\rho(x_i, y_i, z_i) = \psi^*\psi = |\psi|^2$$

So the probability of finding the particle inside a small volume element ΔV_i is given by

$$P_i = \psi^*(x_i, y_i, z_i)\psi(x_i, y_i, z_i)\Delta V_i$$

If we know for certain that the particle is somewhere within a fixed space, we can divide up the space into a large number of tiny volume elements, and if we add up the probabilities of finding the particle in each of the volume elements we should get 1 as the total:

$$\Sigma_i P_i = \sum_i \psi^*(x_i, y_i, z_i)\psi(x_i, y_i, z_i)\Delta V_i = 1$$

Taking the limit as the size of each volume element goes to zero:

$$\int \psi^*(x_i, y_i, z_i)\psi(x_i, y_i, z_i)dV = 1 \tag{9.22}$$

A wave function that satisfies this equation is said to be normalized. We will next obtain a relationship between the wave function ψ and the creation and annihilation operators.

9.12.3 *Field Operators and Probability Amplitude*

Let us divide the cavity into a very large number of microscopic volume elements ΔV_i. We consider the case where there is only one particle inside the cavity. Since this particle can be found somewhere within the cavity, clearly the probability of finding it somewhere inside the cavity is 1.

The probability of finding the particle inside the element ΔV_i is

$$\psi^*(x_i, y_i, z_i)\psi(x_i, y_i, z_i)\Delta V_i$$

so that

$$\sum_i \psi^*(x_i, y_i, z_i)\psi(x_i, y_i, z_i)\Delta V_i = 1.$$

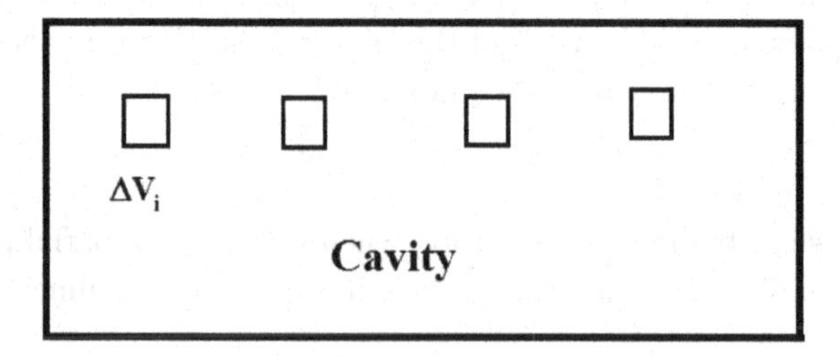

The schematic diagram shows the cavity with four volume elements (greatly exaggerated in size for clarity), one of them marked ΔV_i. Suppose we detect a photon inside this volume element ΔV_i. If the particle is detected within the element ΔV_i then it cannot simultaneously be anywhere else within the cavity. The quantum mechanical way of representing this fact is to say that the photon has been annihilated from the cavity as a whole and then created within the small volume element ΔV_i.

We may write the amplitude for this process as $\langle \Delta V_i | a^\dagger a | 1 \rangle$.

This is a complex number of the form $\langle \psi | \phi \rangle$. Suppose we were to represent these quantum mechanical states by matrices. So let $|\psi\rangle = \begin{bmatrix} a \\ b \end{bmatrix}$ and let $|\phi\rangle = \begin{bmatrix} c \\ d \end{bmatrix}$.

So $\langle \psi | \phi \rangle = \begin{bmatrix} a^* & b^* \end{bmatrix} \begin{bmatrix} c \\ d \end{bmatrix} = a^* c + b^* d$.

It is easy to see that $\langle \phi | \psi \rangle = \begin{bmatrix} c^* & d^* \end{bmatrix} \begin{bmatrix} a \\ b \end{bmatrix} = c^* a + d^* b$.

Thus $\langle \psi | \phi \rangle^* = \langle \phi | \psi \rangle$ and so we can write $|\langle \psi | \phi \rangle|^2 = \langle \psi | \phi \rangle \langle \phi | \psi \rangle$.

Replacing $|\phi\rangle$ by $|\Delta V_i\rangle$ and $|\psi\rangle$ by $a^\dagger a|1\rangle$, we may write

$$|\langle \Delta V_i | a^\dagger a | 1 \rangle|^2 = \langle 1 | a^\dagger a | \Delta V_i \rangle \langle \Delta V_i | a^\dagger a | 1 \rangle$$

So

$$\langle 1 | a^\dagger a | \Delta V_i \rangle \langle \Delta V_i | a^\dagger a | 1 \rangle = \psi^*(x_i, y_i, z_i) \psi(x_i, y_i, z_i) \Delta V_i$$

This equation can be satisfied if we make

$$\psi(x_i, y_i, z_i) = \frac{\langle \Delta V_i | a^\dagger a | 1 \rangle}{\sqrt{\Delta V_i}}$$

So

$$\rho(x_i, y_i z_i) = |\psi(x_i, y_i, z_i)|^2 = \frac{|\langle \Delta V_i | a^\dagger a | 1 \rangle|^2}{\Delta V_i}$$

9.12.4 *Wave Function of a Plane Wave Photon*

In Section 5.3.5 we came across Eq. (5.36) which describes the plane wave solution for the electric field in the electromagnetic wave equation:

$$\mathbf{E} = \mathbf{E}_0 \cos(\mathbf{k} \cdot \mathbf{r} - \omega t) \tag{9.23}$$

The amplitude of this wave is E_0 and its energy density is proportional to E_0^2. The quantum mechanical wave function that describes the photon field should also satisfy the wave equation, but this function need not be real. Let us begin with the most general complex solution of the electromagnetic wave equation:

$$\mathbf{E} = \mathbf{E}_0 \exp i(\mathbf{k} \cdot \mathbf{r} - \omega t) \tag{9.24}$$

For a wave propagating in the x direction, we write the magnitude of the electric field as

$$E = E_0 \exp i(k_x x - \omega t) \tag{9.25}$$

where $\omega = k_x c$ and E_0 is real.

Here E is a complex number. $|E|^2 = E_0^2$, a real number. So, we can call E the wave function ψ (which may not be normalized), since $|\psi|^2$ is proportional to the probability density and hence to the number of photons per unit volume. Thus ψ can be chose to be equal to E to within a normalizing factor.

The energy of a quantum is $\mathrm{E} = h\nu = \hbar\omega$, and the momentum — along the x direction — is $p_x = h/\lambda = h\nu/c = \hbar k_x$. From Eq. (9.25) we obtain

$$\frac{\partial\psi}{\partial x} = ik_x\psi = (ip_x/\hbar)\psi \tag{9.26}$$

and

$$\frac{\partial\psi}{\partial t} = -i\omega\psi = -(i\mathrm{E}/\hbar)\psi \tag{9.27}$$

And so we obtain the equations

$$-i\hbar\frac{\partial\psi}{\partial x} = p_x\psi \tag{9.28}$$

and

$$i\hbar\frac{\partial\psi}{\partial t} = \mathrm{E}\psi \tag{9.29}$$

Equations (9.24) and (9.25) represent plane waves with constant ω and k_x, i.e. all the photons present in this wave have constant energy $\mathrm{E} = \hbar\omega$ and momentum $p_x = \hbar k_x$. So in Eq. (9.28) $-i\hbar\frac{\partial}{\partial x}$ is an operator, p_x is an eigenvalue, and ψ is the corresponding eigenfunction. In Eq. (9.29) $i\hbar\frac{\partial}{\partial t}$ is the operator, E the eigenvalue, and ψ the eigenfunction.

Thus we have derived two operators which act upon wave functions, viz. the momentum and the energy operators. In Cartesian coordinates these operators take the form $-i\hbar\frac{\partial}{\partial x}$, $-i\hbar\frac{\partial}{\partial y}$, $-i\hbar\frac{\partial}{\partial z}$, (or $-i\hbar\nabla$), $i\hbar\frac{\partial}{\partial t}$. Cartesian coordinates are suitable for plane waves. We can also write these momentum and energy operators for spherical and cylindrical coordinates, which would be appropriate for spherical and cylindrical waves, respectively.

9.13 Operators and Eigenvalues

9.13.1 *Physical Observables*

An operator is a mathematical entity that has a special relationship to the observed physical quantity that it represents. Suppose we want to measure a physical quantity A of a quantum mechanical system, say an atom. To do this experimentally we would set up the required apparatus to make this measurement. Quantum mechanics gives us a method of theoretically calculating the value of this observable quantity. And for this we need to find the wave function of the system ψ first. Once we have obtained ψ as a (complex) function of the coordinates, we can calculate the physical quantity mathematically.

The probability of finding the particle inside a small volume ΔV_i containing the point (x_i, y_i, z_i) is $\psi(x_i, y_i, z_i)^*\psi(x_i, y_i, z_i)\Delta V_i$. We first ask the question: if the particle were found in this volume element, what would be the measured value of variable A? Let us call this value A_i. We define the *expected value of* A as the average value we obtain for a measurement of A when carried out over a very large number of *independent* experiments. The word *independent* is extremely important. Each measurement is independent of the previous one. Since there is a finite probability of finding the particle inside any volume ΔV_i, and this probability is in general different from that of finding the particle in another volume ΔV_j, the expected value of the observable A is obtained by weighting the value of A in some volume ΔV_i by the probability of finding it inside ΔV_i. A is a function of the coordinates, and so the value of A will not be uniform over the entire region. By the laws of probability, the average measured value of A, which is also the theoretically calculated expected value of A, written as $\langle A \rangle$, is

$$\langle A \rangle = \frac{\int \psi^*(x, y, z)\hat{A}\psi(x, y, z)dV}{\int \psi^*(x, y, z)\psi(x, y, z)dV} \tag{9.30}$$

If the wave function ψ is normalized the denominator is 1. This definition is therefore valid even when ψ may not be normalized. We shall derive this equation in the following subsection.

9.13.2 *Application to Electrons*

We have seen that in classical electrodynamics the fields are generated by sources, which are stationary and moving charges, and the fields in turn operate on charges. At the quantum level, the electromagnetic field exists as photons which interact with charged particles such as electrons, protons, muons, etc. We have seen examples of interaction of photons with electrons in the Compton effect and the photoelectric effect. When photons interact with atoms the interaction is primarily with the atomic electrons.

The position, the momentum, and the energy of an electron are measured through the interaction of the electron with the electromagnetic field, i.e. through the interaction of the electron with one or more photons. (The mass of the electron is too small for direct measurement of the interaction of an electron with a gravitational field.) Hence, it is a plausible hypothesis that the operators for momentum and energy should have the same form for electrons and photons.

So, if $\psi(x, y, z)$ is an electron wave function, the momentum operator is $-i\hbar\frac{\partial}{\partial x}$, and the energy operator is $i\hbar\frac{\partial}{\partial t}$.

Now, a photon always travels at the speed c, but an electron always travels slower than light. For a slow moving electron we can use the Newtonian formula for kinetic energy, which is a good approximation for speeds much slower than light. We could then construct a kinetic energy operator thus:

$$\hat{T} = \frac{\hat{p}^2}{2m} = -\frac{\hbar^2}{2m}\left[\frac{\partial^2}{\partial x^2} + \frac{\partial^2}{\partial y^2} + \frac{\partial^2}{\partial z^2}\right] = -\frac{\hbar^2}{2m}\nabla^2 \tag{9.31}$$

where the hat over the letter signifies that the symbol is an operator.

The operator for the position is the position coordinate itself. So $\hat{x} = x$, $\hat{y} = y$, $\hat{z} = z$, and likewise for functions of the coordinates. For conservative forces the potential energy is a function of the coordinates: $V(x, y, z)$.

So, the Newtonian equation $T + V = E$ has the quantum counterpart:

$$-\frac{\hbar^2}{2m}\nabla^2\psi + V(x, y, z)\psi = i\hbar\frac{\partial\psi}{\partial t} \tag{9.32}$$

This is the original Schrödinger wave equation.

9.13.3 *Eigenvalues and Eigenfunctions*

In a stable atom, or other stable configurations where the total energy does not change with time, E is a constant in the equation

$$i\hbar\frac{\partial\psi}{\partial t} = E\psi \tag{9.33}$$

And so we obtain the time independent Schrödinger wave equation

$$-\frac{\hbar^2}{2m}\nabla^2\psi + V(x, y, z)\psi = E\psi \tag{9.34}$$

The left side of the equation is often abbreviated as $H\psi$ and so the equation is written as

$$H\psi = E\psi \tag{9.35}$$

H is the total energy operator, and is called the Hamiltonian of the quantum mechanical system. In this equation E is the eigenvalue of H and ψ is an eigenfunction of H.

This equation can be applied to many situations, not just an electron in an atom. We need to first know what is the expression for the potential energy V in that situation. We do not know ψ and we do not know E. But knowing V helps us to solve the equation and obtain the values of E that will satisfy the equation.

The Hamiltonian operator is Hermitian, and it has real eigenvalues. Energy is always real. It can be positive or negative. For an electron orbiting a nucleus, the total energy is negative, because the potential energy is negative, and is greater in magnitude than the kinetic energy.

In general, a system with constant energy can have more than one energy eigenvalue, with corresponding eigenfunctions. Suppose a system has 3 eigenvalues for a given Hamiltonian. Then we would have three solutions to the Schrödinger wave equation, which can be written as $H\psi_1 = E_1\psi_1$, $H\psi_2 = E_2\psi_2$ and $H\psi_3 = E_3\psi_3$.

9.13.4 *Electron in a General State*

In general, an electron may not be in an eigenstate. Let us consider the following example:

Suppose the energy of the ground state or lowest energy state of an electron in a hydrogen atom is E_1, and the energy of the first excited state is E_2. So if the electron is initially in the excited state, and drops from the excited state to the ground state, it would emit a photon of frequency ν such that

$$h\nu = E_2 - E_1$$

Conversely, if an electron which is in its ground state is hit with a photon frequency ν, this electron could absorb the photon and its own energy level would rise from E_1 to E_2.

Now, let us consider a container of hydrogen gas which is irradiated with light of frequency ν. The photons of this light beam would be absorbed by some atoms and the electrons of these atoms would then jump to the first excited state. Other atoms will not absorb a photon, and these would remain in the ground state.

So if we have a container of hydrogen which has been irradiated with some photons of frequency ν, then a given atom could at any given time be either in its ground state or in its first excited state. We are talking about the electron's energy state when we talk about the energy state of the atom.

If we make a measurement of the energy of the atom, we would obtain either E_1 or E_2. Suppose we now perform independent measurements on a large number of atoms which are in identical situations. Suppose we find that 36% of the time we obtain E_1 and 64% of the time we obtain E_2. So the probability of finding the atom in the ground state is 0.36 and the probability of finding it in the excited state is 0.64. So what will be the average value of the energy — which we write as $\langle E \rangle$ — that we measured over a thousand independent measurements on identically prepared atoms? The answer is

$$\langle E \rangle = 0.36E_1 + 0.64E_2 \tag{9.36}$$

Prior to making a measurement of the energy of an atom, we can theoretically set up the wave function of the atomic electron. Since we know it may be in one of the two energy states, but we cannot know which it is until we actually perform a measurement, we can definitely say the electron wave function is not an eigenfunction of the energy operator (Hamiltonian). Rather, we write it as a linear combination of the eigenfunctions as

$$\psi = a\phi_1 + b\phi_2 \tag{9.37}$$

When we carried out the measurements we found 36% to have energy E_1 and the remaining to have energy E_2. Could we have predicted this outcome prior to doing the measurements? Yes, if we knew the wave function ψ. We could then have applied the Hamiltonian operator to the wave function ψ:

$$H\psi = aH\phi_1 + bH\phi_2 \tag{9.38}$$

Now, ϕ_1 is an eigenfunction of H with eigenvalue E_1 and ϕ_2 is an eigenfunction of H with eigenvalue E_2. So $H\phi_1 = E_1\phi_1$ and $H\phi_2 = E_2\phi_2$. Hence

$$H\psi = aE_1\phi_1 + bE_2\phi_2 \tag{9.39}$$

Moreover, these two eigenstates are normalized. This means that $\int \phi_1^*\phi_1 dV = 1$ and $\int \phi_2^*\phi_2 dV = 1$.

Eigenstates are also orthogonal. If the particle is definitely in one of these eigenstates, then it is impossible for it to be found in the other state. This implies that $\int \phi_1^*\phi_2 dV = 0$ and $\int \phi_2^*\phi_1 dV = 0$.

Let us multiply both sides of Eq. (9.39) by $\psi^* = a^*\phi_1^* + b^*\phi_2^*$. We obtain

$$\begin{aligned}\psi^* H\psi &= (a^*\phi_1^* + b^*\phi_2^*)(aE_1\phi_1 + bE_2\phi_2)\\ &= |a|^2E_1|\phi_1|^2 + |b|^2E_2|\phi_2|^2 + a^*bE_2\phi_1^*\phi_2 + ab^*E_1\phi_2^*\phi_1\end{aligned}$$

Integrating over the entire region, we get

$$\begin{aligned}\int \psi^* H\psi dV = |a|^2E_1 \int \phi_1^*\phi_1 dV + a^*bE_2 \int \phi_1^*\phi_2 dV + ab^*E_1\\ \int \phi_2^*\phi_1 dV + |b|^2E_2 \int \phi_2^*\phi_2 dV = |a|^2E_1 + |b|^2E_2\end{aligned} \tag{9.40}$$

Hence

$$\int \psi^* H\psi dV = \langle\psi|H|\psi\rangle = \langle E\rangle = |a|^2E_1 + |b|^2E_2 \tag{9.41}$$

Since we know the values of a and b this equation gives us the expected value of the energy $\langle E\rangle$. And if we know E_1 and E_2 we could also calculate $|a|^2$ and $|b|^2$, knowing that the total probability $|a|^2 + |b|^2 = 1$.

For an arbitrary observable A represented by the operator $\hat{A}$, the expected value is given by

$$\langle A\rangle = \int \psi^*\hat{A}\psi dV \tag{9.42}$$

where ψ is a normalized function. Thus Eq. (9.30) is proved.

9.14 Operator Algebra

The bra and ket notation is a handy way of writing wave functions. If $|\phi_1\rangle$ and $|\phi_2\rangle$ are two orthonormal eigenfunctions of the same operator, then $\langle\phi_1|\phi_1\rangle \equiv \int \phi_1(x,y,z)^*\phi_1(x,y,z)dV = 1$, and $\langle\phi_1|\phi_2\rangle \equiv \int \phi_1(x,y,z)^*\phi_2(x,y,z)dV = 0$.

A hermitian matrix is defined as a square matrix that is identical with the complex conjugate of its transpose:

$$H = (H^T)^* \tag{9.43}$$

The complex conjugate of the transpose of a matrix is called the adjoint of the matrix. So a hermitian matrix is equal to its own adjoint, and so it is also called a self-adjoint matrix. The same language is carried over to quantum mechanical operators. Consider two arbitrary states

$$|\psi\rangle = a_1|\phi_1\rangle + a_2|\phi_2\rangle + a_3|\phi_3\rangle \tag{9.44}$$

and

$$|\chi\rangle = b_1|\phi_1\rangle + b_2|\phi_2\rangle + b_3|\phi_3\rangle \tag{9.45}$$

where the $|\phi_i\rangle$ are orthonormal eigenstates of some hermitian operator H. Hermitian operators, as we have learned, have real eigenvalues. It can be proved easily that for any hermitian operator H

$$\langle\chi|H|\psi\rangle = \langle\psi|H|\chi\rangle^* \tag{9.46}$$

where χ and ψ are two different possible states of the same system.

Exercise:
Prove Eq. (9.46).

These spaces that we have considered above are finite-dimensional vector spaces. But we often deal with Hilbert spaces that have an uncountably infinite number of dimensions. The simplest example is the position of a particle. If we take the dimensions of a particle to be zero, the particle can occupy a continuous range of positions in any direction. So the position wave function $\psi(x,y,z)$ is a Hilbert space vector with an uncountably infinite number of components. The inner product of two such vectors is evaluated as

$$\langle\phi|\psi\rangle = \int \phi^*\psi dV \tag{9.47}$$

A normalized vector obeys the rule

$$\langle\psi|\psi\rangle \equiv \int \psi^*(x,y,z)\psi(x,y,z)dV = 1 \tag{9.48}$$

The expected value of an operator $\hat{A}$ becomes

$$\langle A\rangle = \langle\psi|A|\psi\rangle = \int \psi^*(x,y,z)A\psi(x,y,z)dV \tag{9.49}$$

We saw earlier that for a discrete space, a hermitian operator A obeys the relationship

$$\langle\chi|A|\psi\rangle = \langle\psi|A|\chi\rangle^* \tag{9.50}$$

Extending this relationship to continuous space, a hermitian operator $\hat{A}$ has the property

$$\int \chi^*(x,y,z)A\psi(x,y,z)dV = \left(\int \psi^*(x,y,z)A\chi(x,y,z)dV\right)^* \tag{9.51}$$

Since the context shows that A is an operator, we have dispensed with the hat over the symbol.

9.14.1 *Momentum Operator is Hermitian*

The operator for linear momentum $\hat{p}_x = -i\hbar\frac{\partial}{\partial x}$ can be shown to be hermitian:

$$\chi^*\frac{\partial\psi}{\partial x} = \frac{\partial}{\partial x}(\chi^*\psi) - \frac{\partial\chi^*}{\partial x}\psi$$

Therefore

$$\int \chi^*\frac{\partial\psi}{\partial x}dV = \int \frac{\partial}{\partial x}(\chi^*\psi)dV - \int \frac{\partial\chi^*}{\partial x}\psi dV \tag{9.52}$$

The first term on the right side of the equation becomes

$$\int dydz\,[\chi^*\psi]_a^b$$

The wave functions are assumed to go to zero at the boundaries, which includes the integration limits $x = a$ and $x = b$. Thus this integral vanishes, and so we get

$$\int \chi^*\left(-i\hbar\frac{\partial\psi}{\partial x}\right)dV = \int \psi\left(i\hbar\frac{\partial\chi^*}{\partial x}\right)dV = \left(\int \psi^*\left(-i\hbar\frac{\partial\chi}{\partial x}\right)dV\right)^* \tag{9.53}$$

Hence $\hat{p}_x \equiv -i\hbar\frac{\partial}{\partial x}$ is a hermitian operator.

We saw earlier that a hermitian operator has real eigenvalues. Clearly, the measured momentum of a particle is always a real number, even though the operator itself seems to have an imaginary form.

9.14.2 *Commutation of Operators*

If two operators $\hat{A}$ and $\hat{B}$ have the relationship that $\hat{A}\hat{B} = \hat{B}\hat{A}$ then we say that they commute. Not all pairs of operators commute. The annihilation operator a and the creation operator $a^\dagger$ do not commute. $a^\dagger a|0\rangle = 0$. $aa^\dagger|0\rangle = a|1\rangle = |0\rangle$. Hence $a^\dagger a \neq aa^\dagger$. But the operators $\hat{N} = a^\dagger a$ and $\hat{M} = aa^\dagger$ commute.

Commutation of operators has important applications in quantum physics. If two observable quantities can be measured accurately (barring experimental errors) then the operators that represent these observable quantities commute with each other. Conversely, if two operators commute, then the observables they represent can be measured simultaneously, and so one measurement will not jeopardize the other.

For example, let us say a system is represented by the quantum state $|\alpha\rangle$ which is an eigenstate of both the operators $\hat{A}$ and $\hat{B}$ with respective eigenvalues a and b, which means that the observables represented by $\hat{A}$ and $\hat{B}$ can be measured simultaneously with arbitrary accuracy. When we measure the observable $\hat{A}$ the result is a and when we measure the observable $\hat{B}$ the result is b.

Let us express the foregoing processes in quantum mechanical language.

The initial state of the system is $|\alpha\rangle$. Now we subject the system (which may be a single particle moving in space, or an electron orbiting a nucleus, or a proton moving in an electromagnetic field) to the measurement of a variable, such as the momentum of the particle. This interaction of the experimental apparatus with the system is expressed as $\hat{A}|\alpha\rangle$. This also means the state of the system has become $\hat{A}|\alpha\rangle$. So $\hat{A}|\alpha\rangle = a|\alpha\rangle$. Since $\langle\alpha|a|\alpha\rangle = a$, this means the outcome of the measurement is the value a. If instead we were the measure the observable $\hat{B}$ we would get $\hat{B}|\alpha\rangle = b|\alpha\rangle$. If we were to measure $\hat{B}$ first and subsequently measure $\hat{A}$ we would get $\hat{A}\hat{B}|\alpha\rangle = \hat{A}b|\alpha\rangle = ab|\alpha\rangle$. If we change the order of our measurements and measure $\hat{A}$ first and then $\hat{B}$ we would get $\hat{B}\hat{A}|\alpha\rangle = \hat{B}a|\alpha\rangle = ba|\alpha\rangle$. So $\hat{A}\hat{B}|\alpha\rangle = \hat{B}\hat{A}|\alpha\rangle$. Thus $\hat{A}$ and $\hat{B}$ commute.

We have shown that if two observables can be measured with arbitrary accuracy, then their corresponding operators must commute. The converse is also true, that if two operators commute, their corresponding observables can be measured simultaneously with arbitrary accuracy.

The opposite is also true, that if two operators do not commute, the corresponding physical observables cannot be measured simultaneously, and conversely.

Heisenberg showed that the minimum error that occurs in measuring any two physical variables is related to the question of whether or not the operators representing these variables commute. If they commute, it is possible to measure both variables with arbitrary accuracy. If they do not commute, then a simultaneous measurement of both variables will result in a minimum uncertainty in each variable. The product of these uncertainties will be of the order of $\hbar$. This is the uncertainty principle. Thus the uncertainty principle is related to the commutation of the operators.

As a general rule, the physical quantities that correspond to non-commuting operators are related to each other. So the operator for the x position — which we may write as $\hat{x}$ — does not commute with the operator for the x component of momentum $\hat{p}_x$ but the $\hat{x}$ operator commutes with the operator for the y component of momentum $\hat{p}_y$. This means that we can measure the x position and the y component of momentum simultaneously with arbitrary accuracy, but we cannot do the same for the x position and the x component of the momentum.

We had earlier shown that for the photon field there is a zero point energy for all possible frequencies, because the total energy of a field is given by

$$E = \hbar\omega\left(n + \frac{1}{2}\right) \tag{9.54}$$

There is a zero point energy $\frac{1}{2}\hbar\omega$ for every possible frequency, which leads to an infinite zero point energy. We explained that this is a consequence of the uncertainty principle. We will now show how this result can also be understood in the context of non-commutation of operators.

We recall from an earlier chapter that there is a close similarity between the mathematical structures of a harmonic oscillator and the fields in a propagating electromagnetic wave. We will now see that this structural connection extends to quantum mechanics.

A ball of mass m attached to a spring of force constant k executes simple harmonic motion according to the equation

$$m\frac{d^2x}{dt^2} + kx = 0 \tag{9.55}$$

The general solution is

$$x(t) = ae^{i\omega t} + be^{-i\omega t} \tag{9.56}$$

where the angular frequency $\omega = \sqrt{\frac{k}{m}}$.

The total energy of this spring-ball system is a constant, equal to

$$\frac{p^2}{2m} + \frac{1}{2}m\omega^2 x^2 \tag{9.57}$$

where $p = p_x$, the momentum of the ball (in the x direction).

Let us now define the term $a = \sqrt{\frac{m\omega}{2}}\left(x + \frac{ip}{m\omega}\right)$ and its complex conjugate $a^* = \sqrt{\frac{m\omega}{2}}\left(x - \frac{ip}{m\omega}\right)$.

So

$$\omega a^* a = \omega a a^* = \frac{p^2}{2m} + \frac{1}{2}m\omega^2 x^2 \tag{9.58}$$

In the quantum version of this problem, the momentum and the position become operators. The term a becomes an operator, and a^* is replaced by $a^\dagger$. And we can no longer assume that a and $a^\dagger$ commute.

The commutator of two operators A and B is in general an operator, written as $[A, B] \equiv AB - BA$. We will now evaluate $[x, p_x]$. Suppose the system is described by the wave function ψ. A measurement of the variable A is described by the operation $A\psi$. A measurement of B is described by $B\psi$. If A is measured first, and B is measured next, the state becomes $BA\psi$. If the order is reversed, the state becomes $AB\psi$. If $[A, B] = 0$ the order of measurement is irrelevant. One measurement does not interfere with the other. But if $[A, B] \neq 0$, then the order does make a difference. So, we will now determine whether the state $[x, p_x]\psi$ is zero or has a finite value.

$$xp_x\psi - p_x x\psi = -i\hbar x\frac{\partial}{\partial x}\psi + i\hbar\frac{\partial}{\partial x}(x\psi) \tag{9.59}$$

We notice that in the second term on the right the momentum operator acts upon the state $(x\psi)$, because the measurement of ψ by the position operator has rendered it to a different state, viz. $x\psi$.

And so, we obtain

$$[x, p_x]\psi = i\hbar\psi \tag{9.60}$$

Since this is true for any function ψ, we can write the operator equation as

$$[x, p_x] = i\hbar$$

We will use this result in the following calculation for the total energy of a quantum harmonic oscillator. The Hamiltonian of the oscillator is

$$H = \frac{\hat{p}^2}{2m} + \frac{1}{2}m\omega^2 x^2 \tag{9.61}$$

Let us now employ the quantum mechanical operator a defined as

$$a = \sqrt{\frac{m\omega}{2\hbar}}\left(x + \frac{i\hat{p}}{m\omega}\right) \tag{9.62}$$

and its adjoint $a^\dagger$ as

$$a^\dagger = \sqrt{\frac{m\omega}{2\hbar}}\left(x - \frac{i\hat{p}}{m\omega}\right) \tag{9.63}$$

There is a subtle point here that needs careful attention. In Eq. (9.63) we have used the same symbol for the momentum as in Eq. (9.62). This is because $\hat{p}$ is a hermitian operator. So $\hat{p}^\dagger = \hat{p}$. This may appear counterintuitive, since $p = -i\hbar\frac{\partial}{\partial x}$, and thus one might be tempted to write $\hat{p}^\dagger = -\hat{p}$, which would be incorrect. This is one of the peculiarities of quantum mechanics.

$$\omega a^\dagger a = \frac{m\omega^2}{2\hbar}\left(x - \frac{ip}{m\omega}\right)\left(x + \frac{ip}{m\omega}\right) \tag{9.64}$$

$$\hbar\omega a^\dagger a = \frac{p^2}{2m} + \frac{1}{2}m\omega^2 x^2 + \frac{i\omega}{2}[x,p] = H - \frac{1}{2}\hbar\omega \tag{9.65}$$

And thus

$$H = \hbar\omega\left(a^\dagger a + \frac{1}{2}\right) \tag{9.66}$$

H is evidently a hermitian operator, and it corresponds to the energy of a harmonic oscillator. Recalling earlier discussions, we can identify $a^\dagger a$ as the number operator $\hat{N}$, and so we see that a quantized harmonic oscillator has non-zero ground state energy:

$$H = \hbar\omega\left(\hat{N} + \frac{1}{2}\right) \tag{9.67}$$

When we model the electromagnetic field as a series of harmonic oscillators, we see that even in the absence of photons ($N = 0$) the vacuum has zero point energy for every possible frequency.

Chapter 10

Quantum Angular Momentum and the Field

10.1 Quantum Mechanics of a Stable Hydrogen Atom

A hydrogen atom can be modeled crudely as a composite particle consisting of an electron-nucleus pair. The potential energy of the electron-nucleus system in a hydrogen atom — with the center of the coordinate system at the nucleus — and the coordinates of the electron (r, θ, φ) is

$$V = -\frac{e^2}{4\pi\epsilon_0 r} \tag{10.1}$$

The time independent Schrödinger wave equation for an atomic state having total energy E is therefore:

$$-\frac{\hbar^2}{2m}\nabla^2\psi - \frac{e^2}{4\pi\epsilon_0 r}\psi = E\psi \tag{10.2}$$

Since the potential has spherical symmetry, we seek solutions in spherical polar coordinates. The Schrödinger equation then becomes

$$\begin{aligned} &-\frac{\hbar^2}{2m}\left[\frac{1}{r^2}\frac{\partial}{\partial r}\left(r^2\frac{\partial\psi}{\partial r}\right) + \frac{1}{r^2\sin\theta}\frac{\partial}{\partial\theta}\left(\sin\theta\frac{\partial\psi}{\partial\theta}\right) + \frac{1}{r^2\sin^2\theta}\frac{\partial^2\psi}{\partial\varphi^2}\right] \\ &-\frac{e^2}{4\pi\epsilon_0 r}\psi = E\psi \end{aligned} \tag{10.3}$$

This second order differential equation can be solved by the method of separation of variables. Most standard textbooks on quantum mechanics and atomic physics provide the detailed solutions. Here we will supply the results.

We assume a solution of the form $\psi = R(r)Y(\theta, \varphi)$ where R is the radial function and Y the angular function. Substituting and carrying out the mathematical manipulations we obtain the solutions

$$Y_\ell^m(\theta,\varphi) = \epsilon\sqrt{\frac{(2\ell+1)(\ell-|m|)!}{4\pi(\ell+|m|)!}}e^{im\varphi}P_\ell^m(\cos\theta) \tag{10.4}$$

where ℓ is a non-negative integer, and $|m| \leq \ell$. The P_ℓ^m are the Associated Legendre functions. The factor $\epsilon = (-1)^m$ for $m \geq 0$ and $\epsilon = 1$ for $m \leq 0$. These functions are called spherical harmonics, and are orthonormal for ℓ and m:

$$\int_0^{2\pi}\int_0^{\pi}[Y_\ell^m(\theta,\varphi)]^*[Y_{\ell'}^{m'}(\theta,\varphi)]\sin\theta d\theta d\varphi = \delta_{\ell\ell'}\delta_{mm'} \tag{10.5}$$

The complete solution can be written as

$$\psi_{n\ell m} = \sqrt{\left(\frac{2}{na}\right)^3\frac{(n-\ell-1)!}{2n[(n+\ell)!]^3}}e^{-\frac{r}{na}}\left(\frac{2r}{na}\right)^\ell\left[L_{n-\ell-1}^{2\ell+1}(2r/na)\right]Y_\ell^m(\theta,\varphi) \tag{10.6}$$

The constant a is called the Bohr radius, defined as

$$a = \frac{4\pi\epsilon_0\hbar^2}{me^2} = 0.529 \times 10^{-10}\ \text{m}$$

The associated Laguerre polynomial $L_{n+\ell}^{2\ell+1}$ is evaluated as

$$L_{n-\ell-1}^{2\ell+1}(\rho) = (-1)^{2\ell+1}\left(\frac{d}{d\rho}\right)^{2\ell+1}\left[e^\rho\left(\frac{d}{d\rho}\right)^{n+\ell}(e^{-\rho}\rho^{n+\ell})\right] \tag{10.7}$$

The numbers n, ℓ and m are called the quantum numbers of the hydrogen atom. n is a natural number, ℓ is a non-negative whole number less than n, and m can be positive, zero or negative, and as noted above, $|m| \leq \ell$. The solutions of the Schrödinger wave equation enable us to obtain the possible values of the total energy of the atom, in terms of the quantum number n, which is also called the principal quantum number:

$$\langle\psi_{n\ell m}|H|\psi_{n\ell m}\rangle = E_n = -\left[\frac{m}{2\hbar^2}\left(\frac{e^2}{4\pi\epsilon_0}\right)^2\right]\frac{1}{n^2} \tag{10.8}$$

Here the symbol m is the mass of the electron, not a quantum number. We notice that the energy levels are independent of the values of the quantum numbers ℓ and m. States with different quantum numbers but the same energy are said to be *degenerate*.

This is found to be in perfect agreement with the value calculated on the basis of the Bohr-de Broglie model of the atom without applying wave mechanics. And we saw earlier that there is excellent agreement between the theoretical and experimental values.

One very important conclusion is that Coulomb's law is valid at the level of the atom, and can be assumed to be true for quantum mechanical calculations at these atomic distances.

10.2 Quantization of Angular Momentum

We saw earlier that in the Bohr-de Broglie model of the atom the angular momentum of an electron orbiting the nucleus in an atom is quantized, i.e. the magnitude of the angular momentum is a natural number multiple of $\hbar$, so that the only possible values of the electron angular momentum are $\hbar$, $2\hbar$, $3\hbar$, $4\hbar$, etc. So, not only can the angular momentum not take on any intermediate value between these, but it also cannot be zero. We shall see that this last restriction is lifted when we apply quantum mechanics to the atom. Predictions made on the basis of the old quantum theory had to make way for the more accurate results of quantum mechanics.

The angular momentum of a particle about an axis is defined as

$$\mathbf{L} = \mathbf{r} \times \mathbf{p} \tag{10.9}$$

where $\mathbf{p}$ is the momentum of the particle and $\mathbf{r}$ the displacement vector from the nearest point on the axis to the particle itself. We saw that Bohr had derived the energy levels of a hydrogen atom by applying Newton's laws of motion to the atomic electron. This gives us confidence to apply Eq. (10.9) to the orbital angular momentum of an atomic electron.

We choose the position of the nucleus — which being much more massive than the electron may be assumed to be stationary for our calculations — as the origin of the coordinate system, and so $\mathbf{r}$ is the position vector of the electron. Since the linear momentum is replaced by a quantum operator, we shall replace the angular momentum also by a corresponding quantum operator. And there are three of them, corresponding to the three coordinate axes.

We write down the three components of the angular momentum operator as

$$\hat{L}_x = y\hat{p}_z - z\hat{p}_y; \;\; \hat{L}_y = z\hat{p}_x - x\hat{p}_z; \;\; \hat{L}_z = x\hat{p}_y - y\hat{p}_x \tag{10.10}$$

The operators can be expressed in vector form as

$$\hat{\mathbf{L}} = -i\hbar\mathbf{r} \times \nabla \tag{10.11}$$

We could write out each component operator equation explicitly in spherical polar coordinates, but we are particularly interested in the z component — not because there is any essential difference between the z axis and the other axes, but because conventionally we take the z axis to be the direction of an external magnetic field. The expression for angular momentum becomes particularly simple.

$$\begin{aligned}\hat{L}_z = \hat{k} \cdot [-i\hbar(\mathbf{r} \times \nabla)] &= -i\hbar\hat{k} \cdot \left[\hat{e}_r r \times \left(\hat{e}_r \frac{\partial}{\partial r} + \hat{e}_\theta \frac{1}{r}\frac{\partial}{\partial \theta} + \hat{e}_\varphi \frac{1}{r\sin\theta}\frac{\partial}{\partial \varphi}\right)\right] \\ &= -i\hbar\hat{k} \cdot \left(\hat{e}_\varphi \frac{\partial}{\partial \theta} - \hat{e}_\theta \frac{1}{\sin\theta}\frac{\partial}{\partial \varphi}\right)\end{aligned} \tag{10.12}$$

Since $\hat{k} \cdot \hat{e}_\varphi = 0$ and $\hat{k} \cdot \hat{e}_\theta = -\sin\theta$, we obtain

$$\hat{L}_z = -i\hbar\frac{\partial}{\partial \varphi} \tag{10.13}$$

Since the electron carries a charge, the motion of the electron is equivalent to a circulating current, which is therefore a magnetic dipole, with its dipole moment vector coaxial with its angular momentum vector. So, the z component of the angular momentum is measured by applying a magnetic field along the z direction. Now, the wave function for the atomic electron contains only one factor which is a function of φ, which is $e^{im\varphi}$, and so

$$\hat{L}_z\psi = m\hbar\psi \tag{10.14}$$

Now, for a given value of ℓ, $|m| \le \ell$. So, for $\ell = 3$, m can take on the values $-3, -2, -1, 0, 1, 2, 3$. What this means is that for a given atom with $\ell = 3$ a measurement of the z component of the angular momentum would yield any one of these values $-3\hbar$, $-2\hbar$, $-\hbar$, 0, $+\hbar$, $+2\hbar$, $+3\hbar$. If we do a large number of independent measurements of L_z the average or expected value is 0 (assuming that the positive and negative values are equally likely). By applying the operator twice we get the eigenvalue equation

$$\hat{L^2}_z\psi = m^2\hbar^2\psi \tag{10.15}$$

So, if we were to measure L_z^2 for $\ell = 3$, we would get one of the values 0, $\hbar^2$, $4\hbar^2$, $9\hbar^2$. The expected value of L_z^2 is clearly positive. Now, for each value of ℓ, there are $2\ell+1$ possible values of m. If we grant equal probability to all these values, then the expected value of L_z^2 would be

$$\hbar^2 \frac{(9+4+1+0+1+4+9)}{2\times 3+1} = 4\hbar^2.$$

In general, the expected value of L_z^2 for any value of ℓ is given by

$$\langle L_z^2\rangle = 2\hbar^2 \frac{(1^2+2^2+3^2+...\ell^2)}{2\ell+1} = \hbar^2 \frac{\ell(\ell+1)}{3} \tag{10.16}$$

Of course, we could have placed the external measuring field in any direction, and we could just as well have measured L_x or L_y. We can therefore write:

$$\langle L_x^2\rangle = 2\hbar^2 \frac{(1^2+2^2+3^2+...\ell^2)}{2\ell+1} = \hbar^2 \frac{\ell(\ell+1)}{3} \tag{10.17}$$

and

$$\langle L_y^2\rangle = 2\hbar^2 \frac{(1^2+2^2+3^2+...\ell^2)}{2\ell+1} = \hbar^2 \frac{\ell(\ell+1)}{3} \tag{10.18}$$

The operators corresponding to different components of angular momentum do not commute. We can show that

$$[\hat{L}_x, \hat{L}_y] = i\hbar\hat{L}_z;\ \ [\hat{L}_y, \hat{L}_z] = i\hbar\hat{L}_x;\ \ [\hat{L}_z, \hat{L}_x] = i\hbar\hat{L}_y \tag{10.19}$$

Exercises:

1. Prove that $1^2+2^2+3^2+...n^2 = \frac{n(n+1)(2n+1)}{6}$. Hint: Use mathematical induction. First show that it is true for small values of n such as 1 and 2 (basis of induction). Then show that if it is true for some number k, it must also be true for $k+1$ (induction transition or step).
2. Prove the relations of Eq. (10.19).

From Eqs. (10.19) it follows that it is not possible to measure two different components of the angular momentum simultaneously.

We can define L^2 as the square of the total angular momentum:

$$\hat{L}^2 = \hat{L}_x^2 + \hat{L}_y^2 + \hat{L}_z^2 \tag{10.20}$$

We can show that

$$[\hat{L}^2, L_x] = [\hat{L}^2, L_y] = [\hat{L}^2, L_z] = 0 \tag{10.21}$$

which means we can simultaneously measure the total angular momentum and any *one* component.

What is the expected value of L^2? From Eq. (10.20) we obtain

$$\langle L^2 \rangle = \langle L_x^2 \rangle + \langle L_y^2 \rangle + \langle L_z^2 \rangle = \hbar^2 \ell(\ell + 1) \tag{10.22}$$

And so the expected value of the total angular momentum, defined as $\sqrt{\langle L^2 \rangle}$, becomes

$$\langle L \rangle = \hbar \sqrt{\ell(\ell + 1)} \tag{10.23}$$

Our derivation may give the impression that $\hat{L}^2$ is not a measurable quantity, and that we can obtain $\langle L^2 \rangle$ only through repeated independent measurements of L_x^2, L_y^2 and L_z^2. However, because $\hat{L}^2$ commutes separately with each of the $\hat{L}_i$ component operators, $\hat{L}^2$ is a hermitian operator that has real measurable eigenvalues. So if $\psi_{n\ell m}$ is an eigenfunction of $\hat{L}_z$, it must also be an eigenfunction of $\hat{L}^2$. And since $\langle \psi | \hat{L}^2 | \psi \rangle = \hbar^2 \ell(\ell + 1)$, it follows that

$$\hat{L}^2 \psi_{n\ell m} = \hbar^2 \ell(\ell + 1) \psi_{n\ell m} \tag{10.24}$$

As stated earlier, the Laguerre functions of Eq. (10.7) require that $\ell < n$. This means that for $n = 1$, ℓ must be 0. So, for the lowest energy state, the angular momentum is zero, and not $\hbar$, as expected in the Bohr-de Broglie model of the atom.

Exercises:

1. Prove Eq. (10.21).
2. Derive an explicit expression for $\hat{L}^2$ and show that it has eigenvalues $\ell(\ell + 1)$.

10.2.1 *Potential and Kinetic Energy*

The kinetic energy of a rotating rigid body is obtained in classical mechanics as

$$T = \frac{L^2}{2I} \tag{10.25}$$

where I is the moment of inertia of the body about its axis of rotation. Thus the quantity L^2 is proportional to the kinetic energy of the atomic electron.

Now, the magnetic moment of a circulating current I moving in a circle of area A has the magnitude $\mu = IA$. The potential energy of a magnet with moment $\boldsymbol{\mu}$ in a magnetic field $\mathbf{B}$ is $U = -\boldsymbol{\mu} \cdot \mathbf{B}$. And for an orbiting electron, its magnetic moment is proportional to its angular momentum. Thus, when we measure L_z, we are measuring the potential energy of the electron in the magnetic field.

10.2.2 *Angle between Orbital Angular Momentum and Measuring Field*

The total angular momentum of an electron in an atom is $\sqrt{\ell(\ell+1)}\hbar$ where $\ell = 0, 1, 2, 3$, etc. is called the *orbital quantum number* of the electron. Suppose ℓ were to have the value 2. Then the total angular momentum is $\sqrt{6}\hbar = 2.45\hbar$ and the possible components of this angular momentum are $2\hbar$, $\hbar$, 0, $-\hbar$ and $-2\hbar$. The coefficient of $\hbar$ in these numbers is called the *magnetic quantum number* because it becomes significant when the atom is placed in a magnetic field. The magnetic quantum number is expressed by the number m which is a positive, negative or zero integer.

The magnetic quantum number is manifested in the presence of a magnetic field. The angular momentum of the orbiting electron can only take on certain directions relative to the magnetic field. If we call the direction of the magnetic field the z axis, then the angular momentum can only take on these possible angles relative to the z axis such that

$$\cos\theta = \frac{L_z}{L} = \frac{m}{\sqrt{\ell(\ell+1)}} \tag{10.26}$$

For the case where $\ell = 2$, the possible values of θ are derived as follows: For positive values of m we get $\cos\theta = 2/2.45$ which makes $\theta = 35.3^0$, $\cos\theta = 1/2.45$ for which $\theta = 65.9^0$ and $\cos\theta = 0$ for which $\theta = 90^0$. The corresponding angles are the same for the negative values of m.

10.3 Spin

In classical mechanics every object has its proper identity, implying that each particle carries a label distinguishing it from every other particle. So every electron has its own "selfhood" and a particular history including its various trajectories in the course of time. In quantum mechanics two identical particles — such as two electrons or two protons — do not have distinctive labels or identities. Thus it is meaningless to talk about the history of any one such particle. This distinction has important implications for the statistics of identical particles in quantum mechanics.

Now, let us consider two identical particles located at two points on the x axis equidistant from the origin. Let us say one particle is at $x = +a$ and the other at $x = -a$. Now, suppose the system of the two particles is rotated counterclockwise by 180^0 or π radians. This new state is indistinguishable from the original as far as a physical measurement is concerned, since there is no way of physically distinguishing the identical particles. But we cannot assume that the quantum mechanical wave function is unchanged. One particle was initially at $\varphi = 0$ and after the rotation it was at $\varphi = \pi$. Suppose we were to describe the state of the pair of particles by the wave function ψ. While we expect that $|\psi(\varphi+\pi)|^2 = |\psi(\varphi)|^2$, we cannot assume *a priori* that $\psi(\varphi) = \psi(\varphi+\pi)$. Next, let us rotate the system once more through π radians in the counterclockwise direction. Now the system is exactly in the original state before the rotations began, and therefore we expect $\psi(\varphi+2\pi) = \psi(\varphi)$.

Let us define a rotation operator R such that

$$R\psi(\varphi) = \psi(\varphi+\pi) \tag{10.27}$$

Since this state is physically indistinguishable from the unrotated state, we must have

$$|\psi(\varphi+\pi)|^2 = |\psi(\varphi)|^2 \tag{10.28}$$

So, we can write

$$\psi(\varphi+\pi) = \lambda\psi(\varphi) \tag{10.29}$$

where λ is a complex number such that $|\lambda|^2 = 1$. If we were to rotate the system by another π radians continuing in the counterclockwise direction, we would be back to the original configuration, which is both physically and quantum mechanically indistinguishable from the original. So

$$RR\psi(\varphi) = R^2\psi(\varphi) = \lambda^2\psi(\varphi) = \psi(\varphi) \tag{10.30}$$

Thus $\lambda^2 = 1$, and so $\lambda = \pm 1$.

So, depending on the value of λ, upon a rotation through π, the wave function either changes sign, or it does not.

A rotation through π is also equivalent to an exchange of particles. So, if $\lambda = -1$, the wave function changes sign when the particles are exchanged, and if $\lambda = 1$, the wave function does not change sign under a particle exchange.

Now, if the two identical particles occupy exactly the same position, and have the same quantum numbers, such as two electrons in an atom bigger than hydrogen, the system would be physically and mathematically unchanged under an exchange of particles. Which would mean that the wave function should not change. This has important consequences. If $\lambda = -1$, the wave function can remain unchanged only if it is zero. So, two identical particles for which $\lambda = -1$ cannot occupy the same position.

Particles with $\lambda = -1$ are called Fermi-Dirac particles or fermions. Particles with $\lambda = +1$ are called Bose-Einstein particles or bosons. So, two identical fermions cannot occupy the same position if they have the same quantum numbers, i.e. two identical fermions cannot have identical wave functions. This is called the Pauli exclusion principle.

An electron in an atom is described by radial (r), polar (θ) and azimuthal (φ) coordinates. The z component of angular momentum is proportional to the function $e^{im\varphi}$. We know that the total angular momentum of a system of bodies is conserved, as long as there is no external torque on the system. So, if an atomic electron has an intrinsic angular momentum due to some sort of rotation about its axis in addition to its orbital angular momentum, then the vector sum of these two will always be conserved when there is no external torque on the atom. So, we should look for an intrinsic or spin angular momentum wave function proportional to $e^{is\varphi}$ where s is some real number.

If the wave function is rotated through 2π, the wave function remains unchanged (boson) or changes sign (fermion).

So $e^{is(\varphi+2\pi)} = e^{is\varphi}$ (boson) and $e^{is(\varphi+2\pi)} = -e^{is\varphi}$ (fermion).

Therefore $e^{2is\pi} = +1$ for a boson, and so $s = 0, 1, 2, 3...$, and $e^{2is\pi} = -1$ for a fermion, and hence $s = 1/2, 3/2, 5/2, ...$

We obtain the intrinsic angular momentum (about the z axis) by the equation

$$-i\hbar\frac{\partial}{\partial\varphi}e^{is\varphi} = \hbar s e^{is\varphi} \tag{10.31}$$

Thus the intrinsic angular momentum is $\hbar/2$, $3\hbar/2$, $5\hbar/2...$ for a fermion, and 0, $\hbar$, $2\hbar...$ for a boson.

The number s is called the spin quantum number or simply the spin of the particle. The spin angular momentum $s\hbar$ is written as S, and its components as S_x, S_y, S_z.

For a fermion of spin half, $S_x = \hbar/2$ or $-\hbar/2$, and the same goes for S_y and S_z.

For a fermion of spin $\frac{3}{2}$, the z component of the spin angular momentum could take on the four values $\frac{3\hbar}{2}, \frac{\hbar}{2}, -\frac{\hbar}{2}, -\frac{3\hbar}{2}$. The z component is written as $m_s\hbar$, and m_s can take on $2s+1$ values. The number $2s+1$ is the spin degeneracy of the particle. If S is the total spin angular momentum, then

$$\langle S^2 \rangle = \langle S_x^2 \rangle + \langle S_y^2 \rangle + \langle S_z^2 \rangle = s(s+1)\hbar^2 \tag{10.32}$$

so that $\langle S \rangle = \sqrt{s(s+1)}\hbar$.

Exercise:
Prove Eq. (10.32).

10.3.1 *Stern-Gerlach Experiment*

An electron is negatively charged, and since it has spin angular momentum, it also has an intrinsic magnetic moment $\boldsymbol{\mu}$. The potential energy of such a magnet in a magnetic field $\mathbf{B}$ is given by $-\boldsymbol{\mu} \cdot \mathbf{B}$. Since the spin of an electron is $\frac{1}{2}$, the component of spin along a magnetic field can take on only two values: $\pm\frac{1}{2}$. Based on this fact an experiment can be performed that demonstrates that an electron spin can only be parallel or antiparallel to a magnetic field. This is done by sending a stream of electrons through a magnetic field. If the direction of the spins is entirely random before going through the magnetic field, there is a 50 per cent probability that the spin gets aligned parallel to the field and a 50 per cent probability that the spin gets aligned antiparallel to the field. Let a stream of electrons be sent along the x axis, and let us apply a magnetic field in the z direction.

As the electrons pass through the magnetic field they become oriented in opposite directions, but they cannot be distinguished as long as the field is uniform. Since the two states have different energies, depending on the sign of $-\boldsymbol{\mu} \cdot \mathbf{B}$, we could use their difference in energy to split the beam according to the spin state. A method for splitting the beam in this manner is to introduce a non-uniformity in the magnetic field, by giving the field a gradient along the z direction. This means that the value of B is not uniform but increases as we move upwards and decreases as we move downwards. Particles tend to move from a position of higher potential energy to one of lower potential energy. Since the potential energy of the electron is proportional to B, if the energy is positive, it would move in the direction of decreasing B, and in the opposite direction for negative energy. The incoming electron beam is thus split into $2s+1=2$ streams, with each stream having a different potential energy.

Such experiments carried out by Stern and Gerlach showed that the electron beam was split into two streams as it passed through the apparatus. Classical physics predicts the electron beam to fan out in both directions with electrons deflected through a continuous range of angles, because in classical physics the electron spin can have any orientation relative to a magnetic field. But the experiment showed only two clearly demarcated streams, thereby demonstrating that the spin had only two orientations relative to the magnetic field. Thus the Stern-Gerlach experiment provides experimental proof for the quantum theory of electron spin, and at the same time confirms that the electron has spin $\frac{1}{2}$.

Since an electron is a fermion, it obeys the Pauli exclusion principle. This becomes significant in any atom having more than one electron, of which helium with two electrons is the simplest example. In the ground state of the helium atom the principal quantum number $n=1$. In any atom, for $n=1$, the only possible value of the orbital quantum number ℓ is zero, since $\ell \leq n-1$. (The electron cannot be pictured as orbiting the nucleus but as oscillating in a straight line with the nucleus at the mean position of rest. This is a crude picture, because at the quantum level one can no longer talk about trajectories.) There are two electrons in the ground state of the helium atom, each having $n=1$ and $\ell=0$. But since the two electrons of the atom cannot have all the same quantum numbers, the electrons must differ in their spin quantum number m_s which is the magnetic spin quantum number, taking the values $-\frac{1}{2}$ and $+\frac{1}{2}$. Thus the two electrons must have opposite spins. If we were to measure their spins simultaneously in any direction, say the x direction, if the spin of one electron is along the positive x direction, the other electron will be found to have spin along the negative x direction.

The next atom Lithium, has three electrons. Two electrons have $n=1$ and $\ell=0$, and they have opposite spins. The third electron cannot have $n=1$ and $\ell=0$, because it is impossible to have three electrons with different spin directions. And since ℓ is always less than n, the third electron has $n=2$ and $\ell=0$.

Thus the periodic table of elements is built up with the Pauli exclusion principle as the guiding rule.

10.3.2 *Spin and Polarization of the Photon*

Experiments on the spectra of atoms have shown that when an atom emits or absorbs a photon, the atom must undergo a change in orbital quantum number by 1, i.e. the only permitted transitions are those for which $\Delta\ell = \pm 1$. The change of angular momentum is accounted for by the absorbed or emitted photon. Thus the photon must have spin 1, and the measured intrinsic angular momentum in any direction must be $\pm\hbar$.

Since the photon is a boson, it is not constrained by the Pauli principle. There is no reason why any number of photons cannot be placed in the same quantum state. And this is why lasers are possible. In a laser (Light Amplification by Stimulated Emission of Radiation) a very large number of photons are generated in the same quantum state, thus providing great coherence and intensity.

Since the photon has spin 1, we would expect that a measurement of the photon angular momentum should yield $\hbar, 0$ or $-\hbar$ units in any direction. But the photon has never been measured with zero angular momentum. Thus its degeneracy is 2, and not 3, as one might expect from the formula $2s+1$. One explanation for this anomaly is that the photon travels at speed c and has zero rest mass.

A plane polarized electromagnetic wave has its electric and magnetic fields oscillating in planes perpendicular to each other, and to the direction of propagation. The propagation is along the Poynting Vector $\epsilon_0 c^2 \mathbf{E} \times \mathbf{B}$, which is evidently perpendicular to both $\mathbf{E}$ and $\mathbf{B}$.

If we superpose two identical plane waves with the electric fields perpendicular to each other, we would get a wave which is a resultant of these component waves. For a particular phase difference of the fields, the resultant electric field will circulate in the clockwise direction in a plane perpendicular to the propagation of the wave. Such a wave is said to be circularly polarized with *negative helicity.* For a different phase difference, the wave will be circularly polarized in the counterclockwise direction, and thus with *positive helicity.* The helicity Λ of a particle is defined as the component of its spin in the direction of motion. So, if the momentum of the photon is $\mathbf{p}$, and its spin $\mathbf{s}$, its helicity equals

$$\Lambda = \frac{\mathbf{s} \cdot \mathbf{p}}{p} \tag{10.33}$$

A circular polarized wave has spin angular momentum $+\hbar$ or $-\hbar$ along the axis of propagation. So $\mathbf{s}$ is either parallel or antiparallel to $\mathbf{p}$. Since $S = 1$ for a photon, its helicity $\Lambda = \pm 1$.

The state of a right circular polarized photon (with negative helicity) can be represented by $|R\rangle$ and one with positive helicity by $|L\rangle$. A plane polarized wave can be converted to a circular polarized wave by passing through suitable filters. A plane polarized wave can be considered as a superposition of two opposite circular polarizations, and a circular polarized wave is a superposition of two perpendicular plane polarizations. So, a normalized photon wave — one photon within the region — is expressible in either plane polarized ($|x\rangle$ or $|y\rangle$) or circular polarized ($|L\rangle$ or $|R\rangle$)forms, which have definite relations to each other:

$$\begin{aligned}|R\rangle &= \frac{|x\rangle + |y\rangle}{\sqrt{2}}\\ |L\rangle &= \frac{|x\rangle - |y\rangle}{\sqrt{2}}\end{aligned} \tag{10.34}$$

Equations (10.34) can be solved to obtain the inverse equations

$$\begin{aligned}|x\rangle &= \frac{|R\rangle + |L\rangle}{\sqrt{2}}\\ |y\rangle &= \frac{|R\rangle - |L\rangle}{\sqrt{2}}\end{aligned} \tag{10.35}$$

The states $|x\rangle$ and $|y\rangle$ form an orthonormal basis. $\langle x|x\rangle = 1$, $\langle x|y\rangle = 0$, etc. And an arbitrary polarization state can be written as $|\psi\rangle = a|x\rangle + b|y\rangle$. Alternatively, we can also choose the circularly polarized states as bases and write the same polarization state as $|\psi\rangle = c|R\rangle + d|L\rangle$.

A linear polarized wave can be converted to a circular polarized wave by passing it through a suitable crystal called a half-wave plate. The conversion process is represented quantum mechanically by the operators

$$\begin{aligned}\hat{C}_R &\equiv |R\rangle\langle x| + |R\rangle\langle y|\\ \hat{C}_L &\equiv |L\rangle\langle x| + |L\rangle\langle y|\end{aligned} \tag{10.36}$$

Exercises:
1. Evaluate (i) $\hat{C}_R|x\rangle$ (ii) $\hat{C}_L|y\rangle$.
2. Evaluate (i) $\langle R|x\rangle$ (ii) $\langle L|y\rangle$.
3. Evaluate $\hat{C}_R\hat{C}_L$. Interpret this physically.
4. Derive operators $\hat{P}_x$ and $\hat{P}_y$ for the conversion of circular polarization to plane polarization.
5. Derive an expression for $\hat{C}_R^\dagger$. Show that $\hat{C}_R^\dagger\hat{C}_R$ is the identity operator $\hat{I}$, so that $\langle x|\hat{C}_R^\dagger\hat{C}_R|x\rangle = 1$, and $\langle y|\hat{C}_R^\dagger\hat{C}_R|y\rangle = 1$.

10.4 The Pauli Spin Matrices

10.4.1 *Wave Function and Electron Spin*

An electron can be described by a spatial wave function $\psi(x, y, z)$ in terms of the coordinates (x, y, z). But this wave function does not tell us the spin state of the electron. Since an electron has spin $\frac{1}{2}$, this spin can be oriented either parallel or antiparallel to a magnetic field.

Thus, in order to describe the state of the electron, it is not sufficient simply to provide the probability amplitude for finding the electron at some point in space — for that is what $\psi(x, y, z)$ is — but we should also provide the probability amplitude for finding the electron with spin up or down along some axis. When we say *finding*, we mean measuring the spin of the electron using a magnetic field. Conventionally we use the z axis as the direction of the field.

In quantum mechanical language, we write the spin state of an electron as

$$|\phi\rangle \quad = \quad |+\rangle \qquad \text{or} \qquad |-\rangle$$

where $|+\rangle$ is the state in which the electron has spin up and $|-\rangle$ the state with spin down. If both these are normalized, and if both these states (spin up or spin down) are equally likely — i.e. we have not made any measurement on whether the spin is up or down — then the quantum mechanical spin state of the electron may be written as this normalized wave function

$$|\phi_1\rangle = \frac{|+\rangle + |-\rangle}{\sqrt{2}} \tag{10.37}$$

This state is a linear combination of equally weighted amplitudes for spin up and spin down. Of course, the plus sign between the two state functions is not the only possibility. A different linear combination would yield the following wave function

$$|\phi_2\rangle = \frac{|+\rangle - |-\rangle}{\sqrt{2}} \tag{10.38}$$

The most general spin state with an arbitrary linear combination of up and down states is written as

$$|\psi\rangle = a|+\rangle + b|-\rangle \tag{10.39}$$

where a and b are complex numbers and $|a|^2 + |b|^2 = 1$. The probability of finding spin up is $|a|^2$ and that of finding spin down is $|b|^2$.

Exercise:
Assuming that $|+\rangle$ and $|-\rangle$ are orthonormal, show that $|\phi_1\rangle$ and $|\phi_2\rangle$ are orthonormal.

A fuller description of the electron state is obtained by combining the spin and the space states. The standard procedure is to form the simple product. So, if the spatial wave function is ψ, the total wave function would be something like

$$\left(\frac{|+\rangle + |-\rangle}{\sqrt{2}}\right)\psi$$

We have some familiarity with the matrix representation of quantum states. In this situation each of the spin state vectors can be written as a column matrix. A common convention is the following:

$$|+\rangle = \begin{bmatrix} 1 \\ 0 \end{bmatrix} \tag{10.40}$$

and

$$|-\rangle = \begin{bmatrix} 0 \\ 1 \end{bmatrix} \tag{10.41}$$

So, if an electron is definitely in a state of spin up, it would be described by the wave function

$$\psi(x,y,z)\begin{bmatrix} 1 \\ 0 \end{bmatrix} = \begin{bmatrix} \psi(x,y,z) \\ 0 \end{bmatrix}$$

If the electron is in a state with equal probability of being found with spin up or down, then the wave function could take on the shape

$$\psi\left(\frac{|+\rangle + |-\rangle}{\sqrt{2}}\right) = \frac{1}{\sqrt{2}}\begin{bmatrix} \psi \\ \psi \end{bmatrix}$$

or more generally as

$$\frac{1}{\sqrt{2}}\begin{bmatrix} e^{i\alpha}\psi \\ e^{i\beta}\psi \end{bmatrix}$$

where α and β are real numbers. In general, however, we cannot assume that the electron has equal probability of its spin being up or down. Moreover, the probability of finding the electron with spin up (or down) will in general vary from point to point in space. This is because the electric and magnetic fields may not be uniform in the region. So the most general wave function would have the form

$$\begin{bmatrix} \psi_+(x,y,z) \\ \psi_-(x,y,z) \end{bmatrix}$$

where $\psi_+(x,y,z)$ and $\psi_-(x,y,z)$ are the amplitudes for finding the electron with spin up or down at the point (x,y,z) respectively. But we shall consider the simpler case where the spin does not depend on the spatial coordinates, as in a uniform magnetic field:

$$\psi(x,y,z)\begin{bmatrix} a \\ b \end{bmatrix}$$

where a and b are complex numbers, and $\begin{bmatrix} a \\ b \end{bmatrix}$ is normalized, so that $\begin{bmatrix} a^* & b^* \end{bmatrix}\begin{bmatrix} a \\ b \end{bmatrix} = |a|^2 + |b|^2 = 1.$

If an electron is in the spin state $|\phi\rangle$, the probability amplitude for finding its spin in the up direction is

$$\langle +|\phi\rangle = \begin{bmatrix} 1 & 0 \end{bmatrix}\begin{bmatrix} a \\ b \end{bmatrix} = a \tag{10.42}$$

Thus the probability of finding the electron with spin up is $|a|^2$. Likewise the probability of finding the electron with spin down is $|b|^2$, and $|a|^2 + |b|^2 = 1$.

Exercises:

1. Given a normalized spin wave function $\begin{bmatrix} \frac{1}{2} \\ b \end{bmatrix}$ find the value of $|b|$.

2. Show that $\begin{bmatrix} \cos\theta \\ \sin\theta \end{bmatrix}$ and $\begin{bmatrix} \sin\theta \\ -\cos\theta \end{bmatrix}$ are orthonormal.

Suppose we have an electron in an eigenstate with spin up along the z axis. Now, if the spin angular momentum is measured along the z axis, the state of the electron will not change, and the result of the measurement will be $+\hbar/2$ along the z axis. We could therefore represent the process of measurement by this operator equation:

$$S_z \begin{bmatrix} 1 \\ 0 \end{bmatrix} = \hbar/2 \begin{bmatrix} 1 \\ 0 \end{bmatrix} \tag{10.43}$$

If the spin of the electron were along the negative z direction, the measurement of the spin angular momentum would yield $-\hbar/2$:

$$S_z \begin{bmatrix} 0 \\ 1 \end{bmatrix} = -\hbar/2 \begin{bmatrix} 0 \\ 1 \end{bmatrix} \tag{10.44}$$

This suggests that S_z is a 2×2 matrix:

$$\begin{bmatrix} \hbar/2 & 0 \\ 0 & -\hbar/2 \end{bmatrix} = \hbar/2 \begin{bmatrix} 1 & 0 \\ 0 & -1 \end{bmatrix}$$

That was quite simple. But what about S_x and S_y? These require some more work.

First, since space is isotropic, we expect that S_x and S_y are related to S_z by a simple rotation of axes. We recall that the trace of a matrix is unchanged under a rotation. Since Tr $[S_z] = 0$, it follows that Tr $[S_x] =$ Tr $[S_y] = 0$ as well.

Secondly, since all three operators S_x, S_y, S_z represent observables, they are hermitian.

Thirdly, since the spin angular momentum of an electron is $\hbar/2$ when measured in any direction, $S_x^2 = S_y^2 = S_z^2 = \frac{\hbar^2}{4} I$, where I is the 2×2 unit matrix.

We will begin with the matrix $S_x = \hbar/2 \begin{bmatrix} a & b \\ c & d \end{bmatrix}$.

Since this matrix has 0 trace, $d = -a$. Since it is hermitian, a is real, and $c = b^*$. Next, since $S_x^2 = (\hbar^2/4)I$, $a^2 + |b|^2 = 1$.

So, $S_x = \hbar/2 \begin{bmatrix} a & b \\ b^* & -a \end{bmatrix}$.

Now, the Hilbert space representing electron spin is a two-dimensional space, and the states representing spin in the z direction ($|z+\rangle$ and $|z-\rangle$) are orthonormal, i.e. $\langle z+|z+\rangle = 1$; $\langle z+|z-\rangle = 0$, etc. So, we can take $|z+\rangle$

and $|z-\rangle$ as a pair of unit vectors that span this space. If we now decide to measure the spins along the x direction, we would then find the electron either in the state $|x+\rangle$ or the state $|x-\rangle$. But since we have made $|z+\rangle$ and $|z-\rangle$ the base vectors of this space, we should be able to write the state vectors corresponding to spins along the x axis and the y axis also as linear combinations of $|z+\rangle$ and $|z-\rangle$.

So, we could write

$$|x+\rangle = \frac{|z+\rangle + |z-\rangle}{\sqrt{2}} \tag{10.45}$$

and

$$|x-\rangle = \frac{|z+\rangle - |z-\rangle}{\sqrt{2}} \tag{10.46}$$

(Of course, other linear combinations of $|z+\rangle$ and $|z-\rangle$ are possible, so long as $|x+\rangle$ and $|x-\rangle$ come out orthonormal.) Therefore, Eq. (10.45) becomes

$$|x+\rangle = \frac{1}{\sqrt{2}}\left(\begin{bmatrix}1\\0\end{bmatrix} + \begin{bmatrix}0\\1\end{bmatrix}\right) = \frac{1}{\sqrt{2}}\begin{bmatrix}1\\1\end{bmatrix} \tag{10.47}$$

Since $|x+\rangle$ is an eigenfunction of the operator S_x,

$$S_x|x+\rangle = \hbar/2|x+\rangle \tag{10.48}$$

Writing out the matrix elements explicitly,

$$\begin{bmatrix}a & b\\ b^* & -a\end{bmatrix}\begin{bmatrix}1\\1\end{bmatrix} = \begin{bmatrix}1\\1\end{bmatrix} \tag{10.49}$$

So, we get the equations $a + b = 1$ and $b^* - a = 1$. Adding, we find

$$b^* + b = 2 \tag{10.50}$$

Next, we use the eigenfunction equation

$$S_x|x-\rangle = -|x-\rangle \tag{10.51}$$

which becomes

$$\begin{bmatrix}a & b\\ b^* & -a\end{bmatrix}\begin{bmatrix}1\\-1\end{bmatrix} = -\begin{bmatrix}1\\-1\end{bmatrix} \tag{10.52}$$

And we find that $a - b = -1$ and $b^* + a = 1$. So we get $2a + b^* - b = 0$. $b^* - b$ is wholly imaginary, and a is real. The inference is that $a = 0$, and b is real. And so $b = 1$. The required matrix is

$$S_x = \frac{\hbar}{2}\begin{bmatrix}0 & 1\\ 1 & 0\end{bmatrix} \tag{10.53}$$

Next, we will find an expression for S_y. The eigenstates of this operator are $|y+\rangle$ and $|y-\rangle$. Let us try these two states which are made of equal contributions from $|z+\rangle$ and $|z-\rangle$:

$$|y+\rangle = \frac{|z+\rangle + i|z-\rangle}{\sqrt{2}} \qquad \text{and} \qquad |y-\rangle = \frac{|z+\rangle - i|z-\rangle}{\sqrt{2}}$$

So

$$|y+\rangle = \frac{1}{\sqrt{2}}\begin{bmatrix}1\\ i\end{bmatrix} \qquad \text{and} \qquad |y-\rangle = \frac{1}{\sqrt{2}}\begin{bmatrix}1\\ -i\end{bmatrix}$$

A similar procedure yields the result

$$S_y = \frac{\hbar}{2}\begin{bmatrix}0 & -i\\ i & 0\end{bmatrix} \tag{10.54}$$

Exercise:
Derive Eq. (10.54).

So we have obtained three different hermitian matrices that adequately represent the components of the spin of an electron. Of course, these matrices are by no means unique, and we could have come up with different matrices that serve the same purpose. Each such choice of a set of matrices is called a representation. Our choice is the Pauli representation. The matrices $\begin{bmatrix}0 & 1\\ 1 & 0\end{bmatrix}$, $\begin{bmatrix}0 & -i\\ i & 0\end{bmatrix}$ and $\begin{bmatrix}1 & 0\\ 0 & -1\end{bmatrix}$ are called the *Pauli spin matrices* and are written as σ_x, σ_y and σ_z respectively.

10.4.2 *Trace of a Hermitian Matrix*

Let H be a 2×2 hermitian matrix. Let $X = \begin{bmatrix}x_1\\ x_2\end{bmatrix}$ be an eigenvector of H with eigenvalue λ_1. Let $Y = \begin{bmatrix}y_1\\ y_2\end{bmatrix}$ be the other eigenvector of H with eigenvalue λ_2. These eigenvectors are orthonormal, meaning $X^\dagger Y = Y^\dagger X = 0$ and $X^\dagger X = Y^\dagger Y = 1$.

Now, $HX = \lambda_1 X$ and $HY = \lambda_2 Y$, which can be written in matrix form as

$$\begin{bmatrix}h_{11} & h_{12}\\ h_{21} & h_{22}\end{bmatrix}\begin{bmatrix}x_1\\ x_2\end{bmatrix} = \begin{bmatrix}\lambda_1 x_1\\ \lambda_1 x_2\end{bmatrix} \tag{10.55}$$

$$\begin{bmatrix} h_{11} & h_{12} \\ h_{21} & h_{22} \end{bmatrix} \begin{bmatrix} y_1 \\ y_2 \end{bmatrix} = \begin{bmatrix} \lambda_2 y_1 \\ \lambda_2 y_2 \end{bmatrix} \tag{10.56}$$

Let us construct the matrix U:

$$U = \begin{bmatrix} x_1 & y_1 \\ x_2 & y_2 \end{bmatrix} \tag{10.57}$$

and its adjoint

$$U^\dagger = \begin{bmatrix} x_1^* & x_2^* \\ y_1^* & y_2^* \end{bmatrix} \tag{10.58}$$

From Eqs. (10.55) and (10.56) we obtain

$$HU = \begin{bmatrix} \lambda_1 x_1 & \lambda_2 y_1 \\ \lambda_1 x_2 & \lambda_2 y_2 \end{bmatrix} \tag{10.59}$$

So

$$U^\dagger HU = \begin{bmatrix} \lambda_1(|x_1|^2 + |x_2|^2) & \lambda_2(x_1^* y_1 + x_2^* y_2) \\ \lambda_1(x_1 y_1^* + x_2 y_2^*) & \lambda_2(|y_1|^2 + |y_2|^2) \end{bmatrix} \tag{10.60}$$

The sum of the diagonal elements of a square matrix H is called its *trace* or *spur*, written Tr [H] $\equiv \sum_i h_{ii}$. Consider the product of two square matrices AB, with elements $(AB)_{ij} = \sum_k A_{ik}B_{kj}$. Tr $[AB] = \sum_i (AB)_{ii} = \sum_i \sum_j A_{ij}B_{ji} = \sum_j \sum_i B_{ji}A_{ij} =$ Tr $[BA]$.

It is easy to show that $UU^\dagger = I$ where I is a 2×2 unit square matrix. So Tr $[U^\dagger HU] =$ Tr $[UU^\dagger H] =$ Tr $[H]$.

Since $|x_1|^2 + |x_2|^2 = 1$ and $|y_1|^2 + |y_2|^2 = 1$, Tr$[U^\dagger HU] = \lambda_1 + \lambda_2$. Hence Tr [H] $= \lambda_1 + \lambda_2$. Thus the trace of a hermitian operator is the sum of its eigenvalues.

Since the eigenvalues of an electron spin operator are equal and opposite in sign, the trace of these operators is zero. Tr $[S_x] =$ Tr $[S_y] =$ Tr $[S_z] = 0$. Naturally, the same holds also for the Pauli spin matrices $\sigma_x, \sigma_y, \sigma_z$.

Exercises:

1. Show that $\begin{bmatrix} 1 \\ i \end{bmatrix}$ and $\begin{bmatrix} 1 \\ -i \end{bmatrix}$ are eigenfunctions of S_y corresponding to the two eigenvalues $\hbar/2$ and $-\hbar/2$.
2. Show that $\begin{bmatrix} i \\ 1 \end{bmatrix}$ and $\begin{bmatrix} -i \\ 1 \end{bmatrix}$ are also eigenfunctions of S_y and find their corresponding eigenvalues.
3. Show that $\sigma_x\sigma_y = i\sigma_z$, $\sigma_y\sigma_z = i\sigma_x$ and $\sigma_z\sigma_x = i\sigma_y$.
4. Show that $\sigma_x\sigma_y = -\sigma_y\sigma_x$, $\sigma_y\sigma_z = -\sigma_z\sigma_y$ and $\sigma_z\sigma_x = -\sigma_x\sigma_z$.

10.4.3 *Unitary Matrices*

A square matrix A having the property $AA^\dagger = I$ is said to be a *unitary matrix.* Clearly, for a unitary matrix $A^\dagger = A^{-1}$.

Some unitary matrices can also be hermitian. A unitary hermitian matrix is its own inverse. If $A = A^\dagger$ and $A^\dagger = A^{-1}$ then $AA = I$.

Exercises:

1. Show that σ_x, σ_y and σ_z are Hermitian and unitary matrices.
2. Prove that the eigenvalues of a hermitian unit matrix are either +1 or −1.
3. Show that if λ is the eigenvalue of a unitary matrix, then $\lambda^*\lambda = 1$.

So when we multiply an eigenvector by the corresponding unitary matrix, we get another vector with the same magnitude. In general, a unitary matrix does not change the magnitude of *any* vector, regardless of whether it is an eigenvector or any other vector. This can be proved easily.

Suppose we have a normalized vector $|\psi\rangle$. Since this vector is normalized, it has magnitude 1, i.e. $\langle\psi|\psi\rangle = 1$. Let us multiply this vector by a unitary matrix U and let us call the resulting product the vector $|\phi\rangle$:

$$U|\psi\rangle = |\phi\rangle \qquad \text{and} \qquad \langle\psi|U^\dagger = \langle\phi|$$

$$\langle\phi|\phi\rangle = \langle\psi|U^\dagger U|\psi\rangle = \langle\psi|I|\psi\rangle = \langle\psi|\psi\rangle = 1. \tag{10.61}$$

10.4.4 *Spin Angular Momentum*

Because angular momentum can be thought of as a vector, with three spatial components, it is natural to think of spin as a vector. The Pauli spin matrices are sometimes written as a vector with three components:

$$\boldsymbol{\sigma} = \hat{\imath}\sigma_x + \hat{\jmath}\sigma_y + \hat{k}\sigma_z$$

where $\hat{\imath}, \hat{\jmath}$ and $\hat{k}$ are unit vectors along the x, y and z axes respectively.

So if we call $\mathbf{S} = \frac{\hbar}{2}\boldsymbol{\sigma}$ the spin angular momentum, we can write

$$\mathbf{S} = \hbar/2(\hat{\imath}\sigma_x + \hat{\jmath}\sigma_y + \hat{k}\sigma_z)$$

Let us now define the square of the spin angular momentum $S^2 = \mathbf{S}\cdot\mathbf{S}$

$$S^2 = \mathbf{S}\cdot\mathbf{S} = \tfrac{\hbar^2}{4}(\hat{\imath}\sigma_x + \hat{\jmath}\sigma_y + \hat{k}\sigma_z)\cdot(\hat{\imath}\sigma_x + \hat{\jmath}\sigma_y + \hat{k}\sigma_z) = \tfrac{\hbar^2}{4}(\sigma_x^2 + \sigma_y^2 + \sigma_z^2)$$

Since $\sigma_x^2 = \sigma_y^2 = \sigma_z^2 = 1$, it follows that

$$S^2 = 3\hbar^2/4$$

So we get this interesting result that when we measure the *component* of the spin angular momentum in any direction we obtain $\hbar/2$ but when we measure the magnitude of the total spin angular momentum we obtain $\sqrt{3}\hbar/2$.

10.5 Spin of an Electron

The earth spins on its axis as it flies through space in a wide orbit round the sun. The axis of the earth's rotation is approximately perpendicular to the direction of its orbital motion. This macroscopic object could serve as a model for an atomic electron which has spin and linear motion as it orbits the nucleus. However, when we consider a microscopic particle such as an electron, it is not possible to think of its spin as a rotation about its axis because an electron is really not a hard spinning sphere of charge. Indeed, it is not possible to picture the electron at all, because any attempt at picturing involves a mental construction of its shape and trajectory of motion, neither of which is real when we are dealing with extremely small objects. Nevertheless, there is some merit to thinking of the spin of an electron as a rotation about its axis.

The law of conservation of angular momentum states that the total angular momentum of a system of objects remains constant unless there is an external rotational force acting on the system. It turns out that for an atomic electron only the sum of its orbital angular momentum and its spin angular momentum is conserved, not the spin or the orbital angular momentum separately. So it is possible for the electron spin to undergo a spontaneous flip, but if that should occur there would also be a change in the orbital angular momentum, maintaining the total angular momentum constant. A second argument for considering spin as a sort of rotation comes from ferromagnetism. In classical physics a spinning charge behaves as a magnet with a north and a south pole. The magnetism of iron comes primarily from the spin of electrons.

Thus it is not unscientific to picture the electron as a spinning charge, all the while keeping in mind that in quantum theory a spinning particle is no more than a convenient symbol. In addition to spin an electron could also have linear momentum due to its motion. The component of the spin along the direction of motion is called the *helicity* of the electron. More precisely, for an electron with spin $\boldsymbol{\sigma}$ and linear momentum $\mathbf{p}$ the helicity $= \boldsymbol{\sigma} \cdot \mathbf{p}/p$ where $\boldsymbol{\sigma} = \hat{i}\sigma_x + \hat{j}\sigma_y + \hat{k}\sigma_z$, the vector formed from the Pauli σ matrices. The helicity of an electron is $\pm\frac{1}{2}$.

The quantity $\boldsymbol{\sigma} \cdot \mathbf{p}$ is a matrix. If the components of the momentum are (p_x, p_y, p_z), then we can write $\boldsymbol{\sigma} \cdot \mathbf{p} = \sigma_x p_x + \sigma_y p_y + \sigma_z p_z$.

Exercises:

1. Show that $\boldsymbol{\sigma} \cdot \mathbf{p} = \begin{bmatrix} p_z & p_x - ip_y \\ p_x + ip_y & -p_z \end{bmatrix}$.
2. Show that $(\boldsymbol{\sigma} \cdot \mathbf{p})^2 = p^2 I$ where I is the unit matrix $\begin{bmatrix} 1 & 0 \\ 0 & 1 \end{bmatrix}$.

10.6 Pauli Equation

10.6.1 *Pauli Equation for a Free Electron*

An important result established in the preceding exercise is that $(\boldsymbol{\sigma} \cdot \mathbf{p})^2 = p^2 I$. This suggests a way to incorporate spin into the wave equation for an electron. For the sake of simplicity, we shall limit ourselves to systems which do not change with time. Then the wave equation takes on a particularly simple shape, since the total energy remains constant, and the wave function is therefore an eigenfunction of the energy operator or Hamiltonian.

If an electron is in a state of definite energy, i.e. an energy eigenstate, which we denote as $|\psi\rangle$, the energy operator or Hamiltonian H is related to the eigenstate by the equation

$$H|\psi\rangle = E|\psi\rangle \tag{10.62}$$

For a free electron, $H \equiv \frac{1}{2}mv^2 = \frac{p^2}{2m}$ in the non-relativistic case where v is much less than c. Writing the momentum as an operator, the free electron energy operator becomes

$$H|\psi\rangle \equiv \frac{\hat{p}^2}{2m}|\psi\rangle$$

If the wave function $|\psi\rangle$ includes the spatial and spin components, we may write it as

$$|\psi\rangle = \begin{bmatrix} \psi_1(x, y, z) \\ \psi_2(x, y, z) \end{bmatrix} \tag{10.63}$$

Here the upper term represents the wave function for an electron with spin up, and the lower term for spin down. A wave function consisting of two components corresponding to the two different spin orientations is called a *spinor*. (We could think of this word as an abbreviation of *spin vector*.)

The energy operator acting on the wave function is expressible as

$$H \begin{bmatrix} \psi_1(x, y, z) \\ \psi_2(x, y, z) \end{bmatrix} \equiv \frac{\hat{p}^2}{2m} I \begin{bmatrix} \psi_1(x, y, z) \\ \psi_2(x, y, z) \end{bmatrix} = \frac{(\boldsymbol{\sigma} \cdot \mathbf{p})^2}{2m} \begin{bmatrix} \psi_1(x, y, z) \\ \psi_2(x, y, z) \end{bmatrix}$$

So

$$\frac{(\boldsymbol{\sigma} \cdot \mathbf{p})^2}{2m} \begin{bmatrix} \psi_1(x, y, z) \\ \psi_2(x, y, z) \end{bmatrix} = E \begin{bmatrix} \psi_1(x, y, z) \\ \psi_2(x, y, z) \end{bmatrix} \tag{10.64}$$

Equation (10.64) is called the time independent Pauli equation for a free electron in the absence of electric and magnetic fields. A closer look at the equation reveals that it is actually two different equations written together in matrix form:

$$\frac{(\boldsymbol{\sigma} \cdot \mathbf{p})^2}{2m} \psi_1(x, y, z) = E\psi_1(x, y, z) \tag{10.65}$$

and

$$\frac{(\boldsymbol{\sigma} \cdot \mathbf{p})^2}{2m} \psi_2(x, y, z) = E\psi_2(x, y, z) \tag{10.66}$$

So there is really no advantage to writing the two equations as a single matrix equation for a free electron in the absence of any fields.

10.6.2 *Electron in an Electromagnetic Field*

However, the situation changes if we introduce a magnetic field. A magnetic field is represented either by its field strength vector $\mathbf{B}$ or by its *vector potential* written as $\mathbf{A}$. Both these vectors are equally valid ways of representing the magnetic field, and we use one or the other depending on the physical context. When a particle of charge q moves in a magnetic field, it carries with it a momentum equal to $q\mathbf{A}$ in addition to its mechanical momentum $\mathbf{p}$. In an earlier chapter (Special Relativity and Electromagnetism) we saw that the Lagrangian for a field containing a charge q traveling with velocity $\mathbf{u}$ is given by

$$\mathcal{L} = -mc^2\sqrt{1-\frac{u^2}{c^2}} - q\Phi + q\mathbf{u}\cdot\mathbf{A} \tag{10.67}$$

The canonical momentum is derived from the Lagrangian by taking the derivative with respect to the position coordinate:

$$P_i = \frac{\partial\mathcal{L}}{\partial u_i} = \frac{mu_i}{\sqrt{1-\frac{u^2}{c^2}}} + \frac{q}{c}A_i = p_i + qA_i$$

where $A_1 = A_x, A_2 = A_y, A_3 = A_z$.

In quantum mechanics, the momentum operator in the presence of a field is obtained not from the mechanical momentum, but from the canonical momentum:

$$\hat{P}_x \to -i\hbar\frac{\partial}{\partial x}$$

Hence the mechanical momentum $\hat{p}_x = \hat{P}_x - qA_x = -i\hbar\frac{\partial}{\partial x} - qA_x$.

The presence of the magnetic field will modify the Pauli equation so that it now reads:

$$\frac{1}{2m}[\boldsymbol{\sigma}\cdot(\hat{\mathbf{P}} - q\mathbf{A})]^2|\psi\rangle = E|\psi\rangle \tag{10.68}$$

where $|\psi\rangle$ is the two-component spinor given by Eq. (10.63). Here again $|\psi\rangle$ is an eigenvector of the energy operator. Since the operator is a 2×2 matrix, it has two eigenvalues (which may be equal under certain circumstances) which we shall call E_1 and E_2.

There is an important vector identity involving the Pauli spin matrices:

$$(\boldsymbol{\sigma}\cdot\mathbf{a})(\boldsymbol{\sigma}\cdot\mathbf{b}) = \mathbf{a}\cdot\mathbf{b} + i\boldsymbol{\sigma}\cdot(\mathbf{a}\times\mathbf{b}) \tag{10.69}$$

Setting $\mathbf{a} = \mathbf{b} = \hat{\mathbf{P}} - q\mathbf{A} \equiv -i\hbar\nabla - q\mathbf{A}$, we get

$$[\boldsymbol{\sigma}\cdot(\hat{\mathbf{P}} - q\mathbf{A})]^2 = (\hat{\mathbf{P}} - q\mathbf{A})^2 + i\boldsymbol{\sigma}\cdot[(-i\hbar\nabla - q\mathbf{A})\times(-i\hbar\nabla - q\mathbf{A})]$$

$$[(-i\hbar\nabla - q\mathbf{A})\times(-i\hbar\nabla - q\mathbf{A})]\,\psi = iq\hbar\,[\nabla\times(\mathbf{A}\psi) + \mathbf{A}\times(\nabla\psi)]$$

$$= iq\hbar\,[\psi\,(\nabla\times\mathbf{A}) - \mathbf{A}\times(\nabla\psi) + \mathbf{A}\times(\nabla\psi)] = iq\hbar\mathbf{B}\psi \tag{10.70}$$

So Eq. (10.68) becomes

$$\left(\frac{\mathbf{p}\cdot\mathbf{p}}{2m} - \frac{q\hbar}{2m}\boldsymbol{\sigma}\cdot\mathbf{B}\right)|\psi\rangle = E|\psi\rangle \tag{10.71}$$

In the above equation $\mathbf{p}$ is no longer the momentum operator but simply the mechanical momentum of the charged particle. The Pauli spin matrices appear in a dot product with the magnetic field. The first term inside the brackets is the kinetic energy operator, and the second term is the potential energy operator. The Pauli equation applies to all spin half charged particles. The second term can be compared to the potential energy of a magnet of dipole moment $\boldsymbol{\mu}$ in a field $\mathbf{B}$ which is $U = -\boldsymbol{\mu}\cdot\mathbf{B}$. Thus, the magnetic moment vector of a spin half particle of charge q is given by

$$\boldsymbol{\mu} = \frac{q\hbar}{2m}\boldsymbol{\sigma} \tag{10.72}$$

The spin magnetic moment of an electron μ_B is called the Bohr magneton. $\mu_B = \frac{e\hbar}{2m} = 9.274 \times 10^{-24}$ J/T.

We saw in an earlier chapter that a classical rotating charge q and mass m with angular momentum L has a magnetic moment of magnitude $\mu = \frac{q}{2m}L$. $\frac{q}{2m}$ is called the gyromagnetic ratio. But we see that for an electron $\mu = \frac{e}{m}S$ where S is the angular momentum $\hbar/2$. So for an electron, the gyromagnetic ratio is e/m, twice the expected value. This ratio between the quantum mechanical and the classical gyromagnetic ratios of an electron is called the g factor of the electron. According to this calculation $g = 2$, but because of the interaction of the electron with the electromagnetic field, a correction is required that makes the electron g factor slightly greater than 2.

10.7 Interaction of an Electron with a Photon

The electromagnetic field is generated by charges, and the field can only be detected by charges. At the microscopic level, the simplest interaction between a charge and a field is the collision between a photon and an electron, otherwise known as the Compton Effect. In an earlier chapter we discussed the exchange of energy and momentum between the two particles. In this section we will examine the angular momentum exchange in this process.

Global conservation of angular momentum is required by isotropy of space, i.e. that the laws of physics are valid in every direction. So, in the collision of a photon with an electron the total angular momentum of the system about any axis remains constant.

Consider a propagating electromagnetic wave with linear momentum density $\mathbf{p} = \epsilon_0 c\mathbf{E} \times \mathbf{B}$. Suppose there is a charge q placed at the origin of the coordinate system. The angular momentum density of the field about the origin is

$$\mathbf{L} = \int \mathbf{r} \times \mathbf{p}\, dV \tag{10.73}$$

So

$$\mathbf{L} = \epsilon_0 c \int \mathbf{r} \times (\mathbf{E} \times \mathbf{B}) dV \tag{10.74}$$

Suppose this electromagnetic wave has low intensity. If the charge was initially stationary, then as the wave begins to interact with it, the charge will be displaced in the direction of the force applied by the wave. The magnetic field $\mathbf{B}$ has no effect on a static charge, and so the initial displacement will be in the direction of the electric field. Let us consider an infinitesimal displacement $\Delta\mathbf{r}$ undergone by the charge. Because of this displacement of the charge, the angular momentum of the wave relative to the charge will also change correspondingly:

$$\Delta\mathbf{L} = \epsilon_0 c \int \Delta\mathbf{r} \times (\mathbf{E} \times \mathbf{B}) dV \tag{10.75}$$

Now,

$$\Delta\mathbf{r} \times (\mathbf{E} \times \mathbf{B}) = (\Delta\mathbf{r} \cdot \mathbf{B})\mathbf{E} - (\Delta\mathbf{r} \cdot \mathbf{E})\mathbf{B} \tag{10.76}$$

A stationary charge is not affected by a magnetic field, but it will undergo a displacement along the electric field — in the same direction for a positive charge, and opposite direction for a negative charge. The magnetic field is perpendicular to the electric field. So $\Delta\mathbf{r} \cdot \mathbf{B} = 0$.

Therefore

$$\Delta\mathbf{L} = -\epsilon_0 c \int (\Delta\mathbf{r} \cdot \mathbf{E})\mathbf{B} dV \tag{10.77}$$

Thus the change of angular momentum of the field relative to the point charge will be along the magnetic field $\mathbf{B}$. Global conservation of angular momentum requires that the charge will undergo an equal and opposite change of angular momentum, relative to itself, i.e. about its own axis in a classical description. So, in classical electrodynamics the charge would begin to spin as it enters the field of the electromagnetic wave.

If the charge in question is an electron, we will apply quantum mechanics. The electron has angular momentum component $\pm\frac{\hbar}{2}$ along the direction of the magnetic field $\mathbf{B}$. And angular momentum cannot be exchanged continuously, but only in quanta. The interaction between the electromagnetic wave and the electron is described as an absorption followed by an emission of a photon (or an emission followed by an absorption). In either case there is a transfer of angular momentum $\hbar$ to or from the electron, because the photon has spin 1 and intrinsic angular momentum $\hbar$. Since the spin angular momentum of the electron in any direction can only be $\hbar/2$ or $-\hbar/2$, an absorption or emission of a photon causes the electron spin to flip, so that the angular momentum *change* of the electron is always $\hbar$.

We know that the forces between electrons are mediated by the electromagnetic field. So, the repulsive forces between two electrons can be described in terms of exchange of photons. An electron emits a photon, and thereby experiences a recoil, and this emitted photon is absorbed by the other electron, imparting it a momentum away from the first electron. This intermediate photon is not detectable, and so is called a *virtual* photon. Virtual particles function as catalysts in fundamental processes. Their careers are fleeting, and they come under the realm of the uncertainty principle. The momentum they exchange is the uncertainty in momentum of the real electrons. Likewise the uncertainty in position is the displacement of the virtual photon, etc.

A virtual photon can be absorbed from the vacuum, and returned to the vacuum. Consider the following process:

An electron and a positron are in close proximity to each other. The electron absorbs a virtual photon from the vacuum field, and the positron emits a virtual photon to the vacuum field. There is no net change to the vacuum field. The electron has gained a momentum to the right, and the positron a momentum to the left. The spin of the electron was flipped when it absorbed the photon, and likewise for the positron. The net result is that the two particles approach each other. The virtual process of absorption and emission of photons is repeated, and the charged particles accelerate towards each other.

If two electrons — or two positrons — are brought close to each other, one would emit a virtual photon which would be absorbed by the other particle. The first particle would recoil — say to the left — and the other would receive a momentum to the right. Thus there would be repulsion.

The Compton effect is the scattering of a free electron by a photon. The Feynman diagram of the process is as shown below:

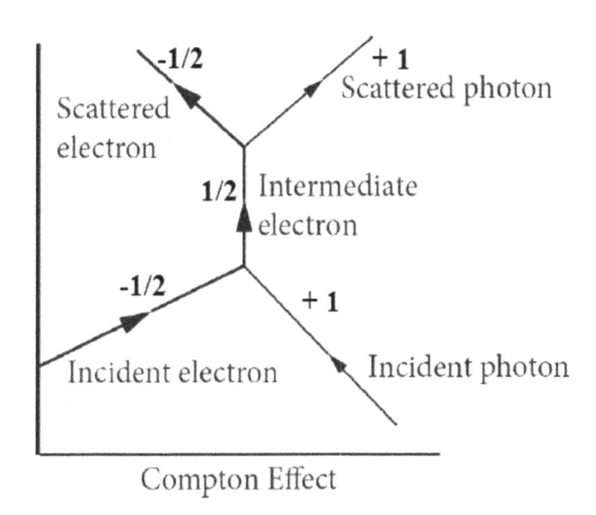

Compton Effect

The incident electron absorbs a photon (incident photon), exists momentarily as an intermediate electron, then continues as a scattered electron after emitting a photon (scattered photon). There are two vertices in this diagram. At each vertex there is conservation of charge, momentum, energy and angular momentum. The absorbed incident photon has the effect of flipping the spin of the electron, so that the intermediate electron has the opposite spin as the incident electron. The intermediate electron emits a photon and thereby its spin flips back to the original.

Conservation of global angular momentum is an important rule in the interaction of elementary particles. Charged particles interact with each other via the electromagnetic field. Most charged elementary particles have spin $\frac{1}{2}$, and the photon has spin 1, with degeneracy 2, i.e. the measured photon angular momentum can only be $\pm\hbar$, and never 0. So the photon is able to mediate the interaction between spin half charged particles, such as electrons, muons and tau particles, as well as their antiparticles having opposite charge. Protons and antiprotons are composite particles constituted of quarks of spin half. Quarks have fractional electric charge values: an up quark (u) has charge $+2e/3$, and a down quark (d) has charge $-e/3$ where the elementary charge e has the magnitude of the electron charge 1.60×10^{-19} C. Having spin half, they can interact with photons. A neutron is constituted of one up quark and two down quarks (udd), and a proton is constituted of two up quarks and one down quark (udu).

Chapter 11

Relativistic Quantum Electrodynamics

11.1 Applying Special Relativity to Pauli's Equation

The Schrödinger wave equation explained the quantum mechanical properties of a non-relativistic particle, and provided the correct energy levels of a hydrogen atom. Pauli's equation incorporated electron spin and correctly yielded the energy eigenvalues in the presence of a magnetic field. However, because it is a non-relativistic equation, it is of limited application. Dirac set out to find an equation that would have all the features of Pauli's equation, but would transcend its limitations. In doing so, he made some interesting discoveries.

The relativistic energy of a particle of rest mass m and momentum of magnitude p is given by $E^2 = p^2c^2 + m^2c^4$. What is interesting about this equation is that the energy and momentum appear only in the second power. This means that for any value of the measured momentum the energy can have two values, one positive, and the other negative:

$$E = \pm\sqrt{p^2c^2 + m^2c^4} \tag{11.1}$$

Negative energies are not uncommon in physical situations. For an atomic electron the potential energy is negative and greater in magnitude than the positive kinetic energy, and so the total energy comes out negative. But Eq. (11.1) shows that E can have negative values even in the limit as $p \to 0$, i.e. for a virtually stationary particle in the absence of a potential. Thus, the rest energy of the particle could also be negative. Dirac realized that he would have to include negative energies as valid possibilities in his relativistic wave equation.

Hence Dirac's wave function for the electron is a vector or row matrix with four elements — positive spin up, positive spin down, negative spin up, negative spin down. Such a matrix is called a *Dirac spinor*:

$$|\psi\rangle = \begin{bmatrix} \psi_1 \\ \psi_2 \\ \psi_3 \\ \psi_4 \end{bmatrix}$$

Each of these four elements is a spatial wave function of the electron. The first (ψ_1) is the wave function of the electron with positive energy spin up, the second (ψ_2) represents positive energy spin down, the third (ψ_3) negative energy spin up, and the fourth (ψ_4) negative energy spin down.

A vector with 4 components requires a 4×4 matrix operator to act upon it. Let us recall that spin was explicitly incorporated into Pauli's non-relativistic equation by rewriting the Hamiltonian as $\frac{(\boldsymbol{\sigma}\cdot\mathbf{p})^2}{2m}$, and for a free electron this is identically equal to $\frac{p^2}{2m}I$. Allowing for positive and negative energies, in the absence of fields — the square of the energy is $E^2 = p^2 + m^2c^4$. So, for a free electron, the Hamiltonian H must satisfy $H^2 = p^2I + m^2c^4I$ where I is a 4×4 unit matrix.

We now define three 4×4 matrices:

$$\boldsymbol{\alpha} \equiv \begin{bmatrix} \mathbf{0} & \boldsymbol{\sigma} \\ \boldsymbol{\sigma} & \mathbf{0} \end{bmatrix}$$

where the symbols $\mathbf{0}$ represent the 2×2 zero matrix $\begin{bmatrix} 0 & 0 \\ 0 & 0 \end{bmatrix}$ and $\boldsymbol{\sigma}$ is the usual 2×2 Pauli spin matrix vector $= \hat{\imath}\sigma_x + \hat{\jmath}\sigma_y + \hat{k}\sigma_z$.

$$\alpha_x = \begin{bmatrix} 0&0&0&1 \\ 0&0&1&0 \\ 0&1&0&0 \\ 1&0&0&0 \end{bmatrix} \quad \alpha_y = \begin{bmatrix} 0&0&0&-i \\ 0&0&i&0 \\ 0&-i&0&0 \\ i&0&0&0 \end{bmatrix} \quad \alpha_z = \begin{bmatrix} 0&0&1&0 \\ 0&0&0&-1 \\ 1&0&0&0 \\ 0&-1&0&0 \end{bmatrix}$$

We also define another 4×4 matrix:

$$\beta = \begin{bmatrix} I & \mathbf{0} \\ \mathbf{0} & -I \end{bmatrix} = \begin{bmatrix} 1&0&0&0 \\ 0&1&0&0 \\ 0&0&-1&0 \\ 0&0&0&-1 \end{bmatrix}$$

We now propose a free particle Hamiltonian:

$$H = c\boldsymbol{\alpha} \cdot \mathbf{p} + \beta mc^2 \tag{11.2}$$

Since $\boldsymbol{\alpha}$ and β are hermitian, the operator H is also hermitian. Now, the energy of the electron is given by $E^2 = p^2 + m^2c^4$, and so the eigenvalues of H are E and $-E$. Since the trace of a hermitian operator is the sum of its eigenvalues, we require that Tr [H] = 0. An examination of the matrices shows that Tr $[\boldsymbol{\alpha}] = 0$, and Tr$[\beta] = 0$, and hence Tr [H] = 0.

We should carry out the following exercises before proceeding further.

Exercises:

1. Show that $\beta^2 = I$ where I is the 4×4 identity matrix.
2. Show that $\beta\boldsymbol{\alpha} = -\boldsymbol{\alpha}\beta$.
3. Using the relations $\sigma_x\sigma_y = -\sigma_y\sigma_x$, etc. show that $(\boldsymbol{\sigma} \cdot \mathbf{p})(\boldsymbol{\sigma} \cdot \mathbf{p}) = p^2 I$ where I is the 2×2 identity matrix.
4. Show that $(\boldsymbol{\alpha} \cdot \mathbf{p})(\boldsymbol{\alpha} \cdot \mathbf{p}) = p^2 I$ where I is the 4×4 identity matrix.

As stated above, for both positive and negative energies, the Hamiltonian H must satisfy $H^2 = (p^2 + m^2c^4)I$ where I is the 4×4 unit matrix. This is easily verified, using the results obtained in the exercise above: $(\boldsymbol{\alpha} \cdot \mathbf{p}) \cdot (\boldsymbol{\alpha} \cdot \mathbf{p}) = p^2 I$ and $\beta\boldsymbol{\alpha} = -\boldsymbol{\alpha}\beta$. And so

$$\left[c\boldsymbol{\alpha} \cdot \mathbf{p} + \beta mc^2\right]\left[c\boldsymbol{\alpha} \cdot \mathbf{p} + \beta mc^2\right] = (p^2 + m^2c^4)I$$

For a spin half particle of charge e in a field described by the potentials $(\phi, \mathbf{A})$ the Dirac Hamiltonian becomes

$$H = c\boldsymbol{\alpha} \cdot (\mathbf{P} - e\mathbf{A}) + \beta mc^2 I + e\phi I \tag{11.3}$$

where $\mathbf{P}$ is the canonical momentum operator $-i\hbar\nabla$. When applied to the four-component spinor, there are four equations, and in general the components cannot be separated, i.e. there is no equation containing only one element ψ_i of the spinor.

It is customary to suppress the unit matrix factor I and so we write the general time-dependent wave equation as

$$[c\boldsymbol{\alpha} \cdot (\mathbf{P} - e\mathbf{A}) + \beta mc^2 + e\phi] \begin{bmatrix} \psi_1 \\ \psi_2 \\ \psi_3 \\ \psi_4 \end{bmatrix} = i\hbar\frac{\partial}{\partial t} \begin{bmatrix} \psi_1 \\ \psi_2 \\ \psi_3 \\ \psi_4 \end{bmatrix} \tag{11.4}$$

which we express in condensed form as

$$H|\psi\rangle = i\hbar\frac{\partial|\psi\rangle}{\partial t} \tag{11.5}$$

For a time independent steady state, with constant total energy, the solution becomes

$$|\psi\rangle = e^{-iEt/\hbar}|u\rangle \tag{11.6}$$

where $|u\rangle$ is a state vector which is independent of the time.

A time independent steady state is a state of constant energy, which could be positive or negative. For each positive energy there are two possible spin states, and likewise there are two negative energy states corresponding to the different spin values. So E is positive for the first two elements of $|u\rangle$, and is negative for the third and fourth elements. It is evident that the $|u_i\rangle$ are orthonormal states.

Let us now consider a free electron in the frame of reference where its momentum $\mathbf{p} = 0$. The Hamiltonian becomes $H = \beta mc^2$, and Dirac's equation takes the form $\beta mc^2|u\rangle = E|u\rangle$. The eigenvalue E takes on the values mc^2 and $-mc^2$, and the corresponding normalized time independent spinor solutions become

$$|u_1\rangle = \begin{bmatrix} 1 \\ 0 \\ 0 \\ 0 \end{bmatrix}, \quad |u_2\rangle = \begin{bmatrix} 0 \\ 1 \\ 0 \\ 0 \end{bmatrix}, \quad |u_3\rangle = \begin{bmatrix} 0 \\ 0 \\ 1 \\ 0 \end{bmatrix}, \quad |u_4\rangle = \begin{bmatrix} 0 \\ 0 \\ 0 \\ 1 \end{bmatrix} \tag{11.7}$$

$|u_1\rangle$ and $|u_2\rangle$ represent solutions of positive energy, and $|u_3\rangle$ and $|u_4\rangle$ represent solutions of negative energy.

11.2 Interpretation of the Negative Energy States

If negative energy states are real, how come every electron does not drop into a negative energy state like an atomic electron dropping from a higher to a lower energy state? If this could happen, there should be no positive energy electron available in the universe, because they would all drop from higher (positive) to lower (negative) energy states. Dirac suggested that the negative energy states were all filled. This would be analogous to an atom in which the lower energy states are occupied by electrons, and by Pauli's exclusion principle the higher energy electrons could not drop to one of these lower energy states, since no two electrons can occupy the same quantum state. So Dirac's hypothesis was that the universe is actually an enormous sea of negative energy electrons of all possible (negative) energies, and so there is no possibility of a positive energy electron dropping to a negative energy state, simply because there is no room available.

But in an atom it would be possible for a higher energy electron to drop to a lower energy state if somehow one of the lower energy electrons got knocked out of the atom, thereby creating a "hole" which could be filled by a higher energy electron.

So Dirac considered the possibility that a negative energy electron could be knocked out of its state and become a positive energy electron, leaving a hole behind. This hole would behave like a positive energy particle. And if a positive energy electron fell into the hole, the hole would disappear.

So let us see what happens when such an electron at rest falls into a hole. Its initial energy is mc^2. As it drops into the hole, its energy becomes $-mc^2$. So the net loss of energy of the electron is $2mc^2$. This energy is radiated away as photons. Since the photons carry momentum, we need two photons to balance energy and momentum. The sum of the energies of the photons would therefore have to be $2mc^2$.

Dirac suggested that a hole in the negative electron sea is equivalent to a particle with the same mass as an electron, but with opposite charge, since the union of an electron and a hole results in photons having no charge.

And so the idea of a *positron* was born. The positron was eventually discovered. Positrons are given off by the nuclei of certain atoms. These positrons collide with electrons and the two particles annihilate to produce a pair of photons. Because these positrons die so quickly after they are generated by the nucleus, they are difficult to detect.

The existence of positrons having identical properties as the electron — but with opposite charge — helps eliminate at least one objection to Dirac's negative energy sea. If all the negative energy electron states are filled with electrons, then the vacuum would not be neutral but would have (infinite) negative charge. However, if we assume that the vacuum is *also* a sea of negative energy positrons, then the net charge of the vacuum would be zero.

Dirac's sea would have a net infinite negative energy. But since the vacuum has infinite positive zero point energy $\frac{1}{2}\hbar\omega$ for every possible frequency ω, one could imagine that these two infinities cancel each other.

11.3 Anomalous Velocity of the Electron

11.3.1 *Dirac's Explanation*

Dirac's equation leads to a surprising result: the electron is always found traveling at the speed of light. A velocity operator $\dot{x}_i$ can be defined for the Dirac electron from the Hamiltonian of Dirac's equation:

$$\dot{x}_i = \frac{\partial H}{\partial p_i} = c\alpha_i \tag{11.8}$$

Each of the alpha matrices is hermitian, and so all the eigenvalues of the alpha matrices are real. Let

$$\alpha_i|\phi_i\rangle = \lambda_i|\phi_i\rangle \tag{11.9}$$

Multiplying both sides by α_i, and noting that $\alpha_i^2 = I$:

$$\alpha_i\alpha_i|\phi_i\rangle = |\phi_i\rangle = \lambda_i^2|\phi_i\rangle \tag{11.10}$$

Therefore, the eigenvalues of α_i are all ± 1, and so each component of the velocity operator has the two eigenvalues $\pm c$.

One could also obtain the eigenvalues of the α operators the "traditional" way, via the following procedure: $\alpha|\phi\rangle = \lambda|\phi\rangle$, which can be written as $(\alpha - \lambda I)\phi = 0$. This is a set of four equations which have solutions only if $|\alpha - \lambda I| = 0$. Thus we obtain the four values of λ_i.

Exercise:
Show from explicit calculation, i.e. by setting the determinant $|\alpha_i - \lambda I| = 0$, that each of the alpha matrices has the four eigenvalues $+1, +1, -1, -1$.

Now, each one of the alpha matrices is hermitian, with real eigenvalues, suggesting that the velocity operator corresponds to a physical observable. It appears that the velocity of the Dirac electron is an observable quantity, and a measurement of this velocity in any direction yields the value $\pm c$, where c is the speed of light. This appears to be in blatant contradiction to the special theory of relativity, according to which a particle with non-zero rest mass cannot be observed traveling at the speed of light.

Dirac provided an explanation for this apparent violation of special relativity:

"To measure the velocity we must measure the position at two slightly different times and then divide the change of position by the time interval. (It will not do to measure the momentum and apply a formula, as the ordinary connexion between velocity and momentum is not valid.) In order that our measured velocity may approximate to the instantaneous velocity, the time interval between the two measurements of position must be very short and hence these measurements must be very accurate. The great accuracy with which the position of the electron is known during the time-interval must give rise, according to the principle of uncertainty, to an almost complete indeterminacy in its momentum. This means that almost all values of the momentum are equally probable, so that the momentum is almost certain to be infinite. An infinite value for a component of momentum corresponds to the value of $\pm c$ for the corresponding component of velocity."[1]

In the last sentence Dirac is alluding to the relationship between velocity and momentum of a relativistic electron

$$\mathbf{v} = \frac{\mathbf{p}c}{\sqrt{m^2c^2 + p^2}} \tag{11.11}$$

whereby every component of velocity becomes $\pm c$ as the corresponding component of momentum goes to $\pm\infty$. But — as Dirac himself acknowledges — the validity of Eq. (11.11) for quantum mechanics is not obvious. Also, Dirac's explanation is that any particle — not just an electron — would have an expected velocity c in any direction. All that is needed is an application of special relativity and the uncertainty principle. But there is a more serious problem with Dirac's argument. According to Dirac's explanation any single measurement of the velocity would yield a value less than c, but the average value of the velocity over a large number of repeated measurements will asymptotically reach the value c. Thus Dirac claims that c is *not an eigenvalue of the velocity operator, but the expected value.* In order to appreciate the importance of this distinction it is helpful to revisit this basic quantum mechanical rule.

Suppose an operator A has the eigenvalues λ_1, λ_2 and λ_3 with corresponding orthonormal eigenfunctions $|\phi_1\rangle, |\phi_2\rangle$ and $|\phi_3\rangle$. So, by definition,

$$A|\phi_i\rangle = \lambda_i|\phi_i\rangle \tag{11.12}$$

Suppose the system is not in a pure eigenstate, but in some random (normalized) state $|\psi\rangle$ related to the eigenstates by

$$|\psi\rangle = a|\phi_1\rangle + b|\phi_2\rangle + c|\phi_3\rangle \tag{11.13}$$

[1]P. A. M. Dirac, *The Principles of Quantum Mechanics*, Fourth Edition (p. 262).

A large number of independent measurements on a state $|\psi\rangle$ will yield the expected value

$$\begin{aligned}\langle\psi|A|\psi\rangle &= (a^*\langle\phi_1| + b^*\langle\phi_2| + c^*\langle\phi_3|)A(a|\phi_1\rangle + b|\phi_2\rangle + c|\phi_3\rangle) \\ &= \lambda_1|a|^2 + \lambda_2|b|^2 + \lambda_3|c|^2 \end{aligned} \tag{11.14}$$

Dirac's claim that c is not an eigenvalue but the expected value of the velocity operator is incorrect. We found that c is an eigenvalue of each of the operators $c\alpha_i$. So we need to find a better explanation.

Now, the velocity operator $\mathbf{v}$ was obtained from the Hamiltonian by applying the formula $v_i = \frac{\partial H}{\partial p_i}$, and this turned out to be a constant, independent of p_i, and therefore valid even when $p_i \to 0$. Thus, even in its rest frame the electron has a measured velocity $v_i = c$ in any direction. Now, in the limit as $\mathbf{p} \to 0$, the eigenvectors (or eigenfunction matrices) of the Hamiltonian βmc^2 are the spinors listed in Eq. (11.7).

Let us now examine if these spinors are also eigenvectors of the velocity operators $c\alpha_i$. It becomes evident that none of them is an eigenvector.

Exercise:
Show that none of the spinors in Eq. (11.7) is an eigenvector of any of the velocity operators $c\alpha_i$.

So, an eigenfunction of the Hamiltonian (definite energy) is not an eigenfunction of velocity. This implies that when the velocity of an electron in its rest frame is found to be c, this electron is not in an eigenstate of the Hamiltonian, i.e. it does not have definite energy, either positive or negative.

Let us consider a linear combination of these eigenvectors, say

$$|\chi_3\rangle = \frac{1}{\sqrt{2}}\left(\begin{bmatrix}1\\0\\0\\0\end{bmatrix} + \begin{bmatrix}0\\0\\1\\0\end{bmatrix}\right) = \begin{bmatrix}\frac{1}{\sqrt{2}}\\0\\\frac{1}{\sqrt{2}}\\0\end{bmatrix} \tag{11.15}$$

The wave function $|\chi_3\rangle$ is a linear combination of two states, one of positive energy, and one of negative energy. Moreover, it is an equally balanced mixture of the two energy eigenstates. So, if the energy of this state is measured, there is a probability of $\frac{1}{2}$ that the energy will be positive (or negative).

It is readily seen that this wave vector $|\chi_3\rangle$ is an eigenvector of the velocity operator $c\alpha_z$. We can construct eigenvectors of the operators $c\alpha_x$ and $c\alpha_y$ as well.

$|\chi_2\rangle = \begin{bmatrix} \frac{1}{\sqrt{2}} \\ 0 \\ 0 \\ \frac{i}{\sqrt{2}} \end{bmatrix}$ is an eigenvector of $c\alpha_y$, and $|\chi_1\rangle = \begin{bmatrix} \frac{1}{\sqrt{2}} \\ 0 \\ 0 \\ \frac{1}{\sqrt{2}} \end{bmatrix}$ is an eigenvector of $c\alpha_x$.

Each of the three spinors $|\chi_1\rangle, |\chi_2\rangle, |\chi_3\rangle$ is an equal mixture of positive and negative energy states. Note that they are not all mutually orthogonal to each other. They are not eigenstates of the same operator.

Corollary 1:
An eigenstate of the velocity operator is an equal mixture of positive and negative energy eigenstates.

So, in order to obtain the value c for the velocity, the electron must be in a state that is a superposition of positive and negative energy states. If an electron is in an energy eigenstate, with either purely positive or purely negative energy, its measured velocity cannot be c. This last sentence also implies that the Hamiltonian and the velocity operators cannot have the same eigenstates, and therefore, the Hamiltonian cannot commute with the velocity operator.

Exercises:
1. Show that $H = \beta mc^2$ does not commute with $c\boldsymbol{\alpha}$.
2. Show that $H = c\boldsymbol{\alpha} \cdot (\mathbf{P} - e\mathbf{A}) + \beta mc^2 I + e\phi I$ does not commute with $c\boldsymbol{\alpha}$.

Consider the spinors $|\phi_1\rangle = \begin{bmatrix} \frac{1}{\sqrt{2}} \\ 0 \\ 0 \\ \frac{i}{\sqrt{2}} \end{bmatrix}$ and $|\phi_2\rangle = \begin{bmatrix} \frac{1}{\sqrt{2}} \\ 0 \\ 0 \\ \frac{-i}{\sqrt{2}} \end{bmatrix}$.

Applying the velocity operator $\hat{v}_y \equiv c\alpha_y$ to these vectors we obtain

$$\hat{v}_y|\phi_1\rangle = c|\phi_1\rangle \tag{11.16}$$

and

$$\hat{v}_y|\phi_2\rangle = -c|\phi_2\rangle \tag{11.17}$$

And it is easily verified that $\langle\phi_1|\phi_1\rangle = \langle\phi_2|\phi_2\rangle = 1$, and $\langle\phi_1|\phi_2\rangle = 0$, and so the two vectors are orthonormal.

Now, consider the vector $|\psi\rangle = \frac{1}{\sqrt{2}}(|\phi_1\rangle + |\phi_2\rangle)$. This is clearly $\begin{bmatrix}1\\0\\0\\0\end{bmatrix}$, which is an eigenvector of the Hamiltonian with positive energy.

Now, the eigenvalue of a velocity operator is $\pm c$. But if the electron is not in an eigenstate of a velocity operator, the measured velocity will not be $\pm c$. We saw above that Dirac suggested that the expected value of the velocity of a particle must be c. Here we will show that is not the case for the most common electron state, a state of pure positive energy. Let us evaluate the expected value of $\hat{v}_y$ in the state $|\psi\rangle$, keeping in mind that we are still in the rest frame of the electron:

$$\langle\psi|\hat{v}_y|\psi\rangle = \frac{1}{\sqrt{2}}(\langle\phi_1| + \langle\phi_2|)\left(\frac{c}{\sqrt{2}}|\phi_1\rangle - \frac{c}{\sqrt{2}}|\phi_2\rangle\right) = \frac{c}{2} - \frac{c}{2} = 0 \quad (11.18)$$

The result is almost trivial. When we measure the velocity of an electron in the positive energy eigenstate, in the reference frame of the electron, the result is zero.

Exercise:
Show that the expected velocity in the x and z directions of an electron in a positive energy state is zero in the rest frame of the electron.

Let us consider an electron in an arbitrary normalized state χ. Let us expand this state vector in terms of the vectors $|\phi_1\rangle$ and $|\phi_2\rangle$ as

$$|\chi\rangle = a|\phi_1\rangle + b|\phi_2\rangle \quad (11.19)$$

Since $|\chi\rangle$ is normalized, $|a|^2 + |b|^2 = 1$. ($0 \leq |a|^2 \leq 1$ and $0 \leq |b|^2 \leq 1$.)

Now, the expected value of $\hat{v}_y$ is given by

$$\langle\chi|\hat{v}_y|\chi\rangle = (a^*\langle\phi_1| + b^*\langle\phi_2|)\hat{v}_y(a|\phi_1\rangle + b|\phi_2\rangle) = c(|a|^2 - |b|^2) \quad (11.20)$$

Since $|a|^2 + |b|^2 = 1$, it follows that $-1 \leq |a|^2 - |b|^2 \leq +1$. Hence, $-c \leq \langle\hat{v}_y\rangle \leq +c$.

The expected value of the velocity of an electron which is in an arbitrary state always has magnitude less than or equal to c. For a pure positive or a pure negative energy state, the coefficients a and b have equal magnitude, and so the expected velocity comes out zero. The expected value becomes

$\pm c$ only for a state with an exactly equal proportion of positive and negative energy states.

A superposition of positive and negative energy states is a valid solution to Dirac equation. Schrödinger labeled the motion of an electron in such a state as zitterbewegung — meaning *jittery motion.* This term not only implies a rapid oscillation between positive and negative energies, but — as we shall see presently — an oscillation in physical space as well.

The phenomenon of zitterbewegung is highly abstract, and can be understood only in the light of quantum mechanics. The following model was developed by experts in quantum mechanics in the twentieth century.

We observe an electron in its rest frame, in the absence of electromagnetic (and gravitational) potentials. According to classical physics this electron would just sit there and not budge. But quantum mechanics shows that the vacuum has energy $\frac{1}{2}\hbar\omega$ for every possible frequency ω. The uncertainty principle allows the electron to give some energy ΔE to the vacuum and retrieve it in a short time Δt (or do the reverse) as long as $\Delta E \Delta t \gtrsim \frac{\hbar}{2}$.

So now, let us picture the following scenario. A free electron of energy mc^2 gives off a quantum of energy $E = 2mc^2$ to the vacuum, thereby becoming an electron of negative energy $-mc^2$. We know that the exclusion principle would prohibit this process because all the negative energy states are already filled, but the uncertainty principle finds a way around this restriction. The vacuum converts the energy it has received to a photon of energy $2mc^2$. This photon knocks out a negative energy electron, thereby creating a hole which is manifested as a positron, and a positive energy electron. In that process the photon is absorbed by the vacuum, and the result is a hole — which is a positron — and also creates a positive energy electron, alongside the original electron which now has negative energy. The negative energy electron now falls into the hole and fills it. All this takes place within the short time interval $\Delta t \sim \hbar/mc^2$ permitted by the uncertainty principle.[2] The net overall result is a positive energy electron, with the same energy as the original positive energy electron, except that this electron is not at the same spatial point as the original electron. The photon that was emitted by the vacuum and the photon that was absorbed by the vacuum have the same energy, and the two photons can be thought

[2]Here we are talking about approximate order of magnitude relationships, and so a factor of 2 is not significant.

of as the same virtual photon. So, in this picture, the original electron interacts with the vacuum and creates a virtual electron-positron pair. The original electron annihilates the virtual positron, and the virtual electron now becomes a real electron.[3]

The spontaneous creation and mutual annihilation of electron-positron pairs is illustrated in the figure below. An electron-positron pair appears out of the vacuum, disappears, and reappears, with the electron and positron having switched places, and the process continues ad infinitum.

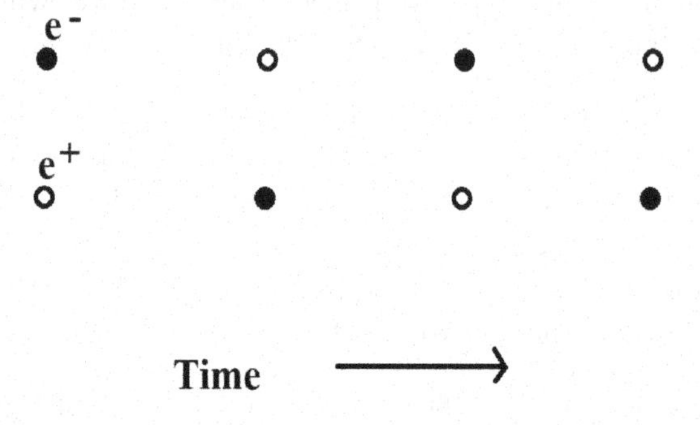

The model of zitterbewegung outlined in the preceding paragraph suggests that zitterbewegung occurs when the free electron interacts with the spontaneous creation of virtual electron-positron pairs by the vacuum. There is a spatial translation that accompanies the oscillation of the electron between positive and negative energies. See the figure below:

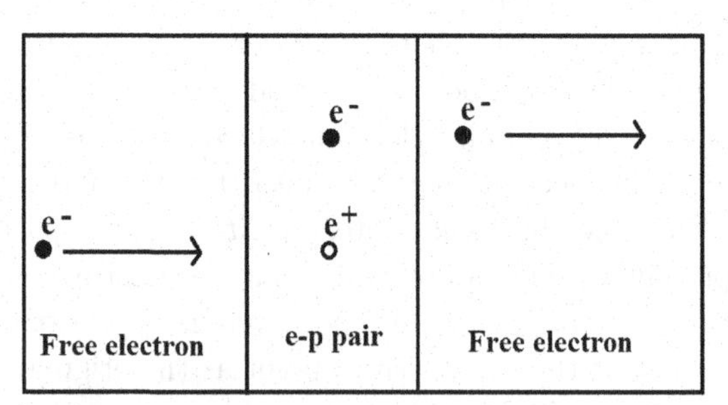

A free electron encounters a virtual electron-positron pair, is annihilated by the positron, and the virtual electron continues as a real free electron

[3]cf. J. J. Sakurai, *Advanced Quantum Mechanics* (Addison-Wesley: Menlo Park, CA, 1967).

The process described above can be expressed using a Feynman diagram, this time with the emission and absorption of virtual photons. Here the electron-positron pair is depicted as two divergent lines flowing from a vertex C into which a photon line flows. Feynman had suggested that a positron can be thought of as an electron traveling backward in time. The two different photon lines represent virtual photons. The photons are drawn from the vacuum and returned to the vacuum. One could also connect the two ends of the photon lines to make it a single curved photon line that joins the two electron vertices.

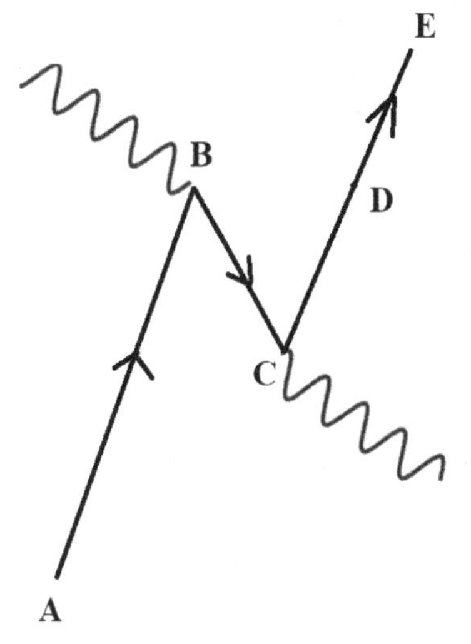

CB = positron, which can be thought of as an electron traveling backward in time from B to C.

AB = free electron; CB = virtual positron;
CD = virtual electon; DE = free electron

An oscillation between positive and negative energies is expressed quantum mechanically as a state which is a linear combination of positive and negative energy states. And such a state has the velocity eigenvalue $\pm c$. So, in the process outlined above, the displacement of the electron from its original position to its final position takes place at the speed of light. And we note that the electron does not travel linearly from one point to the other, but is annihilated at one spot, and re-created in another. The displacement (from B to D in the above diagram) is to be understood as a quantum process, not a classical one.

11.3.2 *Position of an Electron*

The position operator x in Schrödinger's wave mechanics is identical with the position of the particle. The situation is somewhat different in the quantum mechanics of the Dirac wave equation, which incorporates special relativity. The difference will become clearer as we construct the position operator x of the Dirac electron. We follow the procedure taken by Dirac.[4]

We first derive an important formula for the time derivative of an operator A. We will employ the relations $H|\psi\rangle = i\hbar\frac{\partial|\psi\rangle}{\partial t}$ and $\langle\psi|H = -i\hbar\frac{\partial\langle\psi|}{\partial t}$:

$$i\hbar\frac{d\langle A\rangle}{dt} = i\hbar\frac{d\langle\psi|A|\psi\rangle}{dt} = i\hbar\frac{\partial\langle\psi|}{\partial t}A|\psi\rangle + i\hbar\left\langle\psi|\frac{\partial A}{\partial t}|\psi\right\rangle + i\hbar|\langle\psi|A|\frac{\partial|\psi\rangle}{\partial t}$$

$$= -\langle\psi|HA|\psi\rangle + i\hbar\left\langle\frac{\partial A}{\partial t}\right\rangle + \langle\psi|AH|\psi\rangle = \langle\psi|[A,H]|\psi\rangle + i\hbar\left\langle\frac{\partial A}{\partial t}\right\rangle \quad (11.21)$$

If the operator has no explicit time dependence, such as the position operator x or the velocity operator $c\alpha_x$, then the term on the extreme right is zero. Then we would have

$$i\hbar\frac{d\langle A\rangle}{dt} = \langle[A,H]\rangle \quad (11.22)$$

This is true for an arbitrary time-independent operator A and an arbitrary Hamiltonian H, and so we may write

$$i\hbar\frac{dA}{dt} = [A,H] \quad (11.23)$$

Let us apply this equation to one of the alpha operators, say α_x:

$$i\hbar\dot{\alpha}_x = \alpha_x H - H\alpha_x \quad (11.24)$$

Now, for a free particle Hamiltonian $H = c\boldsymbol{\alpha}\cdot\mathbf{p} + \beta mc^2$,

$$\alpha_x H + H\alpha_x = \alpha_x c\alpha_x p_x + c\alpha_x p_x\alpha_x = 2cp_x \quad (11.25)$$

Exercise:
Prove Eq. (11.25).

From (11.24) and (11.25) we obtain

$$i\hbar\dot{\alpha}_x = 2\alpha_x H - 2cp_x \quad (11.26)$$

[4]P. A. M. Dirac, *The Principles of Quantum Mechanics*, Fourth Edition, Oxford, 1967.

and

$$i\hbar\dot{\alpha}_x = -2H\alpha_x + 2cp_x \tag{11.27}$$

Since H and p_x are time independent, when we take the time derivative of Eq. (11.26) we obtain

$$i\hbar\ddot{\alpha}_x = 2\dot{\alpha}_x H \tag{11.28}$$

Upon integration, we get the following expression for $\dot{\alpha}_x$:

$$\dot{\alpha}_x = \dot{\alpha}_x^0 e^{-2iHt/\hbar} \tag{11.29}$$

where $\dot{\alpha}_x^0 = \dot{\alpha}_x(t = 0)$. Since the two factors on the right side of this equation are matrices, their order is important. From Eq. (11.24) we see that $\dot{\alpha}_x$ is hermitian.

Exercise:
Prove that $\dot{\alpha}_x$ is hermitian.

Using Eqs. (11.26) and (11.29), we get the result

$$\alpha_x = \frac{1}{2} i\hbar\dot{\alpha}_x^0 e^{-2iHt/\hbar} H^{-1} + cp_x H^{-1} \tag{11.30}$$

Since $c\alpha_x = \dot{x}$, the time integral of the above equation yields,

$$x = -\frac{1}{4} c\hbar^2 \dot{\alpha}_x^0 e^{-2iHt/\hbar} H^{-2} + c^2 p_x H^{-1} t + a \tag{11.31}$$

where a is a constant of integration.

Using Eq. (11.30),

$$-\frac{1}{4} c\hbar^2 \dot{\alpha}_x^0 e^{-2iHt/\hbar} H^{-2} = \frac{1}{2} ic\hbar(\alpha_x - cp_x H^{-1}) H^{-1} \tag{11.32}$$

And so Eq. (11.31) becomes

$$x = \frac{1}{2} ic\hbar(\alpha_x - cp_x H^{-1}) H^{-1} + c^2 p_x H^{-1} t + a \tag{11.33}$$

This is a 4×4 matrix equation, and so a is a constant matrix. But this equation corresponds to the single variable equation in the macroscopic limit $\hbar \to 0$:

$$x = c^2 p_x H^{-1} t + a \tag{11.34}$$

$p_x = \gamma m v_x$, and $H = \gamma m c^2$, where $\gamma = \frac{1}{\sqrt{1-\frac{v^2}{c^2}}}$. And so we get $x = v_x t + a$, which is the classical equation for displacement at velocity v_x, and the constant number a is the initial position, which can be set to zero. And so, we may set the constant matrix $a = 0$ in our quantum matrix equation:

$$x = \frac{1}{2} ic\hbar(\alpha_x - cp_x H^{-1})H^{-1} + c^2 p_x H^{-1} t \tag{11.35}$$

This is an operator of profound significance. What we provide below is the interpretation that has evolved from the second half of the twentieth century through the present.

Let us now return to the frame where the momentum of the electron $p_x = 0$.

$$x = \frac{1}{2} ic\hbar\alpha_x H^{-1} \tag{11.36}$$

Now, α_x and H are hermitian, and so x is not hermitian. This means x is not an observable, unlike the Schrödinger position operator. But the operator $X^2 \equiv x^\dagger x$ is hermitian:

$$X^2 = \frac{c^2\hbar^2}{4H^2} \tag{11.37}$$

In the rest frame $p = 0$, this operator has eigenvalue $\frac{\hbar^2}{4m^2c^2}$. We observe that this operator X^2 does not distinguish between positive and negative energy states. Therefore, since a velocity eigenstate is a combination of positive and negative energy states, it is also an eigenstate of the operator of Eq. (11.37). So we would expect X^2 to commute with the velocity operators $c\alpha_i$, which indeed it does.

Exercise:
For a free electron in the absence of a field, write out explicitly H^{-1} and H^{-2}, and show that H^{-2} commutes with $\boldsymbol{\alpha}$.

Therefore, we can find states that are simultaneously eigenstates of X^2 and of α_i. For an electron at rest, X^2 has the eigenvalue $\frac{\hbar^2}{4m^2c^2}$. We can then construct the operator $X \equiv \sqrt{X^2}$ which has the eigenvalues $\pm\frac{\hbar}{2mc}$. This suggests that a measurement of the position of the electron in its rest frame yields $\pm\frac{\hbar}{2mc}$. The electron is found at $x = \frac{\hbar}{2mc}$ or at $x = -\frac{\hbar}{2mc}$. So, the electron appears to flit back and forth over a distance of $\frac{\hbar}{mc}$ with each measurement. And each such displacement occurs at the speed of light. Now x itself is not hermitian, and therefore the actual position itself is not an observable, and so we should not imagine that the electron was actually transported at the speed of light through a short distance in the act of measurement. But $X \equiv \sqrt{x^\dagger x}$ is an observable, and this can be measured. We will examine the physical significance of such a measurement presently.

11.4 Electron Velocity and Electron Charge

In classical mechanics a point particle is defined by its mass, and the position of the particle is where the mass is located. And this definition of position does not change if the particle carries a charge. For a charged particle the position of the mass is identical with the position of the charge. The situation is no different in non-relativistic quantum mechanics, where the particle position of an electron has been identified with the "mass point" of the electron. But in the study of the Dirac electron, the coordinate operator $\mathbf{x}$ associated with Dirac's equation is explicitly identified as the "position of the charge",[5] which is not necessarily the position of the mass. And the time derivative of this charge position operator is identified as the velocity operator:

$$\frac{d\mathbf{x}}{dt} = c\boldsymbol{\alpha} \tag{11.38}$$

Defining the velocity operator as the velocity of the charge is physically meaningful, because the charge of a particle is a constant, independent of the velocity, unlike the mass. If this were not the case, the negative charges on the fast moving electrons would not neutralize the positive charge in the nucleus of an atom.

We can therefore identify $q\mathbf{v} \equiv -ec\boldsymbol{\alpha}$ as the electron charge current operator, where $q = -e$ is the negative electron charge. This operator commutes with the velocity operator, thereby indicating that electron charge could be observed being displaced at speed c. And so the phenomenon of zitterbewegung is to be understood as the rapid spatial oscillation of electron charge, even as the energy of the electron fluctuates rapidly between positive and negative values. This zitterbewegung of electron charge has measurable consequences.

[5] A. O. Barut and S. Malin, "Position Operators and localizability of Quantum Systems described by finite and Infinite-Dimensional Wave Equations," *Rev. Mod. Phys.* **40**, 632 (1968).

The hermitian operator $\sqrt{x_i^\dagger x_i}$ measures a distance — not a displacement. The electron charge apparently covers this distance at the speed of light. So, according to this model, the charge disappears at one point in space-time and reappears at another point. Since the displacement of the electron charge takes place at the speed of light, the separation between the two events — the disappearance at one point and the appearance at the other point — is a light-like separation. Since the electromagnetic field itself travels at the speed of light, the electron interacts with its own field. The electron emits a field at point A and when it reappears at point B it catches up with the field it emitted when it was at point A. Thus the electron experiences a potential due to itself at its earlier position. The Liénard-Wiechert potential formula applies to this situation, because this formula takes into account the finite speed of an electromagnetic field. The potential at a point $\mathbf{r}$ at time t due to a charge Q at a point $\mathbf{r}_{Q(t')}$ at time t' is given by

$$\phi(\mathbf{r},t) = \frac{Q}{4\pi\epsilon_0}\int_{-\infty}^{\infty}\frac{1}{|\mathbf{r}-\mathbf{r}_Q(t')|}\delta\left(\frac{|\mathbf{r}-\mathbf{r}_Q(t')|}{c} - t + t'\right)dt' \tag{11.39}$$

Since the zitterbewegung electron is separated from its "twin" by a light-like interval, we could picture "one" electron at $\mathbf{r}$ at time t experiencing the potential of the "other" electron at $\mathbf{r}_Q$ at time t' such that $|\mathbf{r}-\mathbf{r}_Q(t')| = c(t-t') = \frac{\hbar}{mc}$. So in the electron center of mass system the length $|\mathbf{r}-\mathbf{r}_Q(t')|$ is the eigenvalue of the operator $\sqrt{x_i^\dagger x_i}$. The delta function reduces the integral to

$$\phi(\mathbf{r},t) = -\frac{mce}{4\pi\epsilon_0\hbar} \tag{11.40}$$

Since the electron and its zitterbewegung twin have the same negative charge, there is an increase in the potential energy of the electron, which we may call the self-energy of the electron as it interacts with itself via the vacuum field.

Thus zitterbewegung has the effect of increasing the electron energy by an amount equal to

$$-e\phi(\mathbf{r},t) = \frac{mce^2}{4\pi\epsilon_0\hbar} = \frac{mc^2e^2}{4\pi\epsilon_0\hbar c} = mc^2\alpha$$

Here α is the fine structure constant $\frac{e^2}{4\pi\epsilon_0\hbar c} \sim \frac{1}{137}$. The total energy of the electron is therefore $mc^2(1+\alpha)$. Since an increase of energy is manifested as an increase of mass according to Einstein's equation $E = mc^2$, the mass of the electron undergoes a small increase due to its interaction with itself via the electromagnetic field. The augmented mass is called the electromagnetic mass of the electron, and is equal to $m(1 + \alpha)$.

11.4.1 *The Electron as a Spherical Shell*

The electron charge undergoes zitterbewegung in every direction, and so we can picture the electron at any point in time as occupying a point on a shell of radius $\hbar/(2mc)$. The twin electrons at antipodal or diametric points on the shell have the effect of canceling the field inside the shell. So the net electromagnetic energy inside the shell is zero. In quantum mechanical language we say that the expected value of the field energy is zero, i.e. $\langle \frac{\epsilon_0}{2}\mathbf{E}\cdot\mathbf{E}\rangle = 0$. The shell itself carries a net charge of $-e$ since at any instant of time the electron is at one and only point on the shell. But the rapid motion of the charge has the overall effect that this charge is evenly spread out over the entire spherical surface of the shell.

We now use classical electrodynamics to calculate the energy of the space outside the shell. The radius of the shell is $a = \frac{\hbar}{2mc}$. The magnitude of the field at a distance $r > a$ from the center of the shell is

$$E = \frac{e}{4\pi\epsilon_0 r^2} \tag{11.41}$$

The total energy of the field outside the shell is

$$\int_a^\infty \frac{1}{2}\epsilon_0 E^2 dV = \int_a^\infty \left[\frac{\epsilon_0}{2}\frac{e^2}{16\pi^2\epsilon_0^2 r^4}\right] 4\pi r^2 dr = \frac{e^2}{8\pi\epsilon_0 a} = mc^2\alpha \tag{11.42}$$

This is the same as the quantity we derived using the Liénard-Wiechert potential formula. So an electron increases the energy of the space surrounding it by the amount $mc^2\alpha$.

Thus, the zitterbewegung model of the electron solves a very important problem. If the electron is thought of as a point charge of zero radius, then its electromagnetic energy diverges and becomes infinite. So physicists pictured the electron as a tiny charged sphere with a finite radius. But this model ran into several problems. For one thing, a charged sphere experiences an outward force due to the repulsion of the like charges. This outward force must be counteracted by an inward force which ensures that the charge will not explode. That requires an additional theory to account for this binding force. And more importantly, scattering experiments show that the electron must be thought of as a point charge with zero dimensions. But this brings along with it the problem of infinite self-energy. Various mathematical methods have been proposed to deal with this infinity. We have shown that the zitterbewegung model offers a satisfying way of eliminating the infinity.

11.4.2 *Quantization of Maxwell's Equations*

The notion of an electron disappearing at one point and then appearing at a different point is in violation of the equations of classical electrodynamics. For example, the equation of continuity requires that in any volume charge cannot be created or destroyed, and any alteration in charge within a region must be accompanied by a flow of charge out of or into the region:

$$\nabla \cdot \mathbf{J} = -\frac{\partial \rho}{\partial t} \tag{11.43}$$

Applying the divergence theorem to the continuity equation, we obtain:

$$\oiint \mathbf{J} \cdot \hat{n} dS = -\frac{\partial Q}{\partial t} \tag{11.44}$$

The rate of decrease of the charge inside the region is equal to the net flow of current out of the region. Charge cannot simply disappear at one point and reappear at another. But in the process of zitterbewegung a free electron interacts with the vacuum field via spontaneous electron-positron production. The vacuum field is not empty, because it has zero point energy. This energy is capable of generating an electron-positron pair which is immediately annihilated, and another pair generated, with the positions of the electron and positron flipped, and this process continues without end. In the process of zitterbewegung, a free electron interacts with these electron-positron pairs. The free electron annihilates the positron, leaving the electron that was generated. The electron and positron of the pair are not created at the same geometric point, but there is a spatial distance of the order of $\frac{\hbar}{mc}$ between them. (It is conventional to schematically show the electron-positron pair creation and pair-annihilation as taking place at point vertices as shown in the Feynman diagram shown above.) The generation of the electron-positron pair at distinct positions is a violation of the equation of continuity according to classical physics. But in quantum theory one can think of the field and the charges as operators. And these operates act upon the states of the electromagnetic field, represented by $|\Phi\rangle$. So the Maxwell equations are to be understood as relationships between measured or expected values. Thus, the first equation becomes

$$\langle\Phi|\nabla\cdot\mathbf{E}|\Phi\rangle = \left\langle\Phi|\frac{\rho}{\epsilon_0}|\Phi\right\rangle \tag{11.45}$$

So, whereas the instantaneous value of ρ at any point may not be 0, because the field constantly creates and annihilates electron-positron and positron-electron pairs, the expected value is indeed 0. The quantum mechanical equation of continuity becomes $\nabla\cdot\mathbf{J} = -\frac{\partial\langle\rho\rangle}{\partial t}$. In a vacuum $\langle\rho\rangle = 0$ and so $\nabla\cdot\mathbf{J} = 0$. There is no continuous flow of charge in this process.

Special relativity forbids the transport of any massive object at the speed of light c. However, there is no reason in principle why charge cannot be transported at light speed.

The spontaneous appearance and disappearance of virtual electron-positron pairs in the vacuum requires that $\nabla\cdot\mathbf{E}\neq 0$ within a small spatial volume (of dimensions less than of the order of $\frac{\hbar}{mc}$) at any given time even if $\langle\Phi|\nabla\cdot\mathbf{E}|\Phi\rangle = 0$ everywhere. In Maxwellian electrodynamics an electric field propagates at speed c in the absence of charges:

$$\nabla^2\mathbf{E} = \frac{1}{c^2}\frac{\partial^2\mathbf{E}}{\partial t^2} \tag{11.46}$$

In the standard derivation of this equation it is assumed that $\nabla\cdot\mathbf{E} = 0$ everywhere in the space under consideration. But in a quantum description, the vacuum stimulates the constant generation and annihilation of electron-positron pairs. So we cannot assume that $\nabla\cdot\mathbf{E} = 0$, though it is true that $\langle\Phi|\nabla\cdot\mathbf{E}|\Phi\rangle = 0$ everywhere. So, permitting $\nabla\cdot\mathbf{E} = \rho/\epsilon_0$ and taking the divergence of both sides of Eq. (11.46) we obtain

$$\nabla^2\rho = \frac{1}{c^2}\frac{\partial^2\rho}{\partial t^2} \tag{11.47}$$

This is the equation of a wave that travels at speed c. This is not a continuous wave, since the distance of transport of the charge is only about $\frac{\hbar}{mc}$. The electron charge is displaced through this distance at the speed of light. Thus, the quantum theory of Maxwell's equations corroborates the theory of zitterbewegung according to which the electron charge is transported at light speed over a short distance.

11.4.3 *Interaction with Potentials*

Finally, we will show that the electron charge *must* be carried over a short distance at the speed of light in order for electrodynamics to work.

Suppose we have a positively charged body placed in a uniform electric field as shown below:

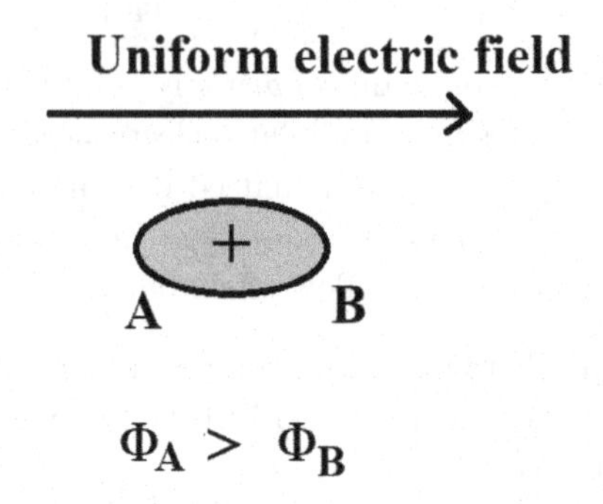

The conductor will pick up a higher potential at the end A, and a lower potential at the end B. Thus it will "sense" the direction of the field and will accelerate in that direction.

Next, suppose a positive point charge is placed in an electric field. The field need not be uniform. In the figure below, the arrow indicates the direction of the field at the position of the charge. In the figure the charge is shown as a black dot for clarity, but it is a point of zero dimension:

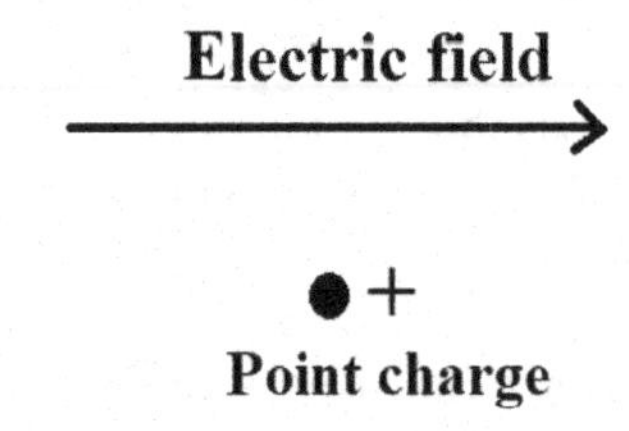

The effect of a field on a charge is measured by the force $\mathbf{F} = q\mathbf{E} = -q\nabla\phi$ acting on the charge, because the force generates a change of momentum. A charge must therefore be able to pick up the gradient of the potential. And in order to pick up the gradient, it should be able to measure the potential at two distinct points. Since a point charge is unable to do this, it follows that *a stationary point charge will not experience a force in an electric field.* This is an important rule, and we state it below more generally, since we know that a stationary charge does not experience a force in a magnetic field.

A stationary *point* charge will not experience a force in an electromagnetic field.

In classical physics, a point charge is considered as the limiting case of a finite object. For a finite object — however small — the gradient of a potential can be defined as $\frac{\Delta\phi}{\Delta x}$ which is meaningful however small Δx becomes, as long as Δx never takes the value 0. But, as we have explained, the point charge of an electron is not generated by shrinking a classical charged sphere to a mathematical point, for such a shrinking would entail an infinite increase of electromagnetic self-energy.

Zitterbewegung — the rapid flitting back and forth — provides a means for a point charge to measure the potential at two different points and thereby measure the electric field $\mathbf{E} = -\nabla\phi$. In order to measure the electric field in any direction a point charge must evaluate the electric potential at two different points in space *simultaneously*. Simultaneity has a natural meaning in Newtonian physics where space and time are distinct. But simultaneity has a different meaning in relativistic mechanics. If by simultaneity we mean that the two events are separated by a light-like interval, then such a definition is relativistically invariant. We recall that a light-like interval remains light-like in any inertial frame of reference.

If the events are separated by a time-like interval, one could always transform via a Lorentz boost to a reference frame where the two points have different time coordinates but the same spatial coordinate, and so the gradient of the potential would not exist at that spatial point in that frame. Thus a time-like interval is ruled out.

If the two events are separated by a space-like interval, they will remain causally independent. The charge would have to be transported faster than light from one point to the other, and we have seen that this is impossible in special relativity. And so a space-like interval is inadmissible.

Hence it is necessary that a point charge should "scan" the space of the electric potential at the speed of light, which remains invariant under Lorentz boosts. And since a light-like interval remains light-like in every frame of reference, including one in which the electron momentum is not zero, the velocity eigenvalue remains $\pm c$ even when the momentum is non-vanishing. So, whereas all our discussion so far has centered around an electron with zero momentum, the zitterbewegung velocity of electron charge is the same even in reference frames where the electron has non-zero momentum.

11.4.3.1 *Experimental Evidence for Zitterbewegung*

Zitterbewegung permits a stationary electron — i.e. one with zero center of mass momentum — to scan the field to determine the gradient of the scalar potential. There is experimental evidence for this. The electrons in an atom experience the Coulomb potential due to the positive nucleus. Because of zitterbewegung, there is a small modification or "correction" to the potential. This correction is experimentally verifiable when the frequency of the emitted spectra are measured with great accuracy. This correction term is called the Darwin correction term. Calculation of this correction is fairly simple.

Let $\delta\mathbf{r}$ be the displacement of the electron charge due to zitterbewegung. So $\delta r \sim \frac{\hbar}{mc}$. This requires a correction to the potential:

$$\langle \delta V \rangle = \langle V(\mathbf{r} + \delta\mathbf{r}) \rangle - \langle V(\mathbf{r}) \rangle \tag{11.48}$$

This correction term can be written as a Taylor expansion up to second order in the δr_i terms:

$$\begin{aligned} \langle V(\mathbf{r} + \delta\mathbf{r}) \rangle - \langle V(\mathbf{r}) \rangle &= \left\langle \delta r \frac{\partial V}{\partial r} + \frac{1}{2} \sum_{ij} \delta r_i \delta r_j \frac{\partial^2 V}{\partial r_i \partial r_j} \right\rangle \\ &= \left\langle \delta r \frac{\partial V}{\partial r} \right\rangle + \left\langle \frac{1}{2} \sum_{ij} \delta r_i \delta r_j \frac{\partial^2 V}{\partial r_i \partial r_j} \right\rangle \end{aligned} \tag{11.49}$$

The net work done by the electrical force on an electron in a complete orbit is zero, since the electrostatic field is a conservative force field. So $\oint \frac{\partial V}{\partial r} dr = 0$, and thus $\left\langle \delta r \frac{\partial V}{\partial r} \right\rangle = 0$.

And so

$$\langle \delta V \rangle = \left\langle \frac{1}{2} \sum_{ij} \delta r_i \delta r_j \frac{\partial^2 V}{\partial r_i \partial r_j} \right\rangle \cong \frac{1}{6} \delta r^2 \nabla^2 V \simeq \frac{\hbar^2}{6m^2c^2} \nabla^2 V$$

where the factor 1/3 takes care of the triple counting of the $\delta r_i \delta r_j$. $\nabla^2 V = -\frac{\rho}{\epsilon_0}$, where ρ is the charge density of the nucleus, responsible for the potential V, as seen by the atomic electron. This charge density can be expressed as $\rho = Ze\psi^*\psi$, where ψ is the electron wave function at the position of the nucleus. We will consider the case of a hydrogen atom for which $Z = 1$. So the change in potential energy due to zitterbewegung becomes

$$-e\langle \delta V \rangle \simeq \frac{\hbar^2}{6m^2c^2} \frac{e^2}{\epsilon_0} \psi^* \psi \tag{11.50}$$

ψ is to be evaluated at the position of the nucleus, which is at $r = 0$. Recall from the previous chapter that an atomic wave function for a single electron is written as a product of a radial function and an angular function:

$$\psi_{n\ell m} = \sqrt{\left(\frac{2}{na}\right)^3 \frac{(n-\ell-1)!}{2n[(n+\ell)!]^3}} e^{-\frac{r}{na}} \left(\frac{2r}{na}\right)^{\ell} \left[L_{n-\ell-1}^{2\ell+1}(2r/na)\right] Y_{\ell}^{m}(\theta,\varphi) \tag{11.51}$$

where the constant a — called the *Bohr radius* — is defined as

$$a = \frac{4\pi\epsilon_0\hbar^2}{me^2}$$

It is evident that at $r = 0$ all the wave functions vanish except those for which $\ell = 0$ (and hence $m = 0$.) For $\ell = m = 0$, the spherical harmonic $Y_0^0 = \frac{1}{\sqrt{4\pi}}$, and, at $r = 0$, the associated Laguerre polynomials are all equal to 1. Hence,

$$\psi^*\psi(0) = \frac{1}{8\pi}\left(\frac{2}{na}\right)^3 \tag{11.52}$$

So the change in energy due to zitterbewegung can be written as

$$\delta U = \frac{\hbar^2}{6m^2c^2}\frac{e^2}{\epsilon_0}\frac{1}{8\pi}\left(\frac{2}{na}\right)^3 \tag{11.53}$$

This is the Darwin term. We see that this correction is inversely proportional to the cube of the principal quantum number n, and becomes less significant for larger values of n.

Exercises:

1. Show that the Darwin term is proportional to nE_n^2, where E_n is the energy of the atom having principal quantum number n.
2. $\delta U = \lambda n E_n^2$. Find the factor λ.

The Darwin term contains only the change of potential energy due to zitterbewegung. This is because in zitterbewegung it is the charge and not the mass that undergoes spatial oscillation. If the mass of the electron were also involved in zitterbewegung, then there would be a contribution from kinetic energy. The absence of a kinetic energy contribution corroborates the theory we have discussed in this chapter.

Chapter 12

Gravity and Electromagnetism

12.1 Dimensions and Their Relationships

Introductory courses in college physics define length, mass and time as the three basic dimensions for constructing all the units in which quantities are expressed in physics. The corresponding symbols are L, M and T. The dimension of length or displacement is L, that of speed or velocity is LT^{-1}, that of acceleration is LT^{-2}, and so on. Since force equals the product of mass and acceleration, the dimension of force is MLT^{-2}. We express these relationships in mathematical symbolism as $[v] = LT^{-1}$, $[a] = LT^{-2}$, $[f] = MLT^{-2}$, etc.

In Newtonian physics, these three dimensions remain independent of one another. A change in one does not affect the other two. A displacement of a spaceship at any velocity relative to an observer will not change the mass of the spaceship or the rate of time flow on the ship as measured by the observer.

The situation is different in special relativity. When B is in uniform motion relative to A, a time interval in B is not the same as in A, and the mass of an object measured by B is not the same as the mass of the same object measured by A. So, the three dimensions of mass, length and time are no longer independent, but both length and mass have a relationship with time. When the position of an object relative to an observer changes with time, its length and its mass — as measured by the observer — also change. The greater the relative velocity, the more pronounced is the change in these measured quantities.

But special relativity does not offer a direct relationship between space and mass. A material body having mass generates a gravitational field in the surrounding space, but the geometry of the space remains Euclidean, just as the presence of a charge creates an electromagnetic field without altering the geometry.

Einstein's general theory of relativity introduced a major change to this picture. According to general relativity, every material object has an effect both on space and time. Space becomes curved, and so even a light ray will follow a curved path close to a massive object like the sun. Time is also altered, and clocks slow down when they are introduced into a gravitational field.

In this chapter we will provide a basic introduction to the general theory of relativity and its significance for understanding the electromagnetic field. We will extend our knowledge of special relativity to include the effects of matter on space. Einstein's hypothesis was that matter has an effect on both space and time, and causes a curvature of the Minkowski spacetime, so that the metric is no longer simply that of flat spacetime $g_{\mu\nu} = (-1, 1, 1, 1)$. Flat spacetime becomes the limiting case of real spacetime in the limit of low gravitation, just as Newtonian dynamics is the limiting case of relativistic dynamics for low velocities.

12.2 Curved Space

The electromagnetic field is embedded in a three-dimensional Euclidean space. Euclidean space is characterized by Euclidean geometry. The sum of three angles of a triangle equals two right angles, and parallel lines — which can be drawn anywhere — will never intersect each other.

Mathematically speaking, other geometries are also possible besides the Euclidean. Consider the surface of a sphere. This is an example of a two-dimensional curved surface. Ordinarily we do not think of the surface of a sphere as a two-dimensional entity. We might think of it as a two-dimensional surface that is curved in the third dimension. And that is because we are accustomed to the three dimensions in relation to our bodies: right-left, front-back, and up-down. And we have learned that though the surface of the earth may appear two-dimensional it actually curves in the third dimension.

Let us now do a little exercise in mathematical abstraction. First, we consider an infinitely long straight line, such as the real number line. Now, suppose there is a point somewhere on this line, which we shall call P. Next, we allow P to travel along this number line in the positive direction. Since P moves only in this direction, the position of P can be specified uniquely by a real number. Next, consider an infinitely long curved line, specifically the sine wave shown below:

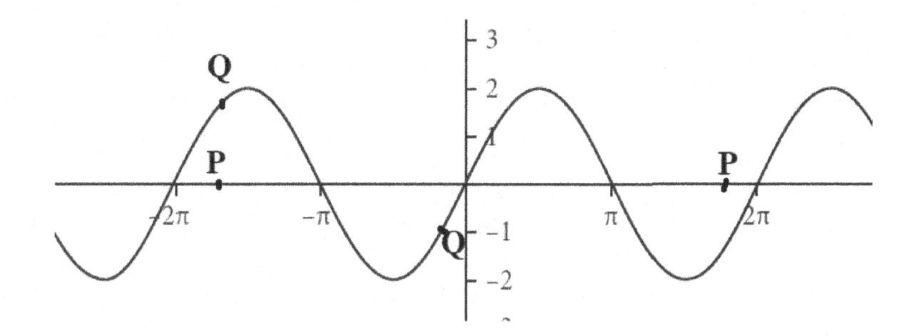

Consider a point Q moving at constant speed along the sine curve. Now, Q is a mathematical point. It has no mass, no right side or left side, no top and bottom, and no front and back. It is not constrained by Newton's laws of motion. The only constraint on it is to move at constant speed along the sine curve. We humans looking at the two paths can be forgiven for imagining that whereas P moves in one dimension, Q moves in two dimensions. But in fact — once the path followed by Q has been chalked out — the motion of Q is entirely one-dimensional. Assuming that it travels at constant speed, the actual distance it has covered at any instant of time is determined solely by a single positive real number representing the time. The positions of P and Q at earlier and later times are shown for illustrative purposes, and it is hard to imagine that Q is moving in a single dimension. But if we label distance along the path as a mathematical dimension, then it makes sense to say that the motion of Q is just as one-dimensional as that of P.

Next, consider a point moving along a closed circle. Such a motion is one-dimensional, but curved and closed. It is one-dimensional, because at any point in time only one parameter is needed to define its position. To think of a circle as one-dimensional requires an effort of mathematical imagination. We will next bring that mathematical imagination to bear on two-dimensional curved surfaces.

The surface of a sphere is an example of a two-dimensional curved surface that is also closed. We shall train our wills to limit our attention to the surface of the sphere, without worrying about its inside and its outside. Indeed, we need to train our imagination into believing there is no inside or outside to this spherical surface, even though this surface does have a radius of curvature. We naturally picture this radius as a line drawn from the center to the surface. But if we prescind from such geometric visualization, this radius of curvature is just a number — a mathematical entity — which can be expressed in terms of quantities that are measurable *on the surface itself.* Indeed, that is how the radius of the earth was measured by Eratosthenes. An idealized but conceptually simple method is the following: Three surveyors meet together at some point, each armed with a large protractor for measuring angles, and a very long piece of string. Each person attaches one end of the string to the person on the right. Then they move away from each other, unraveling the string as they go, moving at equal speeds and thus maintaining equal lengths for the three strings, so that they form an equilateral triangle with the persons at the vertices. They measure the angles between the strings as they move outwards. Initially each angle would be close to 60^0, but as they move further apart the angles would increase. The procedure is continued until each angle is exactly 90^0. Now each string has length equal to one-fourth of the circumference of the earth. So, if the length of a string is ℓ, the radius of the earth is $\frac{2\ell}{\pi}$. This thought experiment illustrates how the curvature of a surface can be measured without leaving the surface.

Whereas radius is a well defined quantity for a spherical shell, there are curved surfaces which do not have a clearly defined radius, such as the surface of an ellipsoid. The shape of the earth can be described as a good approximation to an oblate ellipsoid (or spheroid), with polar radius slightly less than equatorial radius. So, whereas every circle of latitude has a well-defined radius, this is not true for any line of longitude. It is possible to imagine various sorts of curved surfaces which have no well-defined radius, such as that of a potato. A more useful quantity is curvature, which can be defined even when the radius is indefinite. We can think of curvature as the inverse of radius of curvature. A plane surface can be thought of as a spherical surface having infinite radius, and therefore zero curvature. For a more complicated surface the definition of curvature necessarily becomes more abstract. But it is possible to come up with a mathematical definition of curvature applicable to all surfaces. Such a definition must yield 0 for a plane surface, which can be thought of as a spherical surface of infinite radius.

The curvature of a surface such as a sphere or an ellipsoid is positive. A surface shaped like a saddle has negative curvature. A plane surface has zero curvature. A plane surface — of zero curvature — is Euclidean. Euclidean geometry requires that, given any point A and a line not passing through A, one can draw one and only line through A that is parallel to the line outside A. A surface of positive curvature obeys Riemannian geometry. For a Riemannian surface, given any point A and a line not passing through A, it is impossible to draw a line through A that is parallel to (i.e. does not intersect) the line outside A. A surface of negative curvature obeys the geometry of Lobachevsky and Bolyai and is called Bolyai-Lobachevskian (also Lobachevskian or hyperbolic) geometry. In this geometry one can draw more than one straight line through a point A that are parallel to a line outside of A.

In the last paragraph we spoke about straight lines in all three geometries. Our imaginations have been trained to believe that one can draw a straight line only on a Euclidean surface. But when considering alternate geometries, it is necessary to adapt the concept of straightness to the specific geometry of the surface. On a Riemannian surface like the earth's surface, one can define a straight line as a set of points that define the shortest distance between two points. Such a line is also called a geodesic. Every line of longitude on the surface of the earth is a geodesic, but the equator is the only latitude that is a geodesic.

A line — be it straight or curved — is a one-dimensional manifold. A plane or a spherical surface is a two-dimensional manifold. Our physical space is generally thought of as a three-dimensional manifold. The earth itself is a three-dimensional manifold, and its surface is a two-dimensional manifold. We commonly visualize the surface of the earth as a two-dimensional manifold embedded in a three-dimensional manifold.

The notion of embedding is something we implicitly follow when we think of geodesics on the surface of the earth. We do not think of the geodesics as straight lines, but as circles or arcs, because we intuitively embed the manifold of the curved surface in a three-dimensional manifold. But such an embedding is not mathematically required. If we restrict ourselves to a mathematical spherical surface — lacking such features as hills, gorges, volcanoes, etc. — then the surface is simply a two-dimensional manifold with curvature. If we abstain from subconsciously embedding the surface in a three-dimensional manifold, there is no reason why the geodesics should not be thought of as straight lines.

Of course, a manifold need not have uniform curvature at every point. A manifold can be formed by stitching together regions of positive, zero and negative curvature, with the result resembling a fried *papad.* So, curvature is not a property of an entire manifold, but of a point and its immediate neighborhood within the manifold.

12.3 Curvature of Spaces

12.3.1 *Covariant Derivatives*

Let us recall that a vector is a tensor of rank one. We have learned that tensor fields are characterized by the way they transform from one coordinate system to another. A contravariant vector A^μ transforms according to

$$A^{\mu'} = \frac{\partial x^{\mu'}}{\partial x^\mu} A^\mu \tag{12.1}$$

and a covariant vector B_ν transforms according to

$$B_{\nu'} = \frac{\partial x^\nu}{\partial x^{\nu'}} B_\nu \tag{12.2}$$

The gradient of a scalar field ϕ is a covariant vector, and transforms accordingly:

$$\frac{\partial \phi}{\partial x^{\nu'}} = \frac{\partial x^\nu}{\partial x^{\nu'}} \frac{\partial \phi}{\partial x^\nu} \tag{12.3}$$

The above equation can be written in shorthand notation as

$$\partial_{\nu'} \phi = \frac{\partial x^\nu}{\partial x^{\nu'}} \partial_\nu \phi \tag{12.4}$$

Tensors of higher rank have covariant and contravariant indices, and each index transforms according to whether it is above or below, so

$$C^{\mu'}_{\nu'} = \frac{\partial x^{\mu'}}{\partial x^\mu} \frac{\partial x^\nu}{\partial x^{\nu'}} C^\mu_\nu \tag{12.5}$$

We refer to a tensor of contravariant rank ℓ and covariant rank m as a tensor of rank (ℓ, m) or as an (ℓ, m) tensor. So, the tensor C is a $(1,1)$

tensor. A covariant vector is a $(0,1)$ tensor, and a contravariant vector is a $(1,0)$ tensor.

We have seen above that the gradient of a scalar field $(\partial_\mu \phi)$ transforms as a $(0,1)$ tensor. Let us now examine the transformation of a $(1,1)$ tensor of the form $\partial_\mu A^\nu$. We might expect it to transform like any other $(1,1)$ tensor field, but that does not happen, as we will find when we write out the transformation relations explicitly for both the derivative ∂_μ and the vector A^μ:

$$\frac{\partial A^{\nu'}}{\partial x^{\mu'}} = \frac{\partial x^\mu}{\partial x^{\mu'}} \frac{\partial}{\partial x^\mu} \left(\frac{\partial x^{\nu'}}{\partial x^\nu} A^\nu \right) \tag{12.6}$$

$$\frac{\partial A^{\nu'}}{\partial x^{\mu'}} = \frac{\partial x^\mu}{\partial x^{\mu'}} \frac{\partial x^{\nu'}}{\partial x^\nu} \frac{\partial A^\nu}{\partial x^\mu} + \frac{\partial x^\mu}{\partial x^{\mu'}} \frac{\partial^2 x^{\nu'}}{\partial x^\mu \partial x^\nu} A^\nu \tag{12.7}$$

It is evident that the derivative $\frac{\partial A^\nu}{\partial x^\mu}$ does not transform like a tensor, and therefore it is not a tensor. The additional term on the right side is of the form $B^{\nu'}_{\mu'\nu} A^\nu$ (recalling that repeated indices are summed over). We cannot assume that it is a tensor.

The form of Eq. (12.7) suggests that in order to come up with a derivative that is covariant under transformations we could create an expression of the form

$$\nabla_\mu A^\nu \equiv \partial_\mu A^\nu + \Gamma^\nu_{\mu\alpha} A^\alpha \tag{12.8}$$

The ∇_μ on the left is a new symbol which is different from the gradient $\nabla\phi = \partial_\mu \phi$. $\nabla_\mu A^\nu$ is called the *covariant derivative* of A^ν. Since the first term on the right side is evidently not a tensor, the second term on the right side is not a tensor either. But since A^ν itself is a tensor, it follows that the symbol $\Gamma^\nu_{\mu\alpha}$ cannot be a tensor, for then the term would be an inner product of two tensors, and thus a tensor of reduced rank. So $\Gamma^\nu_{\mu\alpha}$ is simply a symbol, and it is commonly called a *Christoffel Symbol.* It is also called a connection, known as the Christoffel connection, or the Levi-Civita connection. We need to find the explicit form of this symbol or connection.

12.3.2 *Christoffel Symbols for Covariant Tensors*

First, let us consider the covariant derivative of a (0, 1) tensor. It is not unreasonable to expect the covariant derivative $\nabla_\mu B_\nu$ to have a similar expansion as the covariant derivative of a (1, 0) tensor. But we need not assume that the expansion will be identical. So, we will assume an expansion of the form

$$\nabla_\mu B_\nu = \partial_\mu B_\nu + \tilde{\Gamma}^\lambda_{\mu\nu} B_\lambda \tag{12.9}$$

The symbol $\tilde{\Gamma}$ is not necessarily equal to Γ. We will now obtain the relationship between these two symbols.

The covariant derivative by definition is a true tensor, meaning that it transforms according to the tensor transformation equations. Now, the contraction of a tensor remains invariant under coordinate transformations. So, the covariant derivative must commute with contractions. By this we mean that the covariant derivative of a contraction of a tensor should equal the contraction of the covariant derivative of the tensor:

$$\nabla_\mu (T^\lambda{}_{\lambda\rho}) = (\nabla T)^{\ \lambda}_{\mu\ \lambda\rho} \tag{12.10}$$

Now, the gradient of a scalar transforms as a (0,1) tensor, and hence it is identical with the covariant derivative of the scalar:

$$\nabla_\mu \phi = \partial_\mu \phi \tag{12.11}$$

Let us consider the scalar $B_\lambda A^\lambda$. We can therefore write

$$\nabla_\mu (B_\lambda A^\lambda) = \partial_\mu (B_\lambda A^\lambda) = (\partial_\mu B_\lambda) A^\lambda + B_\lambda (\partial_\mu A^\lambda) \tag{12.12}$$

The covariant derivative of a (1,1) tensor can be expanded in explicit form, and in our case it becomes

$$\begin{aligned} \nabla_\mu (B_\lambda A^\lambda) &= (\nabla_\mu B_\lambda) A^\lambda + B_\lambda (\nabla_\mu A^\lambda) \\ &= (\partial_\mu B_\lambda) A^\lambda + \tilde{\Gamma}^\sigma_{\mu\lambda} B_\sigma A^\lambda + B_\lambda (\partial_\mu A^\lambda) + B_\lambda \Gamma^\lambda_{\mu\rho} A^\rho \end{aligned} \tag{12.13}$$

Comparing Eqs. (12.12) and (12.13), we obtain the relationship

$$\tilde{\Gamma}^\sigma_{\mu\lambda} B_\sigma A^\lambda = -\Gamma^\lambda_{\mu\rho} B_\lambda A^\rho \tag{12.14}$$

We can relabel dummy indices, and rewrite this equation as

$$\tilde{\Gamma}^\sigma_{\mu\lambda} B_\sigma A^\lambda = -\Gamma^\sigma_{\mu\lambda} B_\sigma A^\lambda \tag{12.15}$$

B and A are arbitrary vectors, and so this equality implies that

$$\tilde{\Gamma}^\sigma_{\mu\lambda} = -\Gamma^\sigma_{\mu\lambda} \tag{12.16}$$

So these are the expressions for the covariant derivatives of different tensors:

$$\nabla_\mu A^\nu = \partial_\mu A^\nu + \Gamma^\nu_{\mu\lambda} A^\lambda \tag{12.17}$$

$$\nabla_\mu B_\nu = \partial_\mu B_\nu - \Gamma^\lambda_{\mu\nu} B_\lambda \tag{12.18}$$

We will next find explicit expressions for these Christoffel symbols.

12.3.3 *Torsion Tensor*

In our study of electromagnetism, we found that the fields generated by static charges can be expressed as $\mathbf{E} = -\nabla\phi$, and therefore $\nabla \times \mathbf{E} \equiv \text{curl}\,\mathbf{E} = 0$. This last equation can also be expressed as

$$\partial_i E_j = \partial_j E_i \tag{12.19}$$

We say that such a field is *irrotational.* Extending the definition to an n-dimensional field, an irrotational vector field obeys $\partial_\mu B_\nu = \partial_\nu B_\mu$. From Eq. (12.18), we obtain the equation

$$\nabla_\mu B_\nu - \nabla_\nu B_\mu = \partial_\mu B_\nu - \partial_\nu B_\mu - (\Gamma^\lambda_{\mu\nu} - \Gamma^\lambda_{\nu\mu})B_\lambda \tag{12.20}$$

If B is an irrotational vector,

$$\nabla_\mu B_\nu - \nabla_\nu B_\mu = -(\Gamma^\lambda_{\mu\nu} - \Gamma^\lambda_{\nu\mu})B_\lambda \tag{12.21}$$

Let us label the term inside the brackets as $T^\lambda_{\mu\nu} \equiv \Gamma^\lambda_{\mu\nu} - \Gamma^\lambda_{\nu\mu}$. On the left side we have the difference of two tensors, and therefore a tensor. So the right side is also a tensor. And since B_λ is a tensor, the expression $T^\lambda_{\mu\nu}$ must be a tensor. It is called the *torsion tensor.* So, Eq. (12.20) can be written as

$$\nabla_\mu B_\nu - \nabla_\nu B_\mu = \partial_\mu B_\nu - \partial_\nu B_\mu - T^\lambda_{\mu\nu} B_\lambda \tag{12.22}$$

The torsion tensor is a measure of the difference between the covariant curl and the regular curl of a vector field. We can set it equal to zero for all physically meaningful coordinate spaces. For such spaces, $\Gamma^\lambda_{\mu\nu} = \Gamma^\lambda_{\nu\mu}$. A Christoffel symbol that is symmetric in its lower indices is said to be *torsion-free.*

12.3.4 *Metric Compatibility*

An important property of physical spaces is that the covariant derivative of the metric tensors $g_{\mu\nu}$ and $g^{\mu\nu}$ should vanish:

$$\nabla_\alpha g_{\mu\nu} = \nabla_\alpha g^{\mu\nu} = 0 \tag{12.23}$$

This property of a connection is called *metric compatibility.* All the manifolds we study in general relativity are metric compatible. Metric compatibility allows a covariant derivative to commute with raising and lowering indices. So, given a vector field A^λ

$$g_{\mu\lambda}\nabla_\rho A^\lambda = \nabla_\rho(g_{\mu\lambda}A^\lambda) = \nabla_\rho A_\mu \tag{12.24}$$

12.3.5 *Expression for the Christoffel Symbol*

We will now show that given the two conditions that the connection should be torsion-free and be metric compatible, there is a unique expression for the Christoffel symbol. And we will derive that expression.

We will first write out the covariant derivative of the metric tensor in three different ways, basically the same equation but using different letters each time:

$$\nabla_\rho g_{\mu\nu} = \partial_\rho g_{\mu\nu} - \Gamma^\lambda_{\rho\mu} g_{\lambda\nu} - \Gamma^\lambda_{\rho\nu} g_{\mu\lambda} = 0$$

$$\nabla_\mu g_{\nu\rho} = \partial_\mu g_{\nu\rho} - \Gamma^\lambda_{\mu\nu} g_{\lambda\rho} - \Gamma^\lambda_{\mu\rho} g_{\nu\lambda} = 0$$

$$\nabla_\nu g_{\rho\mu} = \partial_\nu g_{\rho\mu} - \Gamma^\lambda_{\nu\rho} g_{\lambda\mu} - \Gamma^\lambda_{\nu\mu} g_{\rho\lambda} = 0 \tag{12.25}$$

Subtracting the second and the third equation from the first, and using the symmetry of the connection, we obtain

$$\partial_\rho g_{\mu\nu} - \partial_\mu g_{\nu\rho} - \partial_\nu g_{\rho\mu} + 2\Gamma^\lambda_{\mu\nu} g_{\lambda\rho} = 0 \tag{12.26}$$

Multiplying by $g^{\sigma\rho}$, we obtain the expression for the connection:

$$\Gamma^\sigma_{\mu\nu} = \frac{1}{2} g^{\sigma\rho} (\partial_\mu g_{\nu\rho} + \partial_\nu g_{\rho\mu} - \partial_\rho g_{\mu\nu}) \tag{12.27}$$

For the three-dimensional Cartesian metric ($g_{ij} = \delta_{ij}$) and the Minkowski metric $(-1, 1, 1, 1)$ the Christoffel symbols are all zero. So for such metrics the covariant derivative is equal to the ordinary derivative: $\nabla_\mu A^\nu = \partial_\mu A^\nu$. Let us consider the spherical coordinates ($x^1 = r, x^2 = \theta, x^3 = \varphi$). An infinitesimal distance ds is expressed in this coordinate system as

$$ds^2 = g_{11}(dx^1)^2 + g_{22}(dx^2)^2 + g_{33}(dx^3)^2 = dr^2 + r^2 d\theta^2 + r^2 \sin^2\theta d\varphi^2 \tag{12.28}$$

The non-diagonal metric components are all zero, and the diagonal components are $g_{11} = 1$, $g_{22} = r^2$ and $g_{33} = r^2 \sin^2\theta$. Most of the Christoffel symbols vanish, with a few exceptions.

Exercises:

1. Evaluate all the Christoffel symbols for the metric of Eq. (12.28).
2. Prove that this metric satisfies metric compatibility: $\nabla_\sigma g_{\mu\nu} = 0$.

12.3.6 *Divergence Theorem*

The covariant derivative of a vector field A^μ is given by

$$\nabla_\mu A^\nu = \partial_\mu A^\nu + \Gamma^\nu_{\mu\lambda} A^\lambda \tag{12.29}$$

We now define the covariant divergence of A^μ as

$$\nabla_\mu A^\mu = \partial_\mu A^\mu + \Gamma^\mu_{\mu\lambda} A^\lambda \tag{12.30}$$

Now, it can be shown that $\Gamma^\mu_{\mu\lambda} = \frac{1}{2} g^{\rho\mu} \partial_\lambda g_{\mu\rho}$. (You are asked to show this in the following exercise.) If g is the determinant of the metric tensor $g_{\mu\nu}$, then it is not hard to show that

$$\frac{1}{2} g^{\rho\mu} \partial_\lambda g_{\mu\rho} = \frac{1}{\sqrt{|g|}} \partial_\lambda \sqrt{|g|} \tag{12.31}$$

Exercises:
1. Show that $\Gamma^\mu_{\mu\lambda} = \frac{1}{2} g^{\rho\mu} \partial_\lambda g_{\mu\rho}$.
2. Prove Eq. (12.31).

And so we obtain the following expression for the covariant divergence:

$$\nabla_\mu A^\mu = \frac{1}{\sqrt{|g|}} \partial_\mu (\sqrt{|g|} A^\mu) \tag{12.32}$$

the divergence theorem in three-dimensional space is written as

$$\iiint_V \partial_\mu V^\mu d^3x = \oiint_S V^\mu \hat{n}_\mu d^2x \tag{12.33}$$

where $\hat{n}_\mu$ is a outward unit vector normal to the surface S.

Now, let $V^\mu = \sqrt{|g|} A^\mu$. Let the space have n dimensions. We will replace the three-dimensional volume V by the n-dimensional "volume" Σ, and the $(n-1)$-dimensional "surface" by $\partial\Sigma$. We will use g for the metric inside the "volume" and γ for the metric on the "surface." We will use Eq. (12.32) to replace the divergence by the covariant divergence, and so the covariant divergence theorem becomes

$$\boxed{\int_\Sigma \nabla_\mu A^\mu \sqrt{|g|} d^n x = \int_{\partial\Sigma} \hat{n}_\mu A^\mu \sqrt{|\gamma|} d^{n-1} x} \tag{12.34}$$

12.3.7 *Parallel Transport*

12.3.7.1 *Parallel transport in flat space*

In an introductory course in vector analysis we learned how to add vectors. We learned that a vector is drawn as an arrow with a head and a tail. So, to form the vector sum $\mathbf{A}+\mathbf{B}$ we place the head of $\mathbf{A}$ on the tail of $\mathbf{B}$ and join the tail of $\mathbf{A}$ to the head of $\mathbf{B}$:

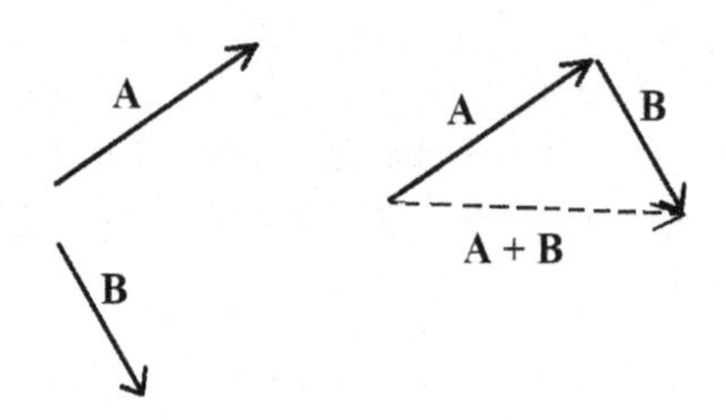

In the process of sliding a vector across the page we took care not to alter the direction or the magnitude of the vector. Since a vector is defined as a quantity having magnitude and direction, the vector remains the same no matter where it is drawn, as long as magnitude and direction do not change. This spatial displacement of a vector is called *parallel transport* of a vector. A vector is not altered through parallel transport.

Parallel transport of vectors was helpful in drawing a triangle of vectors, such as the one shown above, to determine the resultant of two or more forces *acting on a body at the same point.* We later learned that the notion of parallel transport was not useful when we had forces acting at different points on an extended rigid body, producing torque.

12.3.7.2 *Parallel transport in curved space*

On a curved space, such as the surface of a globe, a vector cannot be drawn as an arrow, because a straight arrow will touch the globe at only one point or intersect it at two points. On a curved surface a vector is defined by a set of numbers which differ from point to point. These numbers define the components of the vector that is tangential to the surface at a point. A geometrical depiction of such a vector would require an embedding in a higher space, so that the tangent vector at any point on the surface of the earth is an arrow that is almost entirely outside the surface. In the following discussion, we will give ourselves the license to visualize the vector as an arrow, all the time keeping in mind that the vector itself is defined as a set of numbers at each point on the surface.

On a spherical surface, a straight line — defined as the shortest distance between two points — is a geodesic. So, if we were to parallel transport a vector along a geodesic, the length of the vector should not change, and the angle between the vector and the geodesic (i.e. the tangent to the geodesic) should not change. So, if we start at a point where the vector $\vec{A}$ is tangential to the geodesic (angle between $\vec{A}$ and geodesic is zero), then it should remain tangential to the geodesic at every point. Suppose we start with such a vector pointing north somewhere on the equator, and move along a longitude all the way north to the north pole and continue along the geodesic till we reach the equator on the other side of the globe. By now our vector is directed south. If we then parallel transport it along the equator till we reach the point of origin, our vector would have undergone a change of direction by 180^0.

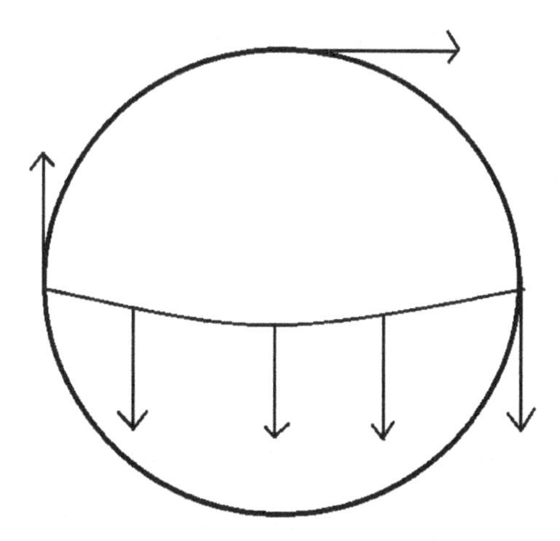

In the above example, the vector changes direction during the parallel transport. We notice that the path taken by the vector was made of two separate geodesics. Along each geodesic the vector did not change direction, but the complete path was not a geodesic, and the vector did undergo a change of direction. In what follows we will discuss parallel transport along arbitrary paths.

Now, a curve in 3-dimensional space can be expressed in parametric form as $x^i = x^i(\lambda)$, etc. where λ is a parameter, and the x^i on the right side stand for (in general different) functions of λ. A familiar example is the projectile, with $\lambda = t$, the time. The path of a projectile is expressible as, e.g. $x = 3.50t; y = 12.3 - 4.90t^2; z = 5.74$. In n-dimensional space, the path of a curve is expressed in concise form as $x^\mu(\lambda)$. In flat Euclidean space, we should be able to parallel transport a vector along any path without changing the vector. But that is not the case with curved space.

For an arbitrary curve $x^\mu(\lambda)$ along which a vector A^μ is parallel transported,

$$\frac{dA^\mu}{d\lambda} = 0 \tag{12.35}$$

The principle of parallel transport can be extended to a (p, q) tensor field such that

$$\frac{d}{d\lambda} A^{\mu_1 \mu_2 \ldots \mu_p}_{\nu_1 \nu_2 \nu_3 \ldots \nu_q} = 0 \tag{12.36}$$

Now, for any tensor A (where we have suppressed the indices for convenience),

$$\frac{dA}{d\lambda} = \frac{dx^\mu}{d\lambda} \frac{\partial A}{\partial x^\mu} \tag{12.37}$$

So, if a tensor A is parallel transported along a curve $x^\mu(\lambda)$, the components of this tensor cannot change along the curve, and so

$$\frac{dx^\mu}{d\lambda} \frac{\partial A}{\partial x^\mu} = 0 \tag{12.38}$$

For a general curved space we upgrade this to a covariant parallel transport by replacing the derivative ∂_μ by the covariant derivative ∇_μ, and we define the condition for parallel transport as:

$$\frac{dx^\mu}{d\lambda} \nabla_\mu A^\nu = \frac{dx^\mu}{d\lambda} \frac{\partial A^\nu}{\partial x^\mu} + \frac{dx^\mu}{d\lambda} \Gamma^\nu_{\mu\rho} A^\rho = 0 \tag{12.39}$$

Noting that $\frac{dx^\mu}{d\lambda} \frac{\partial A^\nu}{\partial x^\mu} = \frac{dA^\nu}{d\lambda}$, we obtain the following equation:

$$\frac{dA^\nu}{d\lambda} + \frac{dx^\mu}{d\lambda} \Gamma^\nu_{\mu\rho} A^\rho = 0 \tag{12.40}$$

This is called the *equation of parallel transport* for a vector.

Note that a vector cannot be parallel transported along an arbitrary curve. Only certain curves $x^\mu(\lambda)$ satisfy the equation of parallel transport.

For a (p, q) tensor we write the equation of parallel transport as

$$\frac{dx^\sigma}{d\lambda}\nabla_\sigma T^{\mu_1\mu_2\ldots\mu_p}{}_{\nu_1\nu_2\ldots\nu_q} = 0 \tag{12.41}$$

For a connection that is metric compatible, i.e. $\nabla_\sigma g_{\mu\nu} = 0$, the equation of parallel transport of the metric tensor is valid for any path $x^\mu(\lambda)$, since

$$\frac{dx^\sigma}{d\lambda}\nabla_\sigma g_{\mu\nu} = 0 \tag{12.42}$$

regardless of the path $x^\mu(\lambda)$. This has an important corollary. The inner product of two vectors $A^\nu B_\nu$ can be expressed as $A^\nu B_\nu = g_{\mu\nu}A^\mu B^\nu$. Suppose two vectors A^μ and B^ν can be parallel transported along some curve $x^\rho(\lambda)$. Then the inner product will automatically be parallel transported along this curve, i.e.

$$\frac{dx^\rho}{d\lambda}\nabla_\rho(g_{\mu\nu}A^\mu B^\nu) = 0 \tag{12.43}$$

Since the inner product of two vectors is preserved, it follows that the norm of a vector $A^\mu A_\mu$ is also preserved in a parallel transport.

Exercise:
Prove Eq. (12.43).

We will now return to the examination of geodesics. A geodesic can be defined as the shortest path between two points, or more accurately, as the path that is shorter (or longer) than all its immediately neighboring non-intersecting paths. But our consideration of parallel transport along the surface of a globe gives us an alternative definition of a geodesic: a geodesic is the path along which the tangent vector is parallel transported:

$$\frac{dx^\rho}{d\lambda}\nabla_\rho\frac{dx^\mu}{d\lambda} = \frac{dx^\rho}{d\lambda}\frac{\partial}{\partial x^\rho}\frac{dx^\mu}{d\lambda} + \Gamma^\mu_{\rho\sigma}\frac{dx^\rho}{d\lambda}\frac{dx^\sigma}{d\lambda} = \frac{d^2x^\mu}{d\lambda^2} + \Gamma^\mu_{\rho\sigma}\frac{dx^\rho}{d\lambda}\frac{dx^\sigma}{d\lambda} = 0 \tag{12.44}$$

A geodesic is a curve $x^\mu(\lambda)$ that satisfies the equation

$$\frac{d^2x^\mu}{d\lambda^2} + \Gamma^\mu_{\rho\sigma}\frac{dx^\rho}{d\lambda}\frac{dx^\sigma}{d\lambda} = 0 \tag{12.45}$$

The geodesic is the shortest or longest path among neighboring non-intersecting paths. A light ray is generated in the laboratory by a laser — a coherent beam of electromagnetic plane waves[1] of high frequency. Such light rays obey Fermat's least time principle, following straight lines through a medium of uniform refractive index, and bending when passing through a boundary between media of different indices. A light ray passing through curved space would follow the shortest path, which is a geodesic. A geodesic is the path of a light ray in any space, whether the curvature be positive, negative or zero.

If the space is flat, with zero curvature, the Christoffel symbols vanish, and the geodesic equation becomes $\frac{d^2x^\mu}{d\lambda^2} = 0$. Integrating, we get the equation of a straight line, $x^\mu = a\lambda + b$.

Suppose we choose to identify the parameter λ with the proper time of a moving particle. By Newton's first law, a particle under the absence of forces would travel along a straight line or a geodesic. Its motion can then be described by the equation

$$\frac{d^2x^\mu}{d\tau^2} + \Gamma^\mu_{\rho\sigma}\frac{dx^\rho}{d\tau}\frac{dx^\sigma}{d\tau} = 0 \tag{12.46}$$

So, for flat space we would get $\frac{d^2x^\mu}{d\tau^2} = 0$, i.e. zero acceleration. If a force were applied to the particle, Newton's second law would be expressible as

$$\frac{d^2x^\mu}{d\tau^2} = F^\mu/m \tag{12.47}$$

If a particle of mass m and having charge q is placed in an electromagnetic field, it experiences acceleration

$$\frac{d^2x^\mu}{d\tau^2} = \frac{q}{m}F^\mu_{\ \nu}\frac{dx^\nu}{d\tau} \tag{12.48}$$

where $F^{\mu\nu}$ is the electromagnetic field tensor. The right-hand side of the equation is the tensor form of the Lorentz force equation in flat space

$$\mathbf{F} = q(\mathbf{E} + \mathbf{v} \times \mathbf{B}) \tag{12.49}$$

Exercise:
Using the expression for the electromagnetic field tensor, derive Eq. (12.48) from the Lorentz force equation.

[1] Lasers can be approximated as plane waves along a distance smaller than the Rayleigh length, which is typically a couple of meters.

Generalizing Eq. (12.48) to curved space, we obtain the expression for the motion of a particle of mass m and charge q in an electromagnetic field $F^{\mu\nu}$:

$$\frac{d^2x^\mu}{d\tau^2} + \Gamma^\mu_{\rho\sigma}\frac{dx^\rho}{d\tau}\frac{dx^\sigma}{d\tau} = \frac{q}{m}F^\mu_{\ \nu}\frac{dx^\nu}{d\tau} \tag{12.50}$$

The behavior of the charged particle depends on the space in which it is moving. For flat space, all the Christoffel symbols vanish, and we get the familiar Eq. (12.48) that is valid for special relativity and electromagnetism.

We have emphasized that charges do not change the curvature of space. Maxwell's equations and the entire classical electrodynamics as well as quantum electrodynamics imply a flat Euclidean space. It is only in Einstein's general theory of relativity that space becomes curved in the presence of matter. But before we reach Einstein's theory we need to obtain a proper expression for the curvature of space.

12.3.8 *Curvature of Space*

A vector can be parallel transported along any closed path in flat space without altering its direction:

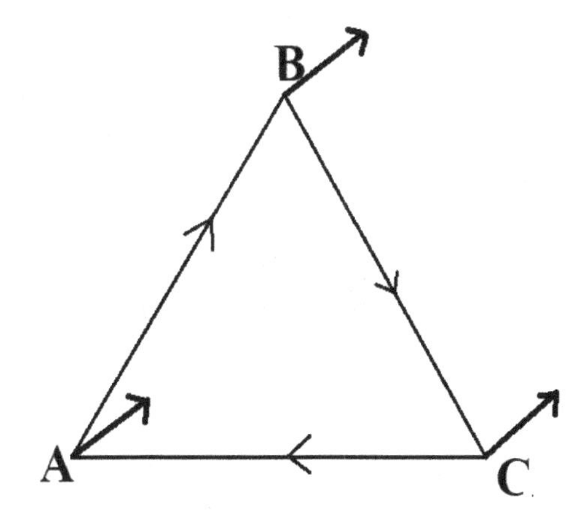

But in curved space, parallel transport along a closed path could result in a change of direction. In the figure below, a vector marked as 1 is parallel transported from A to B to C to A, and ends up as 2, in a different direction.

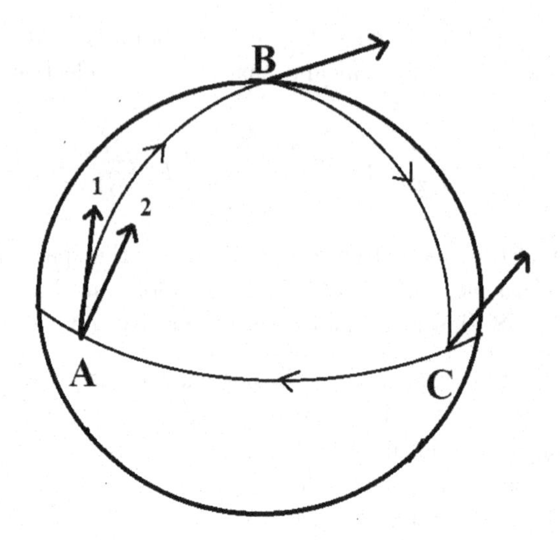

Thus, the deviation of a vector when it is parallel transported along a closed curve is a measure of the curvature of a surface. In the diagram we have shown a sphere with uniform curvature throughout. But an arbitrary surface need not have uniform curvature. Its curvature could vary from point to point. Curvature is therefore a property of a point and its neighborhood, and not of an extended region. Mathematically speaking, the curvature at a point is related to the metric tensor and its derivatives at that point.

Let us consider a parallelogram formed of two infinitesimal vectors dx^μ and dx^ν. Such a parallelogram can be formed exactly in flat space, and these two vectors can have arbitrary magnitudes, but in curved space the shape becomes a parallelogram only in the limit as each vector tends to zero:

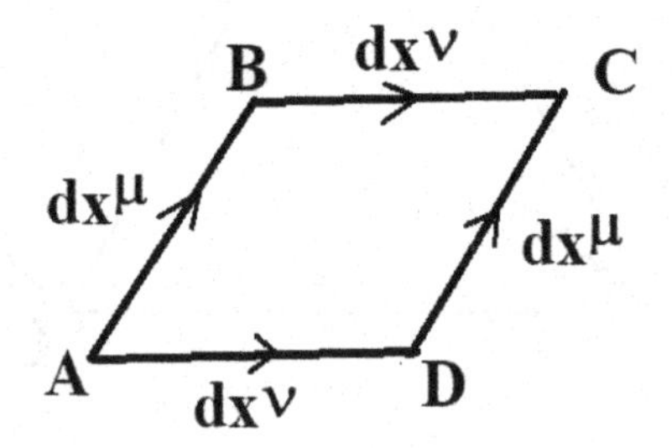

Now, suppose a vector A^ρ is parallel transported from A to C in two different ways: (1) from A to B along dx^μ and then from B to C along dx^ν, and (2) from A to D along dx^ν and then from D to C along dx^μ. If it were flat space, the results of the two transports would be the same. But in curved space, the results would be different. So, curvature of space can be measured by the difference in parallel transport along the two alternative paths. Let us call this difference in the final value of the vector (between the two paths) as δA^ρ. We expect this difference to be proportional to A^ρ itself, and also to the vectors dx^μ and dx^ν. Writing the relationship as a tensor, the constant of proportionality becomes a (1, 3) tensor, which is called the *Riemann tensor* $R^\rho_{\sigma\mu\nu}$:

$$\delta A^\rho = R^\rho_{\sigma\mu\nu} A^\sigma dx^\mu dx^\nu \tag{12.51}$$

We need to find an expression for the Riemann tensor.

Suppose the initial vector at A is A^ρ. When it is parallel transported to B, it becomes $A^\rho + dx^\mu \nabla_\mu A^\rho$. Upon parallel transport to C, it becomes

$$A^\rho + dx^\mu \nabla_\mu A^\rho + dx^\nu \nabla_\nu (A^\rho + dx^\mu \nabla_\mu A^\rho)$$

$$= A^\rho + dx^\mu \nabla_\mu A^\rho + dx^\nu \nabla_\nu A^\rho + dx^\nu dx^\mu \nabla_\nu \nabla_\mu A^\rho \tag{12.52}$$

When the vector is parallel transported from A to D to C, the resulting vector is

$$A^\rho + dx^\nu \nabla_\nu A^\rho + dx^\mu \nabla_\mu A^\rho + dx^\mu dx^\nu \nabla_\mu \nabla_\nu A^\rho \tag{12.53}$$

Subtracting (12.52) from (12.53), we obtain

$$\delta A^\rho = dx^\mu dx^\nu (\nabla_\mu \nabla_\nu - \nabla_\nu \nabla_\mu) A^\rho \tag{12.54}$$

We are interested in the terms on the right side:

$$(\nabla_\mu \nabla_\nu - \nabla_\nu \nabla_\mu) A^\rho = \partial_\mu (\nabla_\nu A^\rho) + \Gamma^\rho_{\mu\lambda} \nabla_\nu A^\lambda - \Gamma^\lambda_{\mu\nu} \nabla_\lambda A^\rho - (\mu \leftrightarrow \nu)$$

$$= \partial_\mu \partial_\nu A^\rho + (\partial_\mu \Gamma^\rho_{\nu\lambda}) A^\lambda + \Gamma^\rho_{\nu\lambda} \partial_\mu A^\lambda + \Gamma^\rho_{\mu\lambda} \partial_\nu A^\lambda + \Gamma^\rho_{\mu\sigma} \Gamma^\sigma_{\nu\lambda} A^\lambda - \Gamma^\lambda_{\mu\nu} \nabla_\lambda A^\rho - (\mu \leftrightarrow \nu)$$

$$= (\partial_\mu \Gamma^\rho_{\nu\sigma} - \partial_\nu \Gamma^\rho_{\mu\sigma} + \Gamma^\rho_{\mu\lambda} \Gamma^\lambda_{\nu\sigma} - \Gamma^\rho_{\nu\lambda} \Gamma^\lambda_{\mu\sigma}) A^\sigma - (\Gamma^\lambda_{\mu\nu} - \Gamma^\lambda_{\nu\mu}) \nabla_\lambda A^\rho$$

We have canceled some terms, and changed some dummy indices. Now, $\Gamma^\lambda_{\mu\nu} - \Gamma^\lambda_{\nu\mu} = T^\lambda_{\mu\nu}$, the torsion tensor, which is zero for all physical spaces of interest to us. Thus, we obtain an expression for the Riemann tensor.

The Riemann Tensor

$$R^\rho_{\sigma\mu\nu} = \partial_\mu \Gamma^\rho_{\nu\sigma} - \partial_\nu \Gamma^\rho_{\mu\sigma} + \Gamma^\rho_{\mu\lambda} \Gamma^\lambda_{\nu\sigma} - \Gamma^\rho_{\nu\lambda} \Gamma^\lambda_{\mu\sigma} \tag{12.55}$$

The Riemann tensor is a measure of the curvature of a surface at any point. We see that the Riemann tensor is a (1,3) tensor, of rank 4. But Eq. (12.54) shows that as far as curvature is concerned, only two indices are significant, since we are ultimately interested only in $\nabla_\mu \nabla_\nu - \nabla_\nu \nabla_\mu$. Hence we need a (0, 2) tensor of the form $R_{\mu\nu}$ to measure curvature. Since the Riemann tensor is (1,3), we will perform a contraction in order to obtain a suitable (0, 2) tensor.

The Riemann tensor has some important properties.

$$R^\rho_{\ \sigma\mu\nu} = -R^\rho_{\ \sigma\nu\mu} \tag{12.56}$$

$$R^\lambda_{\ \lambda\mu\nu} = 0 \tag{12.57}$$

Thus, we can generate a non-vanishing (0, 2) tensor from the Riemann tensor by contracting the top index with either the second or the third lower index. We will choose the second lower index. The resulting (0, 2) tensor is called the Ricci tensor $R_{\mu\nu}$:

$$R^\lambda_{\ \mu\lambda\nu} = R_{\mu\nu} \quad \textbf{Ricci tensor} \tag{12.58}$$

Exercises:

1. Prove Eqs. (12.56) and (12.57).
2. Show that $R_{\mu\nu} = R_{\nu\mu}$.

The trace of the Ricci tensor is called the Ricci scalar or curvature scalar.

Ricci Scalar or Curvature Scalar

$$R = R^\mu_{\ \mu} = g^{\mu\nu} R_{\mu\nu}$$

12.4 Equation for Curved Space

We will now derive Einstein's equation which relates curvature of space to matter and energy. We will follow a Lagrangian approach. We shall set up a Lagrangian of the space as a function of the metric coefficients, and determine the function that minimizes the integral between fixed limits. The variation of the metric is kept zero at these limits. We first consider space without matter.

We need to set up a Lagrangian out of the variables that define curvature. Now, let us recall from Eq. (12.34) that the expression $\nabla_\mu A^\mu \sqrt{|g|}d^n x$ is an invariant or scalar. Since $\nabla_\mu A^\mu$ is manifestly a scalar, it follows that $\sqrt{|g|}d^n x$ is also a scalar. For a metric with a Lorentz signature,[2] g is always negative, and so $|g| = -g$. So $\sqrt{-g}\, d^n x$ is a scalar.

A relativistically invariant action can then be defined as

$$A = \int L\sqrt{-g}\, d^n x \tag{12.59}$$

where the Lagrangian L is a scalar. A suitable candidate for the scalar that describes curvature is the Ricci scalar. And so we define an action for space as

$$A = \int R\sqrt{-g}\, d^n x \tag{12.60}$$

We will first set up the equations of motion for empty space in the absence of matter.

Our equations will follow from the condition that the variation δA is zero:

$$\delta A = \delta \int R\sqrt{-g}\, d^n x = \delta \int g^{\mu\nu} R_{\mu\nu}\sqrt{-g}\, d^n x \tag{12.61}$$

We are seeking an equation for the curvature of space, which is a function of the metric coefficients $g^{\mu\nu}$. So the independent variable is $g^{\mu\nu}$, and we seek the integral for which the variation $\delta A = 0$ for arbitrary small variation $\delta g^{\mu\nu}$. Our objective is therefore to reduce this integral to a form in which the variation $\delta g^{\mu\nu}$ appears as a factor. We expand the integrand using the Leibniz rule, and write the result as the sum of three integrals:

$(\delta A)_1 = \int \delta g^{\mu\nu} R_{\mu\nu}\sqrt{-g}\, d^n x$

$(\delta A)_2 = \int g^{\mu\nu} \delta R_{\mu\nu}\sqrt{-g}\, d^n x$

$(\delta A)_3 = \int R\delta(\sqrt{-g})\, d^n x$

[2]A Lorentz signature for flat space can be $(-1, 1, 1, 1)$ or $(1, -1, -1, -1)$. For curved spaces the elements are no longer ± 1, but the time element always has the opposite sign as the three space elements, and so the determinant is always negative.

The first integral is already in the required form. So we proceed by evaluating $(\delta A)_2$.

We first write out the terms of the Riemann tensor:

$$R^{\rho}_{\mu\lambda\nu} = \partial_\lambda \Gamma^{\rho}_{\nu\mu} + \Gamma^{\rho}_{\lambda\sigma}\Gamma^{\sigma}_{\nu\mu} - (\lambda \leftrightarrow \nu) \tag{12.62}$$

The variations in the Riemann tensor are formed from variations in the Christoffel connections $\delta\Gamma^{\rho}_{\nu\mu}$.

Let us recall that the covariant derivative is a tensor:

$$\nabla_\mu A^\nu = \frac{\partial A^\nu}{\partial x^\mu} + \Gamma^{\nu}_{\mu\sigma} A^\sigma \tag{12.63}$$

A small variation in the connection $\delta\Gamma^{\nu}_{\mu\sigma}$ will affect the covariant derivative, but not the ordinary derivative:

$$\delta(\nabla_\mu A^\nu) = \delta\Gamma^{\nu}_{\mu\sigma} A^\sigma \tag{12.64}$$

Since the left side is a tensor, the right side is also a tensor. And since A^σ is a tensor, it follows that $\delta\Gamma^{\nu}_{\mu\sigma}$ is also a tensor, and hence has a valid covariant derivative.

$$\nabla_\lambda(\delta\Gamma^{\rho}_{\nu\mu}) = \partial_\lambda(\delta\Gamma^{\rho}_{\nu\mu}) + \Gamma^{\rho}_{\lambda\sigma}\delta\Gamma^{\sigma}_{\nu\mu} - \Gamma^{\sigma}_{\lambda\nu}\delta\Gamma^{\rho}_{\sigma\mu} - \Gamma^{\sigma}_{\lambda\mu}\delta\Gamma^{\rho}_{\nu\sigma} \tag{12.65}$$

A variation in the Riemann tensor is obtained from Eq. (12.62):

$$\delta R^{\rho}_{\mu\lambda\nu} = \partial_\lambda(\delta\Gamma^{\rho}_{\nu\mu}) + \delta\Gamma^{\rho}_{\lambda\sigma}\Gamma^{\sigma}_{\nu\mu} + \Gamma^{\rho}_{\lambda\sigma}\delta\Gamma^{\sigma}_{\nu\mu} - (\lambda \leftrightarrow \nu) \tag{12.66}$$

A comparison with Eq. (12.65) shows that

$$\delta R^{\rho}_{\mu\lambda\nu} = \nabla_\lambda(\delta\Gamma^{\rho}_{\nu\mu}) - \nabla_\nu(\delta\Gamma^{\rho}_{\lambda\mu}) \tag{12.67}$$

Contracting the ρ and the λ indices,

$$\delta R_{\mu\nu} = \nabla_\lambda(\delta\Gamma^{\lambda}_{\nu\mu}) - \nabla_\nu(\delta\Gamma^{\lambda}_{\lambda\mu}) \tag{12.68}$$

And so, the second integral becomes

$$(\delta A)_2 = \int d^n x\sqrt{-g}\, g^{\mu\nu}[\nabla_\lambda(\delta\Gamma^{\lambda}_{\nu\mu}) - \nabla_\nu(\delta\Gamma^{\lambda}_{\lambda\mu})] \tag{12.69}$$

Applying metric compatibility ($g^{\mu\nu}\nabla = \nabla g^{\mu\nu}$) and relabeling some dummy indices,

$$(\delta A)_2 = \int d^n x\sqrt{-g}\nabla_\sigma[g^{\mu\nu}(\delta\Gamma^{\sigma}_{\mu\nu}) - g^{\mu\sigma}(\delta\Gamma^{\lambda}_{\lambda\mu})] \tag{12.70}$$

From the divergence theorem of Eq. (12.34) this integral becomes a surface integral over the boundary of the region. We will set the boundary at infinity, where the variation in the metric and its derivatives are zero. Thus the integral vanishes.

We next tackle the third integral $(\delta A)_3$. There is an important theorem that states that if B is a square matrix with non-zero determinant $|B|$, then

$$\ln|B| = \mathrm{Tr}(\ln B) \tag{12.71}$$

We will not prove this, but show that it is true for the metric tensor $g_{\mu\nu}$ which is symmetric in its indices for all physical spaces of interest to us. We will denote the determinant of $g_{\mu\nu}$ by the symbol g. Considering a four-dimensional space, $g = g_{11}g_{22}g_{33}g_{44}$. And Tr $(\ln g_{\mu\nu}) = \ln g_{11} + \ln g_{22} + \ln g_{33} + \ln g_{44} = \ln g$. Hence

$$\ln g = \mathrm{Tr}(\ln g_{\mu\nu}) \tag{12.72}$$

If we introduce a small variation in $\delta g_{\mu\nu}$,

$$\frac{\delta g}{g} = \mathrm{Tr}(g_{\mu\nu}^{-1}\delta g_{\mu\nu}) \tag{12.73}$$

Since $g_{\mu\nu}^{-1} = g^{\mu\nu}$,

$$\delta g = g(g^{\mu\nu}\delta g_{\mu\nu}) \tag{12.74}$$

From $g^{\mu\lambda}g_{\mu\nu} = \delta_\nu^\lambda$ we get $\delta g^{\mu\lambda}g_{\mu\nu} = -g^{\mu\lambda}\delta g_{\mu\nu}$, and thus $g^{\mu\nu}\delta g_{\mu\nu} = -g_{\mu\nu}\delta g^{\mu\nu}$. So

$$\delta g = -g(g_{\mu\nu}\delta g^{\mu\nu}) \tag{12.75}$$

$\delta\sqrt{-g} = -\frac{1}{2\sqrt{-g}}\delta g = \frac{1}{2}\frac{g}{\sqrt{-g}}g_{\mu\nu}\delta g^{\mu\nu} = -\frac{1}{2}\sqrt{-g}g_{\mu\nu}\delta g^{\mu\nu}$.

So

$$(\delta A)_3 = \int R\delta(\sqrt{-g})d^n x = -\frac{1}{2}\int(\delta g^{\mu\nu})Rg_{\mu\nu}\sqrt{-g}d^n x \tag{12.76}$$

Hence, the total integral becomes

$$\delta A = \int(\delta g^{\mu\nu})[R_{\mu\nu} - \frac{1}{2}Rg_{\mu\nu}]\sqrt{-g}d^n x \tag{12.77}$$

This integral vanishes for arbitrary small $\delta g^{\mu\nu}$. Therefore

$$R_{\mu\nu} - \frac{1}{2}Rg_{\mu\nu} = 0 \tag{12.78}$$

$R_{\mu\nu} - \frac{1}{2}Rg_{\mu\nu} \equiv G_{\mu\nu}$ is called the **Einstein tensor**. We have just shown that this tensor vanishes in empty space. There is another important property of this tensor, which we will derive next.

For this derivation we employ the Bianchi identity, which is expressed as a relationship between different components of the fully covariant form of the Riemann tensor: $R_{\rho\lambda\mu\nu} = g_{\rho\sigma}R^{\sigma}_{\lambda\mu\nu}$:

$$\nabla_\lambda R_{\rho\sigma\mu\nu} + \nabla_\rho R_{\sigma\lambda\mu\nu} + \nabla_\sigma R_{\lambda\rho\mu\nu} = 0 \tag{12.79}$$

The proof of this identity is outlined in the box shown on the following page. Multiplying both sides by $g^{\nu\sigma}g^{\mu\lambda}$, and changing indices in the middle term,

$$g^{\nu\sigma}g^{\mu\lambda}(\nabla_\lambda R_{\rho\sigma\mu\nu} - \nabla_\rho R_{\sigma\lambda\nu\mu} + \nabla_\sigma R_{\lambda\rho\mu\nu}) = 0 \tag{12.80}$$

So

$$\nabla^\mu R_{\rho\mu} - \nabla_\rho R + \nabla^\nu R_{\rho\nu} = 0$$

which is also expressible as

$$\nabla^\mu \left(R_{\rho\mu} - \frac{1}{2}Rg_{\rho\mu} \right) = 0 \tag{12.81}$$

The Einstein tensor $G_{\mu\nu} \equiv R_\mu - \frac{1}{2}Rg_{\mu\nu}$ therefore has two important properties: $G_{\mu\nu} = 0$ for empty space without matter, and $\nabla^\mu G_{\mu\nu} = 0$ for all kinds of space.

The Bianchi Identity

We will now derive an important property of the Riemann tensor. To begin, we will write this tensor with all lower indices:

$$R_{\rho\sigma\mu\nu} = g_{\rho\lambda} R^{\lambda}_{\sigma\mu\nu} \tag{12.82}$$

Now, consider a function $f(x)$ which has a maximum at some point $x = a$. So, $f'(a) = 0$, and $f''(a) < 0$. The first derivative is zero, but not the higher derivatives. A physical space without singularities can be approximated as a flat space within a very small neighborhood. So, if we limit ourselves to a small neighborhood of a point in space, the first derivatives of the metric $g_{\mu\nu}$ can be made zero, but not necessarily the higher derivatives. The Christoffel connections are functions of the first derivatives, and hence they vanish, but not the derivatives of these connections. So, within a small space, we approximate the Riemann tensor as

$$R^{\rho}_{\sigma\mu\nu} = \partial_\mu \Gamma^{\rho}_{\nu\sigma} - \partial_\nu \Gamma^{\rho}_{\mu\sigma} \quad \text{(local region)}$$

and

$$R_{\rho\sigma\mu\nu} = g_{\rho\lambda}(\partial_\mu \Gamma^{\lambda}_{\nu\sigma} - \partial_\nu \Gamma^{\lambda}_{\mu\sigma})$$

$$= \frac{1}{2} g_{\rho\lambda} g^{\lambda\tau}(\partial_\mu \partial_\nu g_{\sigma\tau} + \partial_\mu \partial_\sigma g_{\tau\nu} - \partial_\mu \partial_\tau g_{\nu\sigma}) - (\mu \leftrightarrow \nu)$$

$$= \frac{1}{2}(\partial_\mu \partial_\sigma g_{\rho\nu} - \partial_\mu \partial_\rho g_{\nu\sigma}) - (\mu \leftrightarrow \nu)$$

Since the Christoffel symbols vanish in this regime, the covariant derivative becomes equal to the ordinary derivative: $\nabla_\mu \to \partial_\mu$. So

$$\nabla_\lambda R_{\rho\sigma\mu\nu} = \partial_\lambda R_{\rho\sigma\mu\nu} = \frac{1}{2}(\partial_\lambda \partial_\mu \partial_\sigma g_{\rho\nu} - \partial_\lambda \partial_\mu \partial_\rho g_{\nu\sigma}) - (\mu \leftrightarrow \nu)$$

By writing out the terms and adding them, it is seen that

$$\nabla_\lambda R_{\rho\sigma\mu\nu} + \nabla_\rho R_{\sigma\lambda\mu\nu} + \nabla_\sigma R_{\lambda\rho\mu\nu} = 0 \tag{12.83}$$

This is called the **Bianchi identity**. Since it is a tensor equation, it is valid in any coordinate system, not just one that is locally flat (or inertial).

12.4.1 *Space Containing Matter*

We have thus seen that $G_{\mu\nu} \equiv R_{\mu\nu} - \frac{1}{2}Rg_{\mu\nu} = 0$ in a space without matter. Einstein's hypothesis was that the presence of matter led to the curvature of space. We are seeking the complete equation of general relativity, one which relates the curvature of space to the presence of mass.

It may be helpful to recall an analogy from electromagnetism. Experiments by Oersted and others showed that a current flowing along a wire generates a curling electric field. These results were stated as Ampère's law: $\nabla \times \mathbf{B} = \frac{1}{\epsilon_0 c^2}\mathbf{J}$. But this equation is not complete. If we take the divergence of both sides we come up with the result that $\nabla \cdot \mathbf{J} = 0$ in every situation. But this is not true, because currents and charges obey the equation of continuity $\nabla \cdot \mathbf{J} = -\frac{\partial \rho}{\partial t}$. Now, Maxwell's first equation states a relationship between charge and electric field, $\nabla \cdot \mathbf{E} = \frac{\rho}{\epsilon_0}$. And so Maxwell added a term to the right side of Ampère's equation, thereby completing it as: $\nabla \times \mathbf{B} = \frac{1}{\epsilon_0 c^2}\mathbf{J} + \frac{1}{c^2}\frac{\partial \mathbf{E}}{\partial t}$. We will take a parallel route in order to obtain the complete equation relating matter and space.

Our hypothesis is that the presence of matter will transform the equation $G_{\mu\nu} = 0$ into one of the form $G_{\mu\nu} = A_{\mu\nu}$ where $A_{\mu\nu}$ is some (0, 2) tensor that is related to matter, and since $\nabla^\mu G_{\mu\nu} = 0$, it is necessary that $\nabla^\mu A_{\mu\nu} = 0$ as well.

There is a tensor related to matter which satisfies the conditions we are seeking: the energy-momentum tensor or the stress tensor $T^{\mu\nu}$. We construct this tensor in the following manner: First, we assume the conservation of matter, and so we use the equation of continuity

$$\frac{\partial \rho}{\partial t} + \nabla \cdot (\rho \mathbf{v}) = 0 \tag{12.84}$$

where ρ is the mass density and $\mathbf{v}$ the velocity of the matter. This equation can be written as $\partial^\mu J_\mu = 0$ in relativistic form. To make this equation covariant in curved space we write this as $\nabla^\mu J_\mu = 0$. But J_μ is only a rank one tensor, and we are seeking a rank two tensor.

Next, we consider a microscopic region of volume ΔV and choose a point within this region. The total momentum of the matter within this region would be $\rho \mathbf{v} \Delta V$ and the force acting on the region would be

$$\mathbf{f} = \frac{\partial(\rho \mathbf{v})}{\partial t}\Delta V \tag{12.85}$$

If the velocity changes slowly with time, we can use Eq. (12.84) to write the force as

$$\mathbf{f} = \frac{\partial(\rho\mathbf{v})}{\partial t}\Delta V = \frac{\partial \rho}{\partial t}\mathbf{v}\Delta V = -\nabla \cdot (\rho\mathbf{v})\mathbf{v}\Delta V \tag{12.86}$$

If $\mathbf{v}$ also changes gradually with space, we can bring the second $\mathbf{v}$ inside the del:

$$\frac{\partial(\rho\mathbf{v})}{\partial t} = -\nabla \cdot (\rho\mathbf{v}\mathbf{v}) \tag{12.87}$$

The right side is manifestly a tensor, and we can write the left side also in the form of a tensor. We replace $\mathbf{v}$ with the covariant velocity, i.e. $v_\mu = \gamma(c, v_x, v_y, v_z)$ and write the equation as a covariant divergence of a second rank tensor, and relax the restrictions on $\mathbf{v}$:

$$\nabla^\mu T_{\mu\nu} = 0 \tag{12.88}$$

where $T_{0j} = \rho c v_j$ and $T_{ij} = \rho v_i v_j$ where the indices i and j refer to the spatial components (1, 2, 3). The general form is $T_{\mu\nu} = \rho v_\mu v_\nu$.

So, we could incorporate matter into the equation for curvature and write the complete equation as

$$R_{\mu\nu} - \frac{1}{2}Rg_{\mu\nu} = kT_{\mu\nu} \tag{12.89}$$

The constant k can be fixed by comparison with the classical Newton's law of gravitation in the low mass limit. We shall do this in the following subsection.

There are alternate expressions of Einstein's equation of general relativity which we can obtain by taking the trace on both sides:

$$g^{\mu\nu}R_{\mu\nu} - \frac{1}{2}Rg^{\mu\nu}g_{\mu\nu} = kg^{\mu\nu}T_{\mu\nu} \tag{12.90}$$

$$R - \frac{1}{2}R(4) = kT$$

$$R = -kT \tag{12.91}$$

$$R_{\mu\nu} = k\left(T_{\mu\nu} - \frac{1}{2}Tg_{\mu\nu}\right) \tag{12.92}$$

12.4.2 *The Constant k*

Just as for small velocities special relativity can be approximated by Newtonian kinematics, so for small masses general relativity should be approximated by Newtonian gravitational theory. We shall employ this criterion not only to demonstrate the validity of Einstein's equation, but also to derive the value of the constant k in terms of Newton's laws of gravitation.

Let us assume there is a weak static gravitational potential ϕ in some region of space. The acceleration experienced by a test object in this field is therefore $\mathbf{a} = -\nabla\phi$. We will also assume that all speeds in this regime are much smaller than c, so that $\frac{dx}{dt} \ll c$. We label the coordinates (0, 1, 2, 3), so the time coordinate $x^0 = ct$. Then, for $i \neq 0$,

$$\frac{dx^i}{d\tau} \ll \frac{dx^0}{d\tau} \tag{12.93}$$

where $\tau = t\sqrt{1 - \frac{v^2}{c^2}} = t/\gamma$ is the proper time.

A geodesic in this space is expressed by the equation

$$\frac{d^2x^\mu}{d\tau^2} + \Gamma^\mu_{\rho\sigma}\frac{dx^\rho}{d\tau}\frac{dx^\sigma}{d\tau} = 0 \tag{12.94}$$

For small velocities we can neglect all the $\frac{dx^i}{d\tau}$ terms ($i \neq 0$):

$$\frac{d^2x^\mu}{d\tau^2} + \Gamma^\mu_{00}c^2\left(\frac{dt}{d\tau}\right)^2 = 0 \tag{12.95}$$

The field is static, which means that the curvature does not change with time: $\partial_0 g_{\mu\nu} = 0$. So

$$\Gamma^\mu_{00} = \frac{1}{2}g^{\mu\lambda}(\partial_0 g_{\lambda 0} + \partial_0 g_{0\lambda} - \partial_\lambda g_{00}) = -\frac{1}{2}g^{\mu\lambda}\partial_\lambda g_{00} \tag{12.96}$$

Since the mass is small, the curvature is also small. For flat space $g_{00} = -1$. So for a nearly flat space $g_{00} = -1 + h$ where h is small, and $g^{00} = 1/g_{00} = -1 - h$ to a first approximation in h. And so

$$\frac{d^2x^i}{d\tau^2} = \frac{1}{2}c^2 g^{i\lambda}\partial_\lambda(-1 + h)\left(\frac{dt}{d\tau}\right)^2 \tag{12.97}$$

Multiplying both sides by $(\frac{d\tau}{dt})^2$,

$$\frac{d^2x^i}{dt^2} = \frac{1}{2}c^2 g^{i\lambda}\partial_\lambda h = \frac{1}{2}c^2\partial^i h \tag{12.98}$$

Now, the Newtonian acceleration is related to the gravitational potential by

$$\frac{d^2x^i}{dt^2} = -\partial^i\phi \tag{12.99}$$

Hence

$$h = -\frac{2\phi}{c^2} \tag{12.100}$$

Therefore

$$g_{00} = -\left(1 + \frac{2\phi}{c^2}\right) \tag{12.101}$$

For low velocity, all the elements of $T_{\mu\nu}$ can be neglected except $T_{00} = \rho c^2$. And $T = g^{00}T_{00} = -T_{00} = -\rho c^2$ to a first approximation. We showed in the previous subsection that

$$R_{\mu\nu} = k\left(T_{\mu\nu} - \frac{1}{2}Tg_{\mu\nu}\right) \tag{12.102}$$

So, again to a first approximation,

$$R_{00} = k\left[\rho c^2 - \frac{1}{2}(-\rho c^2)(-1)\right] = \frac{1}{2}k\rho c^2 \tag{12.103}$$

Now, $R_{00} = R^{\lambda}_{0\lambda 0}$, but since $R^0_{000} = 0$, the only terms that matter are the R^i_{0i0}. The time derivative is zero, since the fields are static. And we can neglect the terms containing products of Christoffel symbols. So the only significant terms are $R^i_{0j0} = \partial_j\Gamma^i_{00}$. From these we can obtain our required result:

$$R_{00} = R^i_{0i0} = \partial_i\left[\frac{1}{2}g^{i\lambda}(\partial_0 g_{\lambda 0} + \partial_0 g_{0\lambda} - \partial_\lambda g_{00})\right] = -\frac{1}{2}\delta^{ij}\partial_i\partial_j h = -\frac{1}{2}\nabla^2 h \tag{12.104}$$

Now, comparing Eqs. (12.100), (12.103) and (12.104), we obtain

$$\nabla^2\phi = \frac{kc^4}{2}\rho \tag{12.105}$$

This is in the form of the Newtonian equation for gravitational potential

$$\nabla^2\phi = 4\pi G\rho \tag{12.106}$$

Thus, Einstein's equation reduces to Newton's equation in the low mass low velocity regime. This gives us confidence in the theory, and also permits us to obtain the value of k in terms of G:

$$k = \frac{8\pi G}{c^4} \tag{12.107}$$

So, Einstein's equation of general relativity becomes

$$R_{\mu\nu} - \frac{1}{2} R g_{\mu\nu} = \frac{8\pi G}{c^4} T_{\mu\nu} \tag{12.108}$$

This is the fundamental equation of general relativity. This equation basically says that matter (right side) produces curvature in spacetime (left side), and provides the mathematical relationship between the two.

There is an analogy with a corresponding equation from electromagnetism, where $F^{\beta\alpha}$ is the electromagnetic field tensor, and J^β is the four-vector charge-current:

$$\partial_\alpha F^{\beta\alpha} = \frac{J^\beta}{c\epsilon_0} \tag{12.109}$$

Of course, there are differences. This is a tensor equation of rank 1, whereas that of general relativity is a tensor equation of rank 2. This has important consequences at the quantum level. The rank of the tensor is related to the spin of the quantum of the corresponding field.

When the electromagnetic field was quantized, we obtained photons with spin 1. When the gravitational field is quantized, we get gravitons with spin 2.[3] Both photons and gravitons are bosons of zero rest mass, which travel at the same speed c. So, gravitational force is not communicated instantly from one body to another, but propagates at speed c. One of the consequences is that an oscillating mass should radiate gravitational waves. Such waves have been detected.[4]

12.4.3 *A Solution to Einstein's Equation*

An important solution to Einstein's equation is the curvature of spacetime outside a spherical mass such as the sun or some large star. The spherical symmetry of the problem enables us to obtain a simple solution without going through a lot of complicated mathematics. In empty space in the absence of matter a spacetime interval $ds = (cdt, dr, d\theta, d\varphi)$ is expressed by

$$ds^2 = -c^2 dt^2 + dr^2 + r^2 d\theta^2 + r^2 \sin^2\theta d\varphi^2 \tag{12.110}$$

[3] Sean Carroll, *Spacetime and Geometry: An Introduction to General Relativity* (Pearson, 2013), p. 167.

[4] For a lucid account of this discovery, see Janna Levin, *Black Hole Blues and other Songs from Outer Space* (Knopf, 2016).

The metric tensor takes the form

$$g_{\mu\nu} = \begin{bmatrix} -1 & 0 & 0 & 0 \\ 0 & 1 & 0 & 0 \\ 0 & 0 & r^2 & 0 \\ 0 & 0 & 0 & r^2 \sin^2\theta \end{bmatrix} \tag{12.111}$$

The non-zero metric elements are $g_{tt} = -1$, $g_{rr} = 1$, $g_{\theta\theta} = r^2$, $g_{\varphi\varphi} = r^2 \sin^2\theta$. The determinant $g = -r^4 \sin^2\theta$.

Let us now consider the metric in the presence of a mass M at the origin. The gravitational potential at a distance r from this mass is given by

$$\phi = -\frac{GM}{r} \tag{12.112}$$

A comparison with Eq. (12.101) shows that the metric tensor $g_{00} \equiv g_{tt}$ can be expressed to a good approximation for large values of r as

$$g_{tt} = -\left(1 - \frac{2GM}{c^2 r}\right) \tag{12.113}$$

For a spherically symmetric situation the angular metric elements would be unaffected, but the radial element would change. Let the altered radial element be g'_{rr}. Hence the altered determinant becomes $g' = -r^4 \sin^2\theta(1 - \frac{2GM}{c^2 r})g'_{rr}$.

We have seen earlier that the product $\sqrt{-g}\, d^n x$ is a scalar or a Lorentz invariant. Relative velocity or a Lorentz boost alters space and time, but a Lorentz invariant remains the same. The presence of matter has the effect of curving spacetime, but since mass can be considered as another dimension — along with space and time — a distortion of spacetime due to the presence of matter *outside the region* should not alter the invariant $\sqrt{-g}\, d^n x$. For the metric under consideration, $n = 4$. d^4x is not altered by the curvature due to matter. So $\sqrt{-g}$ also remains unaltered. Thus $g' = g$. Therefore

$$g'_{rr} = \left(1 - \frac{2GM}{c^2 r}\right)^{-1} \tag{12.114}$$

So the required metric becomes

$$ds^2 = -\left(1 - \frac{2GM}{c^2 r}\right) c^2 dt^2 + \left(1 - \frac{2GM}{c^2 r}\right)^{-1} dr^2 + r^2 d\theta^2 + r^2 \sin^2\theta d\varphi^2 \tag{12.115}$$

This is called the **Schwarzschild** metric.

We have derived this metric only for the limiting case where $\frac{2GM}{c^2} \ll r$. We will next prove that this equation is valid for all values of r. There are many different functions which take the form $1 - \frac{2GM}{c^2 r}$ in this limit. A simple example is $\left(1 - \frac{2GM}{qc^2 r}\right)^q$ where q is a non-zero real number. As $q \to \infty$, this function becomes $\exp\left(-\frac{2GM}{c^2 r}\right)$. We are seeking a function that is positive for all values of r, thus ensuring that g_{tt} remains negative. Consider $e^{2\alpha(r)}$ where $\alpha(r)$ is real, and $\alpha(r) \to 0$ as $r \to \infty$. Thus $c^2 e^{\alpha(r)}$ will remain positive for all r, and will equal c^2 for large r. The metric can be written as

$$ds^2 = -e^{2\alpha(r)} c^2 dt^2 + e^{-2\alpha(r)} dr^2 + r^2 d\theta^2 + r^2 \sin^2\theta d\varphi^2 \tag{12.116}$$

We are interested only in the region outside the mass, i.e. the region where the Ricci tensor $R_{\mu\nu} = 0$. A zero tensor means every element is zero. Let us choose the element $R_{\theta\theta}$ which we set to zero.

$$R_{\theta\theta} = R^t_{\theta t\theta} + R^r_{\theta r\theta} + R^\theta_{\theta\theta\theta} + R^\varphi_{\theta\varphi\theta} = 0 \tag{12.117}$$

Recalling the expansion of the Riemann tensor

$$R^\rho_{\sigma\mu\nu} = \partial_\mu \Gamma^\rho_{\nu\sigma} - \partial_\nu \Gamma^\rho_{\mu\sigma} + \Gamma^\rho_{\mu\lambda}\Gamma^\lambda_{\nu\sigma} - \Gamma^\rho_{\nu\lambda}\Gamma^\lambda_{\mu\sigma}$$

and of the Christoffel symbol

$$\Gamma^\sigma_{\mu\nu} = \frac{1}{2} g^{\sigma\rho}(\partial_\mu g_{\nu\rho} + \partial_\nu g_{\rho\mu} - \partial_\rho g_{\mu\nu})$$

we obtain for the $\theta\theta$ component of the Ricci tensor:

$$R_{\theta\theta} = -re^{2\alpha(r)}\partial_r \alpha(r) - re^{2\alpha(r)}\partial_r \alpha(r) + 0 + 1 - e^{2\alpha(r)} = 0 \tag{12.118}$$

This simplifies to $e^{2\alpha}(2r\partial_r \alpha + 1) = 1$, which is compacted to $\partial_r(re^{2\alpha}) = 1$, and yields the solution

$$e^{2\alpha} = 1 + \frac{k}{r} \tag{12.119}$$

We know that for large r this must become $1 - \frac{2GM}{c^2 r}$, and hence $k = -\frac{GM}{c^2}$. Thus, the Schwarzschild metric we derived earlier is exact, and is valid for all values of M and r.

Exercise:
Using the metric of Eq. (12.116), obtain expressions for the Riemann tensor components: $R^t_{\theta t\theta}$, $R^r_{\theta r\theta}$, $R^\theta_{\theta\theta\theta}$, and $R^\varphi_{\theta\varphi\theta}$.

The Schwarzschild metric was derived for a static field, one that does not vary with time. But according to *Birkhoff's theorem* the Schwarzschild metric uniquely and accurately describes the spacetime around a spherical body of mass M even when the mass changes with time.

12.5 Some Consequences of General Relativity

12.5.1 *Equivalence Principle*

An important law relating general relativity to Newtonian mechanics and special relativity is called the principle of equivalence, or the equivalence principle.

The equivalence principle — like the second law of thermodynamics — can be stated in different ways. The simplest statement, called *the weak equivalence principle*, is the following:

The inertial mass of a body is identical with its gravitational mass.

Suppose we are in a space ship traveling through zero gravity. Now, suppose the ship enters a gravitational field. All the objects on board will experience the same acceleration. But the same effect could be generated in empty space if the ship turned on its engines and began to accelerate. If the acceleration of the ship was perfectly smooth, it would be impossible to tell whether the ship was accelerating, or it had entered a *uniform* gravitational field.

Einstein's version of the equivalence principle — called the *Einstein Equivalence Principle* — is enunciated as follows:

In small enough regions of spacetime, the laws of physics reduce to those of special relativity; it is impossible to detect the existence of a gravitational field by means of local experiments.

Cavendish obtained Newton's gravitational constant G by measuring the force between two large metal spheres. The Einstein equivalence principle would imply that Cavendish would have gotten the same result whether he did his experiment on earth, or in a space ship accelerating at $g = 9.80\,\text{m/s}^2$, i.e. the gravitational force between the spheres would be the same in both instances. Hence, a more accurate statement, called the *Strong Equivalence Principle*, is stated as:

In small enough regions of spacetime, the laws of physics reduce to those of special relativity; it is impossible to detect the existence of an *external* gravitational field by means of local experiments.

12.5.2 *Doppler Effect due to Gravity*

A four-interval ds is defined as

$$ds^2 = g_{\mu\nu}dx^\mu dx^\nu = -c^2dt^2 + dx^2 + dy^2 + dz^2 \tag{12.120}$$

The proper time τ of a moving body is defined as

$$\begin{aligned}\Delta\tau &= \int dt\sqrt{1-\frac{v^2}{c^2}} = \int \sqrt{dt^2 - (dx^2+dy^2+dz^2)/c^2} \\ &= \frac{1}{c}\int \sqrt{-g_{\mu\nu}dx^\mu dx^\nu}\end{aligned} \tag{12.121}$$

So $d\tau = \frac{1}{c}\sqrt{-g_{\mu\nu}dx^\mu dx^\nu} = \frac{1}{c}\sqrt{-g_{tt}c^2dt^2 - g_{rr}dr^2 - g_{\theta\theta}d\theta^2 - g_{\varphi\varphi}d\varphi^2}$.

Consider a time interval dt at a fixed point in a gravitational field, so that $dr = d\theta = d\varphi = 0$, and $d\tau = \sqrt{-g_{tt}}dt$.

Suppose a pulse of light is directed downward from a point 1 at a height y above the surface of the earth. Point 2 is located on the ground directly below point 1. Let $d\tau_1$ be the proper time duration for the emission of the light beam at frequency ν_1 at point 1, and $d\tau_2$ the proper time for the absorption of the same beam at frequency ν_2 at point 2 on the ground. We expect ν_2 to be different from ν_1 because of the difference in gravitational potential at the two points.

We showed earlier that for a weak gravitational field $g_{tt} = -(1+\frac{2\Phi}{c^2})$.

So, the proper time for the emission of the light wave is

$$d\tau_1 = \sqrt{1+\frac{2\Phi_1}{c^2}}dt \tag{12.122}$$

And the proper time for the absorption of the light wave on the ground is

$$d\tau_2 = \sqrt{1+\frac{2\Phi_2}{c^2}}dt \tag{12.123}$$

Dividing one equation by another,

$$\frac{d\tau_1}{d\tau_2} = \frac{\sqrt{1+\frac{2\Phi_1}{c^2}}}{\sqrt{1+\frac{2\Phi_2}{c^2}}} \tag{12.124}$$

The frequency ν of the wave (ν_1 = emission, ν_2 = absorption) is inversely proportional to the time $d\tau$, and for small values of Φ_1 and Φ_2,

$$\frac{\nu_2}{\nu_1} = 1 + \Phi_1/c^2 - \Phi_2/c^2 \tag{12.125}$$

$$\frac{\nu_2 - \nu_1}{\nu_1} = \frac{\Phi_1 - \Phi_2}{c^2} \tag{12.126}$$

$\Phi = -\frac{GM}{r}$. $r_1 = R + y$, and $r_2 = R$, the radius of the earth. Therefore, for a small vertical displacement y,

$$\frac{\nu_2 - \nu_1}{\nu_1} = \frac{1}{c^2}\left(-\frac{GM}{R+y} + \frac{GM}{R}\right) = \frac{GM}{R^2c^2}y = gy/c^2 \tag{12.127}$$

where $g = GM/R^2$ is the acceleration due to gravity (approximately 9.80 m/s^2). We see from this equation that the frequency of absorption is greater than the frequency of emission.

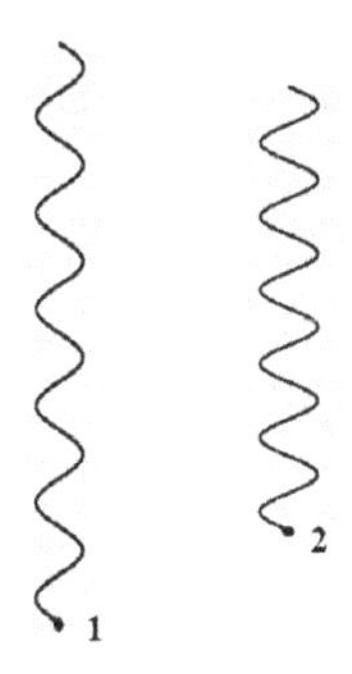

This result that we have just obtained from general relativity agrees with what we get using quantum theory and special relativity.

Suppose a photon of frequency ν_1 is directed downward from a height y above the surface of the earth. It has initial energy $h\nu_1$. By the time it reaches the ground, the force of gravity has done some work done on it, equal to force × displacement = mgy where $m = h\nu_1/c^2$ is the mass of the photon. As the photon travels downward, it accrues kinetic energy by the work energy theorem, and this additional kinetic energy is manifested as an increase of frequency. Considering the photon as a particle of mass $h\nu/c^2$, its final energy

$$h\nu_2 = h\nu_1 + \frac{h\nu_1}{c^2}gy \tag{12.128}$$

which yields the result:

$$\frac{\nu_2 - \nu_1}{\nu_1} = \frac{gy}{c^2} \tag{12.129}$$

There are also other approaches that yield the same result. We can, for instance, apply the principle of equivalence and Lorentz contraction.

We consider an observer standing on the ground, and the light source at a height y above the ground. Applying the principle of equivalence, we could eschew gravitational attraction between the light beam and the earth, and instead restrict ourselves to the mutual acceleration of magnitude g between the two bodies. In this model, the earth is initially at rest relative to the light source. The observer on the earth — initially at rest relative to the source — measures the wavelength of the light emitted by the source, and gets the number λ_1. Then the observer — along with the earth — accelerates towards the source, and measures the wavelength λ_2 when the light source is at the same position as the observer. The relative speed between the source and the observer is now $v = \sqrt{2gy}$. By Lorentz contraction,

$$\lambda_2 = \lambda_1 \sqrt{1 - \frac{v^2}{c^2}} = \lambda_1 \sqrt{1 - \frac{2gy}{c^2}} \tag{12.130}$$

And so, for small values of gy/c^2, and bearing in mind that the speed of light c remains the same,

$$\nu_2 = \frac{\nu_1}{\sqrt{1 - \frac{2gy}{c^2}}} = \nu_1 \left(1 + \frac{gy}{c^2}\right) \tag{12.131}$$

12.5.3 *Deflection of Light by the Sun*

In the above example, we considered a light beam traveling in the direction of a gravitational field. We now consider a light beam traveling perpendicular to the gravitational field of a massive spherical object such as the sun. At a great distance from a source, the wavefront of any wave becomes a plane that is perpendicular to the direction of propagation. Consider an electromagnetic wave passing through the gravitational field of the sun. Gravity has the effect of shortening the wavelength along the direction of the field. The result is that the wavefront changes direction, and so does the propagation. A light wave is therefore deflected upon passing close to the sun.

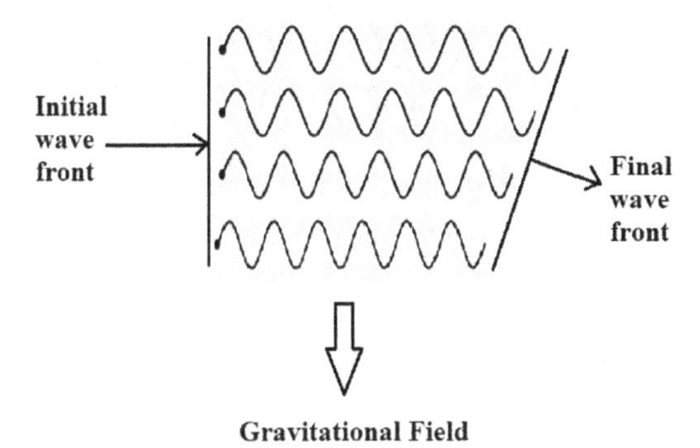

Thus a light ray coming from a distant star will undergo a slight deflection upon passing close to the sun. The total deflection 2θ has been exaggerated in the following diagram:

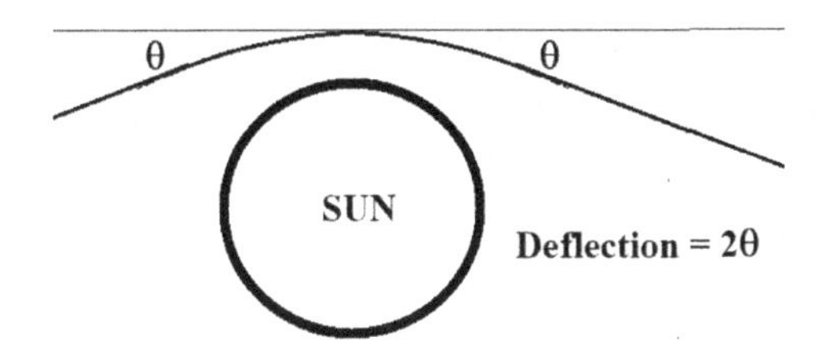

The path taken by the light is a geodesic in this field. The light bends towards the sun, in a radial direction. Let us consider the radial displacement of the path of the light ray, along which dr changes, but $d\theta$ and $d\varphi$ remain constant. A four-displacement ds in this direction is given by

$$ds^2 = g_{tt}c^2dt^2 + g_{rr}dr^2 \tag{12.132}$$

where $g_{tt} = -\left(1 - \frac{2GM}{c^2r}\right)$, $g_{rr} = \left(1 - \frac{2GM}{c^2r}\right)^{-1}$, and M is the mass of the sun. A geodesic along the path of light is a null path, with $ds = 0$. And so, we can write

$$\frac{dr}{dt} = c\sqrt{-\frac{g_{tt}}{g_{rr}}} = c\sqrt{\frac{(1 - \frac{2GM}{c^2r})}{(1 - \frac{2GM}{c^2r})^{-1}}} = c\left(1 - \frac{2GM}{c^2r}\right) \tag{12.133}$$

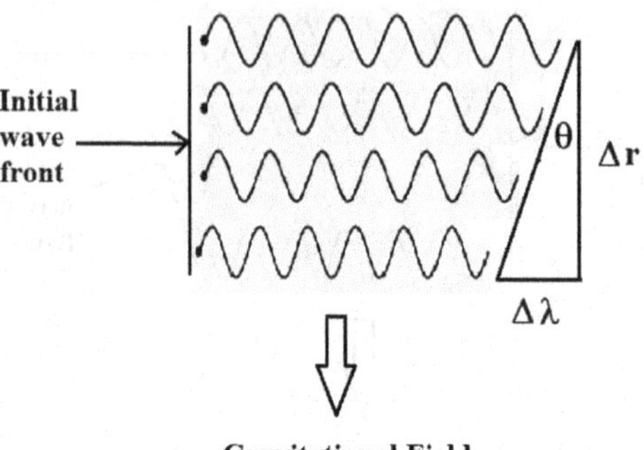

In the triangle shown above,

$$\Delta r = \frac{dr}{dt}\Delta t = c\left(1 - \frac{2GM}{c^2 r}\right)\Delta t \tag{12.134}$$

The horizontal side $\Delta\lambda$ is the shortening in wavelength due to the gravitational field, and this can be expressed in terms of a reduction in wave velocity: $\Delta c = \frac{\Delta\lambda}{\Delta t}$. The apparent reduced speed of the light beam in the field is $\frac{dr}{dt} = (c - \Delta c)$, where $\Delta c = \frac{2GM}{cr}$. In the figure, $\Delta\lambda = \Delta c \Delta t = \frac{2GM}{cr}\Delta t$. All these derivations are valid to first order, since the effect of the sun's gravitational field on the light beam is very small.

The small angle θ can then be calculated from the triangle in the above diagram:

$$\theta = \frac{\Delta\lambda}{\Delta r} = \frac{2GM/cr}{c(1 - \frac{2GM}{c^2 r})} \approx \frac{2GM}{c^2 r} \tag{12.135}$$

Hence, the total deflection of the light ray grazing the sun (mass M, radius R) equals

$$2\theta = \frac{4GM}{c^2 R} \tag{12.136}$$

Putting in the values $G = 6.67 \times 10^{-11}$, $M = 1.99 \times 10^{30}$, $R = 6.96 \times 10^8$, $c = 3.00 \times 10^8$, we obtain $2\theta = 1.75$ arc-seconds. (3600 arc-seconds $= 1^0$, and $57.3^0 = 1$ radian.) This number agrees well with experimental observations.

12.6 Quantization of Gravity

The gravitational field is transmitted at speed c. In analogy with the electromagnetic field, the gravitational field is communicated by quanta called gravitons. A graviton propagates at speed c, and has zero rest mass. The electromagnetic field is a vector field, or a tensor field of rank one, and the corresponding quantum — the photon — is a zero rest mass particle of spin 1. The gravitational field is a tensor field of rank 2, and the corresponding quantum — the graviton — is a zero rest mass particle of spin 2. It has been proved that there cannot be any other field besides gravity that has a quantum of zero rest mass and spin 2. And there is no theoretical or experimental evidence for a massless particle of spin higher than 2.

Now, electrons can be scattered by electrons. Such a scattering is called Møller scattering. Since they do have momentum, and their inertial masses feature in collisions between electrons, we should expect them to have a gravitational force between them. How do we incorporate the gravitational interaction into a Feynman diagram of interacting electrons? Let us try the following diagram:

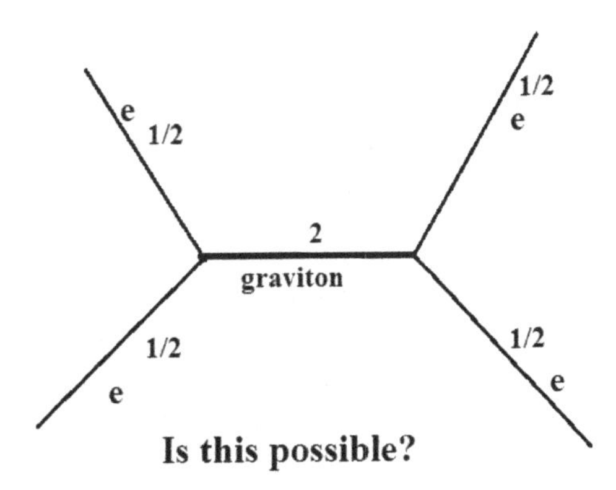

It is hard to picture vertices such as the above. The spin contribution of two electrons at a vertex is either 0 or 1. A single graviton has spin 2, and so it is difficult to see how the above diagram meets conservation of angular momentum at the vertices. A better picture is the following:

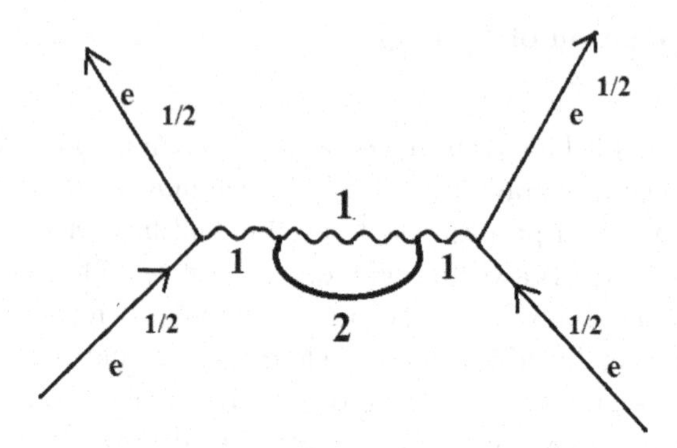

Here we see the gravitational interaction appearing as a correction to the electromagnetic interaction between the electrons. At the first vertex two spin half particles interact with a spin 1 particle. At the second and third vertices two spin 1 particles interact with a spin 2 particle, and at the fourth vertex two spin half particles interact with a spin 1 particle.

The above diagram also describes the fundamental processes occurring in the deflection of the light beam by the sun's gravity. A photon is emitted by an electron in a distant star (first vertex). This photon interacts with a graviton from the sun's gravity (second and third vertices). As a result of this interaction, the photon changes direction, and proceeds till it interacts with an electron in the detecting device (fourth vertex), be it the retina of a human eye, or a photographic film. It is evident that angular momentum is conserved at each of the four vertices. Therefore, the gravitational interaction between two electrons is mediated by the electromagnetic field.

Most of the mass of an atom comes from the nucleus, which is constituted of protons and neutrons. While a neutron has zero charge, it is constituted of three quarks, just as a proton is also constituted of three quarks. Quarks have spin half, and have charges of magnitudes $e/3$ and $2e/3$. Quarks also have spin $1/2$. So, quarks also interact with the gravitons through photons.

Thus both the light electrons, and the heavy protons and neutrons of the nucleus, interact with a gravitational field through the electromagnetic field.

Index

Printed in the United States
by Baker & Taylor Publisher Services